C++ Toolbox for Verified Computing I

R. Hammer M. Hocks
U. Kulisch D. Ratz

# C++ Toolbox for Verified Computing I

## Basic Numerical Problems

Theory, Algorithms, and Programs

With 29 Figures

Springer

Prof. Dr. Ulrich Kulisch
Dr. Rolf Hammer
Dr. Matthias Hocks
Dr. Dietmar Ratz
Institut für Angewandte Mathematik
Universität Karlsruhe
D-76128 Karlsruhe

*Cover figure:* Function of Levy (see also page 337)

Mathematics Subject Classification (1991): 65-01, 65-04, 65F, 65G10, 65H, 65K

ISBN-13:978-3-642-79653-1

Library of Congress Cataloging-in-Publication Data
C++ toolbox for verified computing: theory, algorithms, and programs
R.Hammer [et al.]. p.cm. - Includes bibliographical references and index. Contents:
v. 1. Basic numerical problems.
ISBN-13:978-3-642-79653-1 e-ISBN-13:978-3-642-79651-7
DOI: 10.1007/978-3-642-79651-7
1. C++ (Computer program language) I. Hammer, (Rolf), 1961- .
QA76.73.C153C18 1995 519.4'0285'5133--dc20 95-10173 CIP



Typesetting: Camera ready by authors using $\TeX$
SPIN 10492623 41/3142 - 5 4 3 2 1 0 – Printed on acid-free paper

# Preface

Our aim in writing this book was to provide an extensive set of C++ programs for solving basic numerical problems with *verification of the results.* This *C++ Toolbox for Verified Computing I* is the C++ edition of the *Numerical Toolbox for Verified Computing I.* The programs of the original edition were written in PASCAL–XSC, a PASCAL eXtension for Scientific Computation. Since we published the first edition we have received many requests from readers and users of our tools for a version in C++.

We take the view that C++ is growing in importance in the field of numerical computing. C++ includes C, but as a typed language and due to its modern concepts, it is superior to C. To obtain the degree of efficiency that PASCAL–XSC provides, we used the C–XSC library. C–XSC is a C++ class library for eXtended Scientific Computing. C++ and the C–XSC library are an adequate alternative to special XSC-languages such as PASCAL–XSC or ACRITH–XSC. A shareware version of the C–XSC library and the sources of the toolbox programs are freely available via anonymous ftp or can be ordered against reimbursement of expenses.

The programs of this book do not require a great deal of insight into the features of C++. Particularly, object oriented programming techniques are not required. However, the reader should be familiar with writing programs in a high-level computer language such as PASCAL, C, or FORTRAN. This book is particularly useful for those programmers who have already worked with the PASCAL–XSC edition but have only little knowledge in the C++ language. For those readers our book may be a source of inspiration when switching from PASCAL to C++.

We want to thank the readers of the original version of this book for their overwhelmingly positive comments. We incorporated some minor modifications in our algorithms which take into account the highly valuable suggestions and comments we received from numerous readers, as well as our own experience while using the toolbox. Some errors and misprints in the original edition have now been corrected. Nevertheless, we encourage readers, especially those of this C++ edition, to keep on communicating error reports to us.

Special thanks to Andreas Wiethoff who supported us in all questions concerning C++ and the C–XSC library. Last but not least, we wish to express again our appreciation to all colleagues whose advice helped produce the original edition, particularly those mentioned in the following preface of the PASCAL–XSC edition.

*Karlsruhe, December 1994* *The Authors*

# Preface to the PASCAL–XSC Edition

As suggested by the title of this book *Numerical Toolbox for Verified Computing*, we present an extensive set of sophisticated tools to solve basic numerical problems with a *verification of the results*. We use the features of the scientific computer language PASCAL–XSC to offer modules that can be combined by the reader to his/her individual needs. Our overriding concern is reliability – the *automatic verification of the result* a computer returns for a given problem. All algorithms we present are influenced by this central concern. We must point out that there is no relationship between our methods of *numerical result verification* and the methods of *program verification* to prove the correctness of an implementation for a given algorithm.

This book is the first to offer a general discussion on

- arithmetic and computational reliability,
- analytical mathematics and verification techniques,
- algorithms, and
- (most importantly) actual implementations in the form of working computer routines.

Our task has been to find the right balance among these ingredients for each topic. For some topics, we have placed a little more emphasis on the algorithms. For other topics, where the mathematical prerequisites are universally held, we have tended towards more in-depth discussion of the nature of the computational algorithms, or towards practical questions of implementation. For all topics, we present examples, exercises, and numerical results demonstrating the application of the routines presented.

The different chapters of this volume require different levels of knowledge in numerical analysis. Most numerical toolboxes have, after all, tools at varying levels of complexity. Chapters 2, 3, 4, 5, 6, and 10 are suitable for an advanced undergraduate course on numerical computation for science or engineering majors. Other chapters range from the level of a graduate course to that of a professional reference. An attractive feature of this approach is that you can use the book at increasing levels of sophistication as your experience grows. Even inexperienced readers can use our most advanced routines as black boxes. Having done so, these readers can go back and learn what secrets are inside.

The central theme in this book is that practical methods of numerical computation can be simultaneously efficient, clever, clear, and (most importantly) reliable.

We firmly reject the alternative viewpoint that such computational methods must necessarily be so obscure and complex as to be useful only in "black box" form where you have to believe in any calculated result.

This book introduces many computational verification techniques. We want to teach you to take apart these black boxes and to put them back together again, modifying them to suit your specific needs. We assume that you are mathematically literate, i.e. that you have the normal mathematical preparation associated with an undergraduate degree in a mathematical, computational, or physical science, or engineering, or economics, or a quantitative social science. We assume that you know how to program a computer and that you have some knowledge of scientific computation, numerical analysis, or numerical methods. We do not assume that you have any prior formal knowledge of numerical verification or any familiarity with interval analysis. The necessary concepts are introduced.

Volume 1 of *Numerical Toolbox for Verified Computing* provides algorithms and programs to solve basic numerical problems using automatic result verification techniques.

Part I contains two introductory chapters on the features of the scientific computer language PASCAL–XSC and on the basics and terminology of interval arithmetic. Within these chapters, the important correlation between the arithmetic capability and computational accuracy and mathematical fixed-point theory is also discussed.

Part II addresses one-dimensional problems: evaluation of polynomials and general arithmetic expressions, nonlinear root-finding, automatic differentiation, and optimization. Even though only one-dimensional problems treated in this part, the verification methods sometimes require multi-dimensional features like vector or matrix operations.

In Part III, we present routines to solve multi-dimensional problems such as linear and nonlinear systems of equations, linear and global optimization, and automatic differentiation for gradients, Hessians, and Jacobians.

Further volumes of *Numerical Toolbox for Verified Computing* are in preparation covering computational methods in the field of linear systems of equations for complex, interval, and complex interval coefficients, sparse linear systems, eigenvalue problems, matrix exponential, quadrature, automatic differentiation for Taylor series, initial value, boundary value and eigenvalue problems of ordinary differential equations, and integral equations. Editions of the program source code of this volume in the C++ computer language are also in preparation.

Some of the subjects that we cover in detail are not usually found in standard numerical analysis texts. Although this book is intended primarily as a reference text for anyone wishing to apply, modify, or develop routines to obtain mathematically certain and reliable results, it could also be used as a textbook for an advanced course in scientific computation with automatic result verification.

We express our appreciation to all our colleagues whose comments on our book were constructive and encouraging, and we thank our students for their help in testing our routines, modules, and programs. We are very grateful to Prof. Dr. George Corliss (Marquette University, Milwaukee, USA) who helped to polish the text and the contents. His comments and advice based on his numerical and computational experience greatly improved the presentation of our tools for Verified Computing.

*Karlsruhe, September 1993* *The Authors*

# The computer programs in this book

and a shareware version of the C–XSC library are available in several machine-readable formats. To purchase diskettes in IBM-PC compatible format, use the order form at the end of the book. The programs and the library are also available by anonymous ftp from

`iamk4515.mathematik.uni-karlsruhe.de` (129.13.129.15)

in subdirectory

`pub/toolbox/cxsc`.

Technical questions, corrections, and requests for information on other available formats and software products should be directed to *Numerical Toolbox Software, Institut für Angewandte Mathematik, Universität Karlsruhe, D-76128 Karlsruhe, Germany, e-mail: toolbox@iampc4.mathematik.uni-karlsruhe.de.*

# Table of Contents

# List of Figures

# Chapter 1

# Introduction

This is a reference book for numerical methods with automatic result verification. The methods presented here are practical, reliable, and elegant. We provide theory, algorithmic descriptions, and implementations for methods to solve a number of basic numerical problems in a reliable way. Also, this book can help you to learn how to develop such methods and how to proceed for other problems beyond the scope of the book.

We warn of potential computational problems with standard numerical routines, but we do not claim that our advice is infallible! Therefore, we offer our practical judgment on solving many basic numerical problems with automatic result verification. As you gain experience, you will form your own opinion of how reliable our advice is.

We presume that the reader is able to read computer programs in C++. We use the C–XSC library, a C++ class library for eXtended Scientific Computing, to implement all the "tools" in this *C++ Toolbox for Verified Computing I.* Its wide range of concepts for scientific computation makes C–XSC especially well suited as a specification language for programming with automatic result verification. We discuss the features of C–XSC in Chapter 2.

In the following sections, we first give some hints on the methodology and structure of this book. Then we explain the textual and algorithmic notation we use. We also describe the basic concepts and form of our implementations, and we make some comments on the computational environment. We conclude with a section on the motivation and the necessity of methods with numerical result verification.

## 1.1 Advice for Quick Reading

This book is organized in three parts:

1. the preliminaries,
2. the one-dimensional problems, and
3. the multi-dimensional problems.

The preliminaries give an overview of the extensions of the C++ class library C–XSC and of the properties and advantages of interval arithmetic. The reader with little experience in numerical verification and interval techniques should read the section on interval arithmetic with care and interest. Its introductory character shows what arithmetic improvements can do for the reliability of computed results.

We discuss only basic one- and multi-dimensional problems in parts II and III, respectively. We especially direct the attention of readers unfamiliar with numerical result verification to the following chapters:

- Linear Systems of Equations (Chapter 10),
- Evaluation of Polynomials (Chapter 4),
- Automatic Differentiation (Chapter 5), and
- Nonlinear Equations in One Variable (Chapter 6).

Some other chapters, e.g. optimization, require a little more background knowledge on numerical verification and interval techniques. These algorithms sometimes use the basic algorithms mentioned above (e.g. the linear system solver).

## 1.2 Structure of the Book

In order to make this *Numerical Toolbox* easy and efficient to use, we have chosen some general standards to keep the different chapters of this book uniform. This will help you to find the important information on each topic, e.g. the theoretical background or the implementation of a desired algorithm, without necessarily reading the whole chapter.

Each chapter starts with an introduction to describe the numerical and mathematical problem and to motivate a strategy for its solution. A comprehensive discussion of the theoretical basis necessary to solve the problem and to verify the computational result is given in the section "*Theoretical Background*". Under the heading "*Algorithmic Description*", we present the complete algorithm in a very short, clear, and consistent way.

A very important part of each chapter is the section "*Implementation and Examples*" where an actual implementation for the topic discussed in the chapter is presented. As promised in the title, all routines are designed as "tools" that can be combined in several ways to satisfy your needs in solving a specific numerical problem. The routines are built of different modules so they may be integrated in an existing program, e.g. to overcome a numerical trap at a certain point of your own complex problem-solving algorithm. All algorithms presented in this volume are available on floppy disks and via anonymous ftp (see page ix) to help you use the verification algorithms immediately for testing or for integrating them into existing applications. To demonstrate the effectiveness of each algorithm, we present sample programs that solve practical problems. We point out the superiority of verification methods and precisely defined machine arithmetic over most standard numerical problem-solving strategies.

Most chapters include some "*Exercises*" to encourage you to study the behavior or to improve the algorithms we present. The section "*References and Further Reading*" at the end of each chapter points out to some similar or more sophisticated and specialized algorithms. The appendix contains basic modules which are used by certain verification algorithms.

## 1.3 Typography

We use the following type faces for certain words, names, or paragraphs:

| | |
|---|---|
| *italics* | emphasizes certain words in the text. |
| **boldface** | marks keywords of the pseudo-language (e.g. **if**, **else**, **return**) used in our algorithms. |
| *slanted* | characterizes identifiers of the program listings when they appear in the body of the text. |
| `typewriter` | distinguishes listings and runtime results of programs, including some pretty-printing to improve the readability (see Section 1.5). |

References are always indicated as [*nr*], where the number *nr* corresponds to an entry in the bibliography.

## 1.4 Algorithmic Notation

The algorithms presented in Parts II and III of this book are specified in a pseudo-code similar to the natural language, including mathematical notation (cf. Chapter 3) wherever this is possible. Algorithm 1.1 is a simple example of our algorithmic notation.

**Algorithm 1.1:** DoSomething $(x, y, z, Err)$

1. $a := x + y$;  $b := x - y$;  {Initializations}
2. {Do something special}

   **if** $(a > 0)$ **and** $(b > 1)$ **then**

   $c := \sqrt{a}$;  $d := \ln(a)$;

   $a := -c$;  $b := -d$;

   **else if** $(b = 1)$ **then**

   **return** $Err$ := "Error number 1 occurred!";

   **else**

   **return** $Err$ := "Another error occurred!";
3. **return** $z := a \cdot b \cdot c/d$;

For clarity, the steps of the algorithms are numbered, and the single statements are separated by semicolons. We use indentation and numbering to mark blocks of statements. Within the algorithmic steps, subalgorithms may be used in the form of functions calls.

FunctionName (*ArgumentList*)

*Variable* := FunctionName (*ArgumentList*).

The statements

**return** or **return** *ArgumentList*,

cause an immediate return from the current algorithm, where the second form returns new values for the arguments of *ArgumentList*. Loops have three different forms:

**for** *Index* := *InitialValue* **to** *FinalValue* **do** *Statement(s)*

**repeat** *Statement(s)* **until** *LogicalExpression*

**while** *LogicalExpression* **do** *Statement(s)*.

The statements

$\textbf{exit}_{Index\text{-loop}}$, $\textbf{exit}_{while\text{-loop}}$, and $\textbf{exit}_{repeat\text{-loop}}$

terminate a loop. The statement

$\textbf{next}_{Index}$

terminates the current iteration of the loop by transferring control to the end of the loop specified by *Index*. There, the next iteration of the loop may be initiated, depending on the nature of the loop, its associated increments and tests. When using indexed expressions in mathematical notation, the loop is characterized by a sequence of indices enclosed in parentheses, e.g. $(k = 1, \ldots, n)$.

Conditional statements are

**if** *LogicalExpression* **then** *Statement(s)* **else** *Statement(s)*

with optional **else**-branch.

All comments are enclosed in braces.

## 1.5 Implementation

The authors feel very strongly that you should be able to count on the programs, modules, and code fragments in this book to compile and run as we claim. Therefore, we have compiled and run each one of the code fragments (as parts of entire running programs), modules, and programs. Similarly, the results we give were actually produced by the example programs, except that we show both the printed output and the user-supplied input exactly as it will appear on your screen. The code and results have been included directly, without modification, in this manuscript by a pretty-print program, so you are seeing exactly the code that we ran. We include modules or programs in the text using the typographical style you can see in the sample listings below. For every module, we provide a header file and an implementation file. In the listings, C–XSC include file names, data types, and functions are set in bold type to emphasize that they are provided by the C–XSC library.

```
//----------------------------------------------------------------------------
// File: dosome (header)
// Global functions:
//    DoSomething()        : sample function
//    DoSomethingErrMsg(): delivers an error message text
//----------------------------------------------------------------------------
#include <real.hpp>        // Real arithmetic
#include <rmath.hpp>       // Real standard functions

extern char* DoSomethingErrMsg ( int );
extern void  DoSomething ( real, real, real&, int& );


//----------------------------------------------------------------------------
// File: dosome (implementation)
// Purpose: Sample module
// Global functions:
//    DoSomething()        : sample function
//    DoSomethingErrMsg(): delivers an error message text
//----------------------------------------------------------------------------
#include <dosome.hpp>

const int              // Error codes
  NoError      = 0,
  ErrorOne     = 1,
  AnotherError = 2;

char* DoSomethingErrMsg ( int Err )
{
  static char Msg[80] = "";

  if (Err != NoError) {
    char Hlp[60];

    switch (Err) {
      case ErrorOne:     strcpy(Hlp,"Condition not satisfied"); break;
      case AnotherError: strcpy(Hlp,"Another error occurred");  break;
      default:           strcpy(Hlp,"Code not defined");
    }
    sprintf(Msg,"Error: %s!",Hlp);
  }
  return(Msg);
} // DoSomethingErrMsg

void DoSomething ( real x, real y, real& z, int& Err )
{
  real a, b, c, d;

  a = x + y;  b = x - y;  Err = NoError;     // Initializations

  if ( a > 0.0 &&  b > 1.0)                  // Do something special
    { c = sqrt(a);  d = ln(b);  a = -c;  b = -d; }
```

```
    else if (b == 1.0)
      Err = ErrorOne;
    else
      Err = AnotherError;

    if (Err == NoError)  z = a * b * c / d;    // Else z is undefined
  }
```

We handle all errors and exceptional cases with a function *DoSomethingErrMsg()*. It associates an error message with the error code returned by the routine *DoSomething()*.

## 1.6 Computational Environment

We intend to make the programs in this book as generally useful as possible, not just in terms of the subjects covered, but also in terms of their degree of portability among very different computers. Specifically, we intend that all the programs should work on mainframes, workstations, and personal computers. We selected the C–XSC class library [49] as the implementation environment for our algorithms, because it was designed with this type of portability in mind. A shareware version of the library is included in the software for this book. It is sufficient to compile the modules and programs of this book. The full version of the C–XSC library is available from Numerik Software GmbH [71].

As far as C++ is concerned, we refer the reader to the reference books of Ellis and Stroustrup [16] and Stroustrup [90], [91]. C++ templates and exceptions are not used in our implementations because some of the C++ compilers we employed did not yet support these features.

All our programs have been run and tested on a PC 486 under DOS with Borland C++ 3.1 and 4.0, on Sun SPARCstations under SunOS 4 and SunOS 5 with Gnu C++ 2.5.8, and on HP9000 workstations under HP UX 8.05 with HP C++ 2.0. Our programs contain no known bugs, but we make no claim that they are bug-free or that they are immediately suitable for any particular application. This book has the benefit of about ten years of experience and development in the field of scientific computation with result verification, and we have been very careful in implementing and testing our routines. Comments for improvements of the text and programs are always welcome.

## 1.7 Why Numerical Result Verification?

Floating-point arithmetic is the fast way to perform scientific and engineering calculations. Today, individual floating-point operations on most computers are of maximum accuracy in the sense that the rounded result differs at most by 1 unit in the last place from the exact result. However, after two or more operations, the result can be completely wrong. Computers now carry out up to $10^{11}$ floating-point operations per second. Thus, particular attention must be paid to the reliability of the computed results. In recent years, techniques have been developed in numerical analysis which make it possible for the computer itself to verify the correctness of computed results for numerous problems and applications. Moreover, the computer frequently establishes the existence and uniqueness of the solution in this way. For example, a verified solution of a system of ordinary differential equations is just as valid as a solution obtained by a computer algebra system, which still requires a valid formula evaluation. Furthermore, the numerical routine remains applicable even if the problem does not have a closed-form solution.

In the remainder of this section, we give some historical remarks, some details on the prerequisite knowledge, and some motivation for the methods presented in this book.

### 1.7.1 A Brief History of Computing

Methods of calculation have put their stamp on mathematics throughout history. Progress in connection with numbers has meant progress in mathematics also. The invention of mechanical calculating machines by B. Pascal, G. W. Leibniz, and others, of logarithms and the slide rule, and of many other mechanical and analog computational aids, represent significant milestones in scientific computing.

A great leap in development took place about 50 years ago with the development of the first electronic computers. This technology immediately made it possible to perform arithmetic operations faster than its mechanical or electromechanical predecessors by a factor of about 1000. The great technological gains of this century would have been impossible without modern computation. Today's automobile, airplane, space travel, modern radio and television, and not least the rapid further development of the computer itself were enabled by enormous computational capacity. On the other hand, advances in computer hardware gave massive stimulation to further development of algorithmic and numerical mathematics.

Further development of circuit technology, which again would not have been possible without the computer, has increased the computational capacity of a processor by a factor of about $10^9$, compared to the first electronic computers of the 1950's. Comparison of the numbers $10^3$ and $10^9$ shows that the real computer revolution took place *after* the development of the first electronic computers. Remarkably, there was nothing equivalent to this development on the arithmetical-mathematical side. The enormous advances in computer technology really suggested an attempt to make the computer also more powerful arithmetically. On this, however, mathematicians exerted hardly any influence. In the area of scientific and engineering applications of concern here, the computer is used today in essentially the same way as it was in the middle 1950's. The same four floating-point operations of addition, subtraction, multiplication, and division are still being performed, albeit much faster.

As more and more computations can be done faster and faster, larger scientific problems can be addressed, and it becomes increasingly critical to make scientific computation more trustworthy than the exclusive use of ordinary floating-point arithmetic. This development began about 25 years ago and has gained a considerable following (see, for example, the extensive bibliographies in [70] or [31]). However, methods for validated computation have yet to achieve universal acceptance.

### 1.7.2 Arithmetic on Computers

Usually, electronic computers now have two different number systems built into their hardware: integers and floating-point numbers. Integer arithmetic operates over a limited range of integers. To the extent that the hardware (and also the software) are intact, integer arithmetic and its software extensions operate without error. For example, a rational arithmetic or a multiple-precision arithmetic can be constructed using integer arithmetic. In this way, the computer carries along as many digits as necessary to represent the result of an operation exactly. In number theory or computer algebra, this is the preferred method for arithmetic. Also, most applications in computer science can be carried out error-free with integer arithmetic. Examples are compilation or other forms of translation, and algorithms for searching or sorting.

On the other hand, in numerical or scientific analysis, one works with real numbers. A real number is represented by an infinite decimal or binary fraction. For numerical or scientific computation, real numbers must be approximated by a finite fraction in a computer, ordinarily, by floating-point numbers. Floating-point arithmetic is built into the hardware, and is consequently very fast. However, each floating-point operation is subject to error. Although, each floating-point operation on most modern computers is of maximum accuracy, the result after only a few operations can be completely wrong. We consider two simple examples.

**Example 1.1** Let the two real vectors $x = (10^{20}, 1223, 10^{18}, 10^{15}, 3, -10^{12})$ and $y = (10^{20}, 2, -10^{22}, 10^{13}, 2111, 10^{16})$ be given. Let the scalar product be denoted by $x \cdot y$ (the corresponding components are multiplied, and all the products are summed). This gives (using exact integer arithmetic)

$$x \cdot y = 10^{40} + 2446 - 10^{40} + 10^{28} + 6333 - 10^{28} = 8779.$$

In contrast, floating-point arithmetic on *every* computer (including those with IEEE-arithmetic) gives the value zero for this scalar product. The reason for this is that the summands are of such different orders of magnitude that they cannot be processed correctly in ordinary floating-point format. This catastrophic error occurs even though the data (the components of the vectors) use up less than 5% of the exponent range of small and medium size computers!

**Example 1.2** For the second example, we consider a floating-point system with base 10 and a mantissa length of 5 digits. We want to compute the difference of the two numbers $x = 0.10005 \cdot 10^5$ and $y = 0.99973 \cdot 10^4$ in the presented floating-point arithmetic. This time, both operands are of the same order of magnitude. The computer gives the completely correct result $x - y = 0.77000 \cdot 10^1$. Now, suppose that the two numbers $x$ and $y$ are the results of two previous multiplications. Since we are dealing with 5-digit arithmetic, these products of course have 10 digits. Suppose the unrounded products are

$$x_1 \cdot x_2 = 0.1000548241 \cdot 10^5 \quad \text{and} \quad y_1 \cdot y_2 = 0.9997342213 \cdot 10^4.$$

If we subtract these two numbers, normalize the result, and round it to 5 places, we get $0.81402 \cdot 10^1$, which differs in every digit from the result computed in floating-point arithmetic. In the second case, the value of the expression $x_1 \cdot x_2 - y_1 \cdot y_2$ was computed to the closest 5-digit floating-point number. In contrast, pure floating-point arithmetic with rounding after each individual operation gave a completely wrong result.

One should be very, very careful with the argument, "nothing like that happens in *my* calculations." It is very difficult to know all the data that enter as intermediate results into an hour-long computation on a workstation or a supercomputer. Many computer users appear to be like a wood carver who is forced to carve a work of art with a dull knife, even though it is so easy to sharpen a knife!

In classical error analysis of numerical algorithms, the error of each individual floating-point operation is estimated. It is evident that this is no longer practically possible for an algorithm for which $10^{14}$ floating-point operations are performed in an hour. Thus, an error analysis is not usually performed. Indeed, the fact that the computed result could be wrong is often not taken into consideration.

From the mathematical point of view, the problem of correctness of computed results is of central importance because of the high computational speeds attainable today. The determination of the correctness of computed results is essential in many applications such as simulation or mathematical modeling. One needs an assurance of correct computational results to distinguish between inaccuracies in computation and actual properties of the model. A mathematical model can be developed systematically only if errors entering into the computation can be essentially excluded.

### 1.7.3 Extensions of Ordinary Floating-Point Arithmetic

In the past, scientific computations were carried out using a slide rule or tables of logarithms of numbers and standard functions. The use of logarithms avoids the necessity of keeping track of exponents, as when using a slide rule, but precision is limited in either case. A modern mathematical coprocessor chip can be viewed as automating computation using logarithm tables with a large (but fixed) number of places, including addition and subtraction logarithms. It is so fast, however, that there is no way to observe the cumulative effect of the neglect of small quantities or of cancellation errors.

Early scientific computation can be characterized as multiplication-division intensive. In contrast, accountants and bookkeepers labored with addition and subtraction of long columns of numbers of various magnitudes. Errors in these addition-subtraction intensive computations were not tolerated, and various methods of validation (such as double-entry bookkeeping) were developed to ensure that answers are correct to the penny. In order to handle potentially millions of additions and subtractions accurately, it is evident that the "slide-rule whiz" capabilities of ordinary floating-point arithmetic have to be augmented by the abilities of a patient accountant. Given the state of the art, there is no reason why this cannot be done on a single chip.

Computers were invented to take on complicated work for people. The evident discrepancy between computational power and control of computational errors suggests also turning over the process of error estimation itself to the computer. This has been done successfully for practically all the basic problems of numerical analysis and many applications. To achieve this, the computer arithmetic has to be made more powerful than is ordinarily the case. One thus proceeds from the above observations: In floating-point arithmetic, most errors occur in accumulations, that is, by execution of a sequence of additions and subtractions. On the other hand, multiplication and division in floating-point arithmetic are relatively stable operations. In fixed-point arithmetic, however, accumulation is performed without errors. One only has to provide a fixed-point register in the arithmetic unit which covers the entire floating-point range. If such a hardware register is not available, then it can be simulated in the main memory by software. The resulting loss of speed in many cases is outweighed by the gain in certainty.

If this register is made twice as long as the total dynamic range of the floating-point number system, then dot products of vectors of any finite dimension can always be computed exactly. The products (the summands in the dot product) have double mantissa lengths, and the exponent range is also doubled. For the IEEE arithmetic standard data format, about four times as many bits or bytes are necessary because of its substantially greater exponent range. Even this relatively long fixed-point register presents no problem to modern technology. The possibility of exactly computing the dot product of floating-point vectors of any finite dimension opens a new dimension in numerical analysis. In particular, the optimal dot product proves itself to be an essential instrument in attaining higher computational accuracy.

With the dot product, all operations between floating-point numbers, operations for complex numbers,[1] and in particular, all operations on vectors or matrices with real or complex components can be carried out with maximum accuracy, that is, with only a single rounding in each component by performing the operations in the fixed-point register and then rounding once to store the result back into the appropriate floating-point format. Only a few years ago, just the error-free computation of the product of two floating-point matrices was considered to be harder than the calculation of the eigenvalues of a symmetric matrix! We view the optimal dot product as a basic arithmetic operation. In addition, higher precision floating-point arithmetic can be based on the optimal dot product.

In order to adapt the computer also for automatic error control, its arithmetic must be extended with still another element. All operations on floating-point numbers (addition, subtraction, multiplication, division, and the dot product of floating-point vectors) must be supplied with directed roundings, that is, rounding to the nearest smaller and larger floating-point numbers. An interval arithmetic for real

[1] For complex division, an additional consideration is necessary.

and complex floating-point numbers, as well as for vectors and matrices with real and complex floating-point components can be built with these operations. Intervals bring the continuum into the computer and open a completely new dimension in numerical analysis. An interval is represented in the computer by a pair of floating-point numbers. It describes the continuum of real numbers bounded between these two floating-point numbers. Operations on two intervals in the computer result from operations on two appropriately chosen endpoints of the operand intervals. In this, the computation of the lower endpoint of the result interval is rounded downward, and the computation of the upper endpoint is rounded upward. The result is certain to contain all results of the operation considered applied individually to elements of the first and second operand intervals. The interval evaluation of an arithmetic expression such as a polynomial costs about twice as much as the evaluation of the expression in simple floating-point arithmetic. It provides a superset of the range of the expression or function in the given interval of arguments.

### 1.7.4 Scientific Computation with Automatic Result Verification

With these two arithmetic aids (optimal dot product and interval arithmetic), we are able to perform scientific computations with automatic result verification. We will illustrate this by two simple examples.

**Example 1.3** We first address the question of whether a given real function defined by an arithmetic expression has zeros in a given interval $[x]$. This question cannot be answered with mathematical certainty if only floating-point arithmetic is available. One can evaluate the function say 1000 times in the interval $[x]$. If all computed function values turn out to be positive, then the probability is high that the function does not have any zeros in $[x]$. However, this conclusion is certainly not reliable. Because of roundoff errors, a positive result could be computed for a

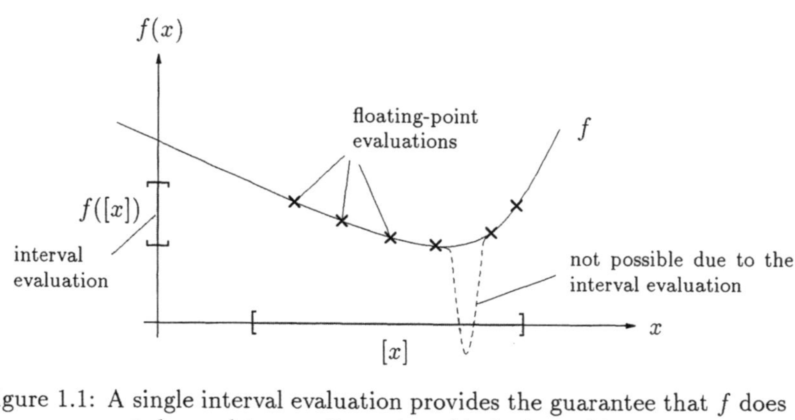


Figure 1.1: A single interval evaluation provides the guarantee that $f$ does not descend to negative values (dashed graph)

negative function value. The function could also descend to a negative value which was not detected due to the choice of the evaluation points. On the other hand, a single interval evaluation of the function may suffice to solve this problem with complete mathematical certainty. If the computed interval result does not contain zero, then the range of values also cannot contain zero, because the computed interval result is a superset of the range. As a consequence, we conclude that the function does not have any zeros in the interval $[x]$. An interval evaluation of the range of the function may fail to validate that the function has no root either because the function really *does* have a root or because of overestimation of the range. In many cases, however, a single interval function evaluation (costing the same as about two floating-point evaluations) provides a guarantee not available from hundreds, thousands, or millions of floating-point evaluations (see Figure 1.1).

**Example 1.4** As a second example, we sketch a method by which one can verify the correctness of the computed solution for a linear system of equations $A \cdot x = b$. First, an approximation for the solution is computed, for example by Gaussian elimination. This approximation is expanded by augmenting it with a small number $\pm\varepsilon$ in each component direction, giving an $n$-dimensional cube $[x]$ with the computed approximation as its midpoint. This cube is just an interval vector consisting of intervals in each coordinate direction. Now, one transforms the system of equations in an appropriate way into a fixed-point form $x = B \cdot x + c$, where one naturally makes use of the fact that one has already computed a presumed approximate solution. For example, one might get $B$ and $c$ from the residual form of $A \cdot x = b$ as discussed in Chapter 10. Then the image $[y] := B \cdot [x] + c$ is computed. If $[y]$ is contained in $[x]$, which can be verified simply by comparison of endpoints, then $[y]$ contains a fixed-point of the equation $x = B \cdot x + c$ by Brouwer's fixed-point theorem (cf. Theorem 3.1). If $[y]$ is strictly contained in $[x]$, so that the endpoints do not touch, then the fixed-point is unique. That is, the computer program has verified computationally that the original matrix $A$ of the system of equations is nonsingular. By iteration using the optimal dot product, the fixed-point can then be determined to arbitrary accuracy in practice. This example illustrates that validation ($Ax = b$ has a unique solution in $[y]$) is distinct from achieving high accuracy (iterating $[x] := B \cdot [x] + c$ until the width of the interval components of $[x]$ are small). In this way, the user obtains a mathematically certain and reliable result.

Examples 1.3 and 1.4 of numerical computation with automatic result verification described here leave open the question of what to do if the verification step fails. For rootfinding (Example 1.3), solving linear systems (Example 1.4), and many other problems, refined methods have been developed which allow valid assertions. A great advantage of automatic result verification is that the computer itself can quickly establish that the computation performed has not led to a correct and usable result. In this case, the program can choose an alternative algorithm or repeat the computation using higher precision.

Similar techniques of automatic result verification can be applied to many other algebraic problem areas, such as the solution of nonlinear systems of equations, the computation of zeros, the calculation of eigenvalues and eigenvectors of matrices, optimization problems, etc. In particular, valid and highly accurate evaluation of arithmetic expressions and functions on the computer is included. These routines also work for problems with uncertain or interval data.

The technique of enclosing function values with high accuracy is an essential aid for scientific computation. Newton's method for the computation of the roots of a system of equations is frequently unstable in a neighborhood of a root, for example when a root results from the mutual cancellation of positive and negative terms. In the immediate vicinity of a root, these terms are only approximately equal and cancellation occurs in ordinary floating-point arithmetic. However, in the case of insufficiently accurate function evaluation, Newton's method can easily overshoot its goal, possibly into the domain of attraction of another root. An iteration process which converges to a solution $x$ in the real numbers thus may not converge to the value $x$ in floating-point arithmetic. The iteration can jump over to an entirely different root. A concluding verification step is thus also advisable for iteration processes.

The techniques sketched here for numerics with automatic result verification use the optimal dot product and interval arithmetic as essential tools in addition to floating-point arithmetic. Interval arithmetic permits computation of guaranteed bounds for the solutions of a problem. One obtains high accuracy by means of the optimal dot product. The combination of both instruments is indeed a breakthrough in numerical analysis. Naive application of interval arithmetic alone always leads to reliable bounds. However, these can be so wide that they are practically useless. This effect was observed 20 years ago by many numerical analysts, leading frequently to the rejection of this useful, necessary and blameless tool. But there are applications, e.g. global optimization methods, which use interval arithmetic with only low accuracy as an essential tool. On the other hand, high accuracy alone is a waste. Validation by interval arithmetic is necessary to finally get results of *high and guaranteed* accuracy.

As already mentioned, interval arithmetic allows us to bound the range of a function over a continuum of points, including those points that are not finitely representable. Hence, interval analysis supplies the prerequisite for solving global problems like the problem of finding *all* zeros of a nonlinear function or the global optimization problem with guarantee, that the global minimum points and the global minimum values have been found.

With automatic differentiation as an essential tool, researchers have also developed methods with automatic result verification for problems of numerical analysis including numerical quadrature or cubature, integral equations, ordinary and partial differential equations. Experimentation with verified solutions of differential equations proves to be very interesting. For example, a "solution" computed by conventional approximating methods can follow completely different trajectories if only two places of accuracy are lacking at a critical location, which cannot be recognized by an ordinary approximating method. In the case of a verified method, the width of the inclusion interval explodes at such a location, indicating that the accuracy must be increased.

In recent years, reliable solutions have been computed for a number of engineering and scientific problems using problem-solving routines with automatic result verification. Frequently, problems for which ordinary solvers and solution methods fail are handled. GAMM (Gesellschaft für Angewandte Mathematik und Mechanik) and IMACS (International Association for Mathematics and Computers in Simulation) have held symposia on the theme "Computer Arithmetic and Scientific Computation with Automatic Result Verification" regularly since 1980. Proceedings of many of these meetings have appeared ([6], [42], [63], [61], [62], [59], [69], [94], [95]). Applications belong to a wide variety of scientific fields including high-speed turbines, filter calculations, plasma flows in a rectangular channel, gear-drive vibrations, high-temperature superconductivity, infiltration of pollution into ground water, inductive drives for buses, periodic solutions of the oregonator (chemical kinetics), geometric modeling, dimensioning of sewers, verified quadrature in chemical transport engineering, nonlinear pile driving, refutation of certain "chaotic" solutions in the three-body problem of celestial mechanics, development of solutions of the Schrödinger equation in wave functions, calculation of magnetohydrodynamic flows for large Hartmann numbers, optical properties of liquid crystal cells, simulation of semiconductor diodes, optimization of VLSI circuits, rotor oscillations, geometric methods for CAD systems, centered beam fracture, and robotics. A documentation of these applications and an extensive bibliography is found in [1].

### 1.7.5 Program Verification versus Numerical Verification

In conclusion, it does not follow in general that if an algorithm has been verified to be correct in the sense of computer science, then a numerical result computed with it will be correct. In this connection, recall the examples above. Here, numerics with automatic result verification is an important additional, simple tool. Naturally, with it alone, one still is not absolutely certain, because the verification step could still lead to a positive result because of a programming error. Verification of the algorithm, compiler, the operating system, and also the computer hardware can never be made superfluous by interval arithmetic. Another useful situation occurs when interval techniques yield a guarantee that the wrong answer has been computed. This demands further debugging. In case of success, numerical result verification at least implies with high probability that all these components have furnished a correct result. Additionally, this result is certainly generally independent of whether program verification has been carried out or not. On the other hand, when the computed result cannot be verified to be correct, this also has a positive value. This result is established quickly and automatically by the computer, without anything having to be done externally. The user can provide program alternatives in this case, for example, choice of different algorithms or methods, or a change to higher precision.

# Part I

# Preliminaries

Chapter 2

# The Features of C–XSC

In this chapter, we present a brief review of the C–XSC library, a C++ class library for eXtended Scientific Computation. For a complete library reference and examples, refer to [49].

C–XSC is a tool for the development of numerical algorithms delivering highly accurate and automatically verified results. It provides a large number of predefined numerical data types, operators, and functions. The types are implemented as C++ classes. Thus, C–XSC allows high-level programming of numerical applications in C++.

The C–XSC library is available for many computers with a C++ compiler translating at least the AT&T language standard 2.0. A shareware version of the library is available by anonymous ftp (see page ). It is sufficient to compile all the modules and programs of this book.

The most important features of C–XSC are:

- Real, complex, interval, and complex interval arithmetic
- Dynamic vectors and matrices
- Subarrays of vectors and matrices
- Dotprecision data types
- Predefined arithmetic operators with highest accuracy
- Standard functions of high accuracy
- Dynamic multiple-precision arithmetic and standard functions
- Rounding control for I/O data
- Error handling
- Library of problem solving routines

The following sections summarize the C–XSC features used in this book. In our program listings, C–XSC include files, data types, and functions are set in bold type to remind you that they are provided by the C–XSC library.

## 2.1 Data Types, Predefined Operators, and Functions

C–XSC provides the following numerical data types:

| | | | |
|---|---|---|---|
| *real* | *complex* | *interval* | *cinterval* |
| *rvector* | *cvector* | *ivector* | *civector* |
| *rmatrix* | *cmatrix* | *imatrix* | *cimatrix* |

Here, the prefix letters *r*, *i*, and *c* are abbreviations for *real*, *interval*, and *complex*. Hence, *cinterval* means complex interval, *cimatrix* denotes complex interval matrices, and *rvector* specifies real vectors. The basic type *real* is based on the IEEE binary floating-point format [4]. There are additional integer array types *intvector* and *intmatrix* which can be used for tag fields or Boolean arrays. For these two types, neither arithmetic nor relational operators are supplied.

Many arithmetic operators are predefined for these types in the library modules (see Section 2.6). All operators deliver results of maximum accuracy. Table 2.1 shows the predefined arithmetic operators in connection with the possible combinations of operand types. The operators of the first row of Table 2.1 are unary operators, i.e. there is no left operand. The operators | and & denote interval hull and intersection, respectively. For vector and matrix types, $*$ denotes the scalar and matrix product, respectively.

Table 2.1: Predefined arithmetic operators

| left operand \ right operand | int<br>real<br>complex | interval<br>cinterval | rvector<br>cvector | ivector<br>civector | rmatrix<br>cmatrix | imatrix<br>cimatrix |
|---|---|---|---|---|---|---|
| *unary* | − | − | − | − | − | − |
| int<br>real<br>complex | +, −, *, /<br>\| | +, −, *, /,<br>\| | * | * | * | * |
| interval<br>cinterval | +, −, *, /,<br>\| | +, −, *, /,<br>\|, & | * | * | * | * |
| rvector<br>cvector | *, / | *, / | +, −, *,<br>\| | +, −, *,<br>\| | | |
| ivector<br>civector | *, / | *, / | +, −, *,<br>\| | +, −, *,<br>\|, & | | |
| rmatrix<br>cmatrix | *, / | *, / | * | * | +, −, *,<br>\| | +, −, *,<br>\| |
| imatrix<br>cimatrix | *, / | *, / | * | * | +, −, *,<br>\| | +, −, *,<br>\|, & |

C–XSC also supplies the assignment operators $\circ$= with $\circ \in \{+, -, *, /, |, \&\}$ for any valid combination of operands in Table 2.1.

In addition to the 4 basic arithmetic operations $+$, $-$, $*$, and $/$, C–XSC provides functions for the corresponding operations with downwardly and upwardly directed rounding. These are the functions:

| | | | |
|---|---|---|---|
| *addd()* | *subd()* | *muld()* | *divd()* |
| *addu()* | *subu()* | *mulu()* | *divu()* |

The suffix characters *d* and *u* indicate the rounding direction downward and upward, respectively.

**Example 2.1** The following sample program computes $(1/3 - 2/7) \cdot 3$ with and without directed rounding operations. Here, the directed rounding operations are used to compute a lower bound of the expression. A lower bound can be obtained by evaluating $(1 \mathbin{\triangledown\!\!\!\!/} 3 \mathbin{\triangledown\!\!\!\!-} 2 \mathbin{\triangle\!\!\!\!/} 7) \mathbin{\triangledown\!\!\!\!*} 3$, where the symbols $\triangledown$ and $\triangle$ denote the downwardly and upwardly directed roundings, respectively (cf. Section 3.5).

```
#include <iostream.h>
#include <real.hpp>

int main ( )
{
  real x, y, z;

  z = (1.0 / 3.0 - 2.0 / 7.0) * 3.0;

  cout << SetPrecision(23,15) << Scientific;
  cout << "z (without directed-rounding operations):" << endl
       << Dec << z << " (Dec.)"
       << Hex << z << " (Hex.)" << endl << endl;

  x = divd(1.0,3.0);  y = divu(2.0,7.0);
  z = subd(x,y);      z = muld(z,3.0);

  cout << "z (lower bound with directed-rounding operations):" << endl
       << Dec << z << " (Dec.)"
       << Hex << z << " (Hex.)" << endl;

  return 0;
}
```

The program generates the following output. Both values of the expression are displayed in decimal and hexadecimal format, respectively:

```
z (without directed-rounding operations):
 1.428571428571428E-001 (Dec.)    +12492492492492e3FC (Hex.)

z (lower bound with directed-rounding operations):
 1.428571428571427E-001 (Dec.)    +1249249249248Ce3FC (Hex.)
```

The decimal values differ only in the last place. However, if you look at the hexadecimal representation, you can see that both values differ by 6 units in the last place of their mantissae. Note, that the last 4 hexadecimal digits represent the exponent of the number.

C–XSC provides only one implicit type casting from *double* to *real*. However, for explicit type casting, C–XSC supplies a large number of type transfer functions. The name of a type transfer function is the name of the type to be generated preceded by an underscore ('_'). For instance, the function *_interval()* can be used to transfer one or two arguments of type *real* into an interval.

```
interval x, y;

x = _interval(1.0,2.0);    // x = [1,2]
y = _interval(1.0);        // y = [1] = [1,1]
```

Type transfer functions also exist for the reversed direction. For the interval types, the functions *Inf()* and *Sup()* provide access to the infimum and supremum component of their argument. For the complex types, the functions *Re()* and *Im()* can be used to get the real and imaginary part of their argument, respectively.

Table 2.2: Predefined relational operators

| right operand / left operand | real | interval | complex | cinterval | rvector | ivector | cvector | civector | rmatrix | imatrix | cmatrix | cimatrix |
|---|---|---|---|---|---|---|---|---|---|---|---|---|
| *unary* | ! | ! | ! | ! | ! | ! | ! | ! | ! | ! | ! | ! |
| real | $V_{\text{all}}$ | $V_{\subset}$ | $V_{\text{eq}}$ | $V_{\subset}$ | | | | | | | | |
| interval | $V_{\supset}$ | $V_{\text{all}}$ | | $V_{\text{all}}$ | | | | | | | | |
| complex | $V_{\text{eq}}$ | | $V_{\text{eq}}$ | $V_{\subset}$ | | | | | | | | |
| cinterval | $V_{\supset}$ | $V_{\text{all}}$ | $V_{\supset}$ | $V_{\text{all}}$ | | | | | | | | |
| rvector | | | | | $V_{\text{all}}$ | $V_{\subset}$ | $V_{\text{eq}}$ | $V_{\subset}$ | | | | |
| ivector | | | | | $V_{\supset}$ | $V_{\text{all}}$ | | $V_{\text{all}}$ | | | | |
| cvector | | | | | $V_{\text{eq}}$ | | $V_{\text{eq}}$ | $V_{\subset}$ | | | | |
| civector | | | | | $V_{\supset}$ | $V_{\text{all}}$ | $V_{\supset}$ | $V_{\text{all}}$ | | | | |
| rmatrix | | | | | | | | | $V_{\text{all}}$ | $V_{\subset}$ | $V_{\text{eq}}$ | $V_{\subset}$ |
| imatrix | | | | | | | | | $V_{\supset}$ | $V_{\text{all}}$ | | $V_{\text{all}}$ |
| cmatrix | | | | | | | | | $V_{\text{eq}}$ | | $V_{\text{eq}}$ | $V_{\subset}$ |
| cimatrix | | | | | | | | | $V_{\supset}$ | $V_{\text{all}}$ | $V_{\supset}$ | $V_{\text{all}}$ |

Table 2.2 shows the predefined relational operators. We use the following notations for some sets of relations

$$V_{\text{all}} = \{==, !=, <=, <, >=, >\} \qquad V_{\text{eq}} = \{==, !=\}$$
$$V_{\subset} = \{==, !=, <=, <\} \qquad V_{\supset} = \{==, !=, >=, >\}.$$

For interval operands, the operators $<=$, $<$, $>=$, and $>$ denote the subset, the proper subset, the superset, and the proper superset relation, respectively.

C–XSC supplies a set of mathematical standard functions, listed in Table 2.3. These functions are available for the scalar data types *real*, *complex*, *interval*, and *cinterval* with generic names. They deliver results of high accuracy for mathematically permissible arguments. The interval standard functions for the types *interval* and *cinterval* enclose the corresponding range of values in guaranteed and sharp bounds.

Table 2.3: Predefined mathematical functions

| Function | Generic Name |
|---|---|
| Sine | sin |
| Cosine | cos |
| Tangent | tan |
| Cotangent | cot |
| Hyperbolic Sine | sinh |
| Hyperbolic Cosine | cosh |
| Hyperbolic Tangent | tanh |
| Hyperbolic Cotangent | coth |
| Square | sqr |
| Integer Power Function | power |
| Exponential Function | exp |
| Power Function | pow |
| Absolute Value | abs |

| Function | Generic Name |
|---|---|
| Arc Sine | asin |
| Arc Cosine | acos |
| Arc Tangent | atan |
| Arc Cotangent | acot |
| Inverse Hyperbolic Sine | asinh |
| Inverse Hyperbolic Cosine | acosh |
| Inverse Hyperbolic Tangent | atanh |
| Inverse Hyperbolic Cotangent | acoth |
| Square Root | sqrt |
| $n$th Root | sqrt |
| Natural Logarithm | ln |

There are some additional mathematical functions which cannot be applied to arguments of arbitrary numerical type. The *sign()* function returns the sign of its real argument. The functions *mid()* and *diam()* compute the midpoint and the diameter of an interval, respectively.

The functions *SetInf()* and *SetSup()* can be used to directly change the value of the infimum and supremum of an arbitrary interval object, respectively. Both functions return an error if you try to generate an irregular interval. An irregular interval is an interval with an upper bound less than its lower bound. However, you can force generating an irregular interval object using the functions *UncheckedSetInf()* and *UncheckedSetSup()*, respectively. The functions *SetRe()* and *SetIm()* can be used to change the value of the real and imaginary part of an object of arbitrary complex type.

## 2.2 Vector and Matrix Handling

The vector and matrix types are realized as dynamic arrays. That is, a user can allocate or deallocate storage space for vectors or matrices *at runtime*. Thus, without recompilation, the same program may use arrays of different sizes. The size is restricted only by the storage of the computer. Furthermore, the memory is used efficiently since the arrays are stored only in their required sizes. When accessing components of the array types, the index range is checked at runtime to provide increased security during programming by avoiding invalid memory accesses.

A vector or matrix can be defined by explicitly specifying its index range enclosed in round braces after the variable name. The parameters for the lower index bounds are optional. By default they are set to 1. The values of the vector or matrix components are undefined.

```
rvector  x(n);                // x[1] ... x[n]
cvector  y(-n,-1);            // y[-n] ... y[-1]
imatrix  A(n,m);              // A[1][1] ... A[n][m]
cimatrix B(0,n-1,5,m+4);      // B[0][5] ... B[n-1][m+4]
```

You can also define a vector or matrix without explicitly indicating the index bounds. In this case a vector of length 1 or a $1 \times 1$ matrix is allocated. You can change its size, as well as the size of any vector or matrix object, at runtime by applying the *Resize()* function. The parameters of the *Resize()* function correspond to the parameters in an explicit vector or matrix definition.

```
rvector  x, y;
imatrix  A, B;

Resize(x,n);              // x[1] ... x[n]
Resize(y,-n,-1);          // y[-n] ... y[-1]
Resize(A,n,m);            // A[1][1] ... A[n][m]
Resize(B,0,n-1,5,m+4);    // B[0][5] ... B[n-1][m+4]
```

If an already allocated array is resized, then the components of the intersection of the old and the new index ranges are assumed. All other components are initialized by zero.

An implicit resizing of a vector or a matrix is also possible during an assignment. If the index bounds of the object on the right-hand side of an assignment do not match those of the left-hand side, then the left-hand side assumes size and index range(s) from the right-hand side.

```
rmatrix  A(n,m), B(n,m), C;

C = A * B;     // C[1][1] ... C[n][m]
```

The size of a vector or a matrix may be determined at any time by calling the functions *Lb()* and *Ub()* for the lower and upper index bounds, respectively. For matrix types, these functions take a second parameter to access the index bounds of the rows (use constant *ROW* or 1) or of the columns (use constant *COL* or 2).

```
for (i = Lb(x); i <= Ub(x); i++) ...

for (i = Lb(A,ROW); i <= Ub(A,ROW); i++)          // Uses predefined constants
  for (j = Lb(A,COL); j <= Ub(A,COL); j++) ...    // ROW and COL

for (i = Lb(A,1); i <= Ub(A,1); i++)              // Uses int constants 1 and 2
  for (j = Lb(A,2); j <= Ub(A,2); j++) ...
```

The function *SetLb()* facilitates setting the lower index bound(s) of a vector or matrix, while preserving the length of the corresponding index range. If *SetLb()* is called to change a lower index bound to the value $l$, then the corresponding upper index bound is changed to $l + u_{\text{old}} - l_{\text{old}}$, where $l_{\text{old}}$ and $u_{\text{old}}$ are the old index bounds. This feature is useful for the implementation of functions that have to deal with arrays of arbitrary index ranges: At the beginning of the function all arrays are changed to standard index range whereas the original index ranges are restored at the end of the function.

```
OldRowLb = Lb(A,ROW);                             // Save lower index bounds
OldColLb = Lb(A,COL);                             // of the rows and columns of A

SetLb(A,ROW,1); SetLb(A,COL,1);                   // Set standard index ranges

for (i = 1 ; i <= Ub(A,ROW); i++)
  for (j = 1 ; j <= Ub(A,COL); j++) ...

SetLb(A,ROW,OldRowLb); SetLb(A,COL,OldColLb);     // Restore old index ranges
```

There are several ways to access the row or column vectors of a matrix. Let `m` an arbitrary matrix object. Since a matrix is an array of vectors, you can access the `i`th row of `m` by `m[i]`. Additionally, C–XSC supplies the equivalent notations `m[Row(i)]` and `Row(m,i)`. Correspondingly, the `i`th column of the matrix `m` can be accessed by the notations `m[Col(i)]` and `Col(m,i)`.

## 2.3 Dot Product Expressions

The implementation of algorithms with automatic result verification or validation in this book makes extensive use of the accurate evaluation of *dot product expressions*, i.e. expressions which can be reduced to a single dot product.

To evaluate this kind of expressions, C–XSC provides the data type *dotprecision* representing a fixed-point format covering the entire range of floating-point products. The dotprecision type allows scalar results — especially sums of floating-point products — to be stored exactly, that is without any rounding error. A dotprecision variable is also called a dotprecision accumulator. To round the exact result of a dotprecision accumulator to a variable of type *real*, the function *rnd()* is supplied. You can prescribe the rounding direction by specifying the optional rounding flags *RND_NEXT*, *RND_DOWN*, or *RND_UP*. The default rounding mode is *RND_NEXT*.

```
real          approx, inf, sup;
interval      encl;
dotprecision  accu;

...                            // Same as: rnd(accu,approx,RND_NEXT);
rnd(accu,approx);              //      or: approx = rnd(accu);
                               //      or: approx = rnd(accu,RND_NEXT);

...
rnd(accu,inf,RND_DOWN);        // Two statements. Use rnd(accu,inf,sup),
rnd(accu,sup,RND_UP);          // alternatively.

encl = _interval(inf,sup);     // Compose an interval
```

Besides the type *dotprecision* for real dot product expressions, C–XSC supplies the data types *cdotprecision*, *idotprecision*, and *cidotprecision* for dot product expressions based on the scalar types *complex*, *interval*, and *cinterval*.

A dotprecision variable can be initialized by assignment of a scalar value. Table 2.4 shows the predefined arithmetic operators for the dotprecision types. Additionally, the assignment operators += and −= are provided for every valid combination of operands in Table 2.4.

Table 2.4: Predefined arithmetic operators for the dotprecision types

| left operand \ right operand | real<br>complex | interval<br>cinterval | dotprecision<br>cdotprecision | idotprecision<br>cidotprecision |
|---|---|---|---|---|
| real<br>complex | | | +, −,<br>\| | +, −,<br>\| |
| interval<br>cinterval | | | +, −,<br>\| | +, −,<br>\|, & |
| dotprecision<br>cdotprecision | +, −,<br>\| | +, −,<br>\| | +, −,<br>\| | +, −,<br>\| |
| idotprecision<br>cidotprecision | +, −,<br>\| | +, −,<br>\|, & | +, −,<br>\| | +, −,<br>\|, & |

The function *accumulate()* can be used to either add the exact product of two scalar operands to a dotprecision accumulator or to add the exact scalar product of two vector operands to a dotprecision accumulator. The overloaded function is available for all reasonable combinations of operand types.

**Example 2.2** The following function computes an optimal enclosure of the defect $b - A \cdot y$ of an approximate solution $y$ of the linear system of equations $A \cdot x = b$. The defect is computed componentwise by exactly evaluating $b_i - A_i \cdot y$ with an accumulator of type *idotprecision*:

```
ivector defect ( rvector b, rmatrix A, rvector y )
{
  idotprecision  accu;
  ivector        encl(Lb(y),Ub(y));

  for (int i = Lb(y); i <= Ub(y); i++) {
    accu = b[i];
    accumulate(accu, -A[i], y);
    encl[i] = rnd(accu);
  }
  return encl;
}
```

## 2.4 Input and Output

The I/O handling of C–XSC is based on the I/O stream concept of C++ using the overloaded stream operators >> and <<. These operators are provided for all numerical data types. Input and output of real floating-point and dotprecision numbers to and from streams is controlled by *manipulators*.

There is an implicit environment which controls how real floating-point and dotprecision variables or expressions are read and written. The most important attributes of that environment and their default values are shown in Table 2.5.

Table 2.5: Attributes of the I/O environment and their default values

| Attribute | Default | Alternatives |
|---|---|---|
| Base | *Dec* | *Dec*, *Hex* |
| Rounding mode | *RndNext* | *RndNext*, *RndDown*, *RndUp* |
| Justification mode | *RightJust* | *RightJust*, *LeftJust* |
| Real format | *width* = 10 | Set with *SetPrecision()* |
| | *digits* = 6 | Set with *SetPrecision()* |
| Output format | *Variable* | *Variable*, *Scientific*, *Fixed* |

The attributes of the I/O environment are changed using manipulators. Syntactically, a manipulator appears as a value that is directed by the >> and << operators to either the input or the output stream. Semantically, a manipulator changes the value of an attribute in the internally maintained I/O environment. A manipulator does not actually perform any insertions or extractions. For example, either of the statements

```
cin  >> RndUp;
cout << RndUp;
```

affects both input *and* output. Either statement changes the I/O environment attribute "Rounding mode" so that variables subsequently read or written are rounded upward.

The current status of the I/O environment can be saved by being pushed onto an internal stack by the manipulator *SaveOpt*. A saved environment is restored by *RestoreOpt*, which effectively pops it off the internal stack.

Some characters in manipulator names are written with upper case letters to avoid ambiguity with existing C++ manipulators. Manipulators are considered as reserved words. Otherwise, a user program might not be able to access them. For example, if a user program declares a variable *RndUp*, then the user variable identifier hides the manipulator identifier. Any occurrence of the identifier *RndUp* refers to the user variable, and not to the manipulator described here. The compiler cannot check for this condition because the user may intend to use a variable of that name and not use the manipulator of the same name.

Most of the manipulators listed in Table 2.5 are self-explanatory. *SetPrecision()* can be used to change the number of digits used for output of a real floating-point number. It takes two parameters, one defining the minimum width for printing the value and a second parameter defining the number of fractional digits. The exact interpretation of both parameters depends on the selection of the actual format (*Variable*, *Scientific*, or *Fixed*). The *Scientific* format results in a scientific notation of the form $[-]d.dd\cdots dd\text{E} \pm eee$, whereas the *Fixed* format generates an exponent-free fixed-point notation of the form $[-]dd\cdots dd.dd\cdots dd$. The *Variable* format generates scientific or fixed-point notation depending on the value being printed. In our sample programs, we use the manipulators *Scientific* and *SetPrecision(23,15)* to get a scientific notation with 15 fractional digits. We use these values because the IEEE binary floating-point format [4] corresponds to about 15 decimal fractional digits of accuracy.

**Example 2.3** The following statements demonstrate how to change the I/O environment by the input and output manipulators of C–XSC.

```
real a, b;

cout << "Enter real values a, b: ";
cout << RndDown; cin  >> a;    // Read 'a' rounded downwards
cout << RndUp;   cin  >> b;    // Read 'b' rounded upwards

cout << SaveOpt;               // Push I/O parameters to stack
cout << Hex;                   // Hexadecimal output format

cout << "a = " << a << endl << "b = " << b << endl;

cout << Dec << Scientific;     // Decimal, scientific output format
cout << SetPrecision(23,15);   // Set field width and fractional digits

cout << "a = " << a << endl << "b = " << b << endl;

cout << RestoreOpt;            // Pop I/O parameters from stack
```

## 2.5 Data Conversion

This section is *critical* to the appropriate use of every routine in this book. The concept is simple: People work in a decimal notation, while computers use a binary representation for numbers. The most common error made by even the most experienced interval guru is to forget that numbers such as 0.1 in input data or as literals in the code are *not* what they seem!

Numerical computations nearly always are executed on non-decimal floating-point systems. The floating-point format usually used is a binary format, so it is inevitable that literal *real* constants must be converted into that data format. This conversion can cause an error. For example, the literal constant 1.1 is not exactly representable in a binary format (see Section 3.7 for details).

To convert literal constants into the internal floating-point format in a controlled way, all C–XSC data types can be read from or written to strings in a manner similar to that used for streams. For this purpose, C–XSC provides a set of overloaded operators `>>` and `<<` that work with strings as they do with streams. The rules for changing the I/O environment by manipulators are applied in exactly the same manner as for the stream operators (see Section 2.4).

**Example 2.4** The following statements can be used to get a correct enclosure of the literal constant 1.1.

```
real     xLow, xUp;
interval x, y;

"1.1" >> RndDown >> xLow;          // Round 1.1 downwardly
"1.1" >> RndUp   >> xUp;           // Round 1.1 upwardly
x = _interval(xLow,xUp);           // Compose an interval

"[1.1,1.1]" >> y;                  // Same result: x == y
```

Note, that `_interval(1.1,1.1)` would result in an interval with identical bounds, because both constants 1.1 are converted into the same value of the internal floating-point format before the interval is generated. That is, $1.1 \notin$ `_interval(1.1,1.1)`.

## 2.6 Predefined Modules

The data types of the C-XSC library are implemented in separate object modules. You can include these types, their operators and functions by *#include* directives for the corresponding header files. The mathematical standard functions are predefined in their own header files. Some header files include other header files. For example, a vector data type cannot be defined without knowing the scalar base type, a matrix data type cannot be defined without knowing the vector component type, and so

Figure 2.1: Hierarchy of the data types and header files of the C-XSC library

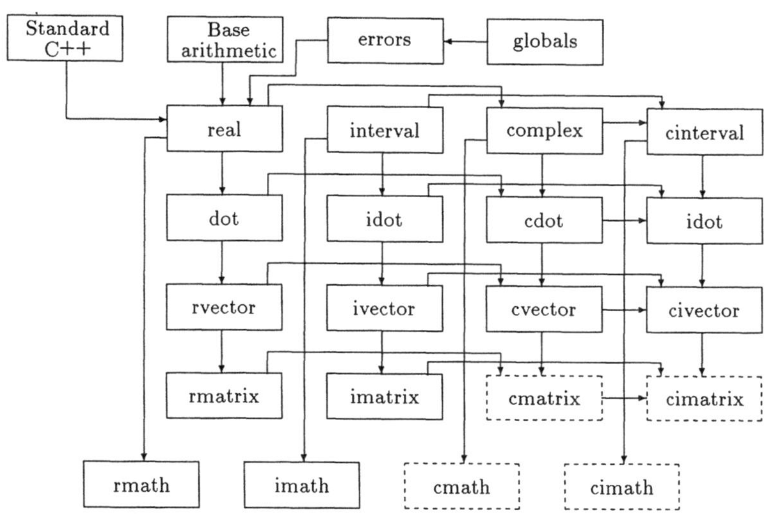

on. Therefore, in a user program, only the highest data type must be defined and included by an *#include* directive. The hierarchy of the C–XSC data types and header files needed to implement the programs of this book is shown in Figure 2.1.

In Figure 2.1 [A]→[B] means that data type $B$ requires data type $A$ or that header file $A$ is included by header file $B$. The dashed boxes are printed for clarity only. They mark data types that are not used in our programs. To include these header files with an *#include* directive in a user program, you must append the file extension *hpp*. On DOS systems each name in Figure 2.1 should be truncated after its 8th character.

## 2.7 C–XSC or Other Libraries?

C–XSC is a powerful tool for solving scientific problems with automatic verification of the results. Due to its implementation as a C++ class library, C–XSC is available for personal computers, workstations, and mainframes.

C++ provides intrinsic safety features such as type checking, type-safe linking of programs, and function prototypes. C–XSC supports additional features for safe programming such as index range checking for vectors and matrices and checking for numerical errors such as overflow, underflow, loss of accuracy, illegal arguments, etc.

All arithmetic operators provided by C–XSC deliver a result of maximum accuracy. The mathematical standard functions deliver results of high accuracy. Moreover, C–XSC provides the possibility to evaluate dot product expressions with maximum accuracy. The evaluation of such expressions is *the* fundamental tool for solving sensitive numerical problems.

However, you need not necessarily use the C–XSC library to implement the algorithms presented in this book. It should be a rather simple task to modify our programs to work with your favorite library, supposed your library provides the efficiency of C–XSC as described in the previous sections. That is, your library must provide

- a real arithmetic and real standard functions,
- an interval arithmetic and interval standard functions,
- the numerical data types introduced in Section 2.1,
- functions to handle these types (vector/matrix handling, I/O handling etc.), and
- a feature for the accurate evaluation of dot product expressions.

The real arithmetic should satisfy the IEEE standard 754 for binary floating-point arithmetic (see [4]).

As far as interval arithmetic is concerned, you can use any library that guarantees correct enclosures for every arithmetic operator or standard function it provides. That is, these interval operators and functions *must* deliver true enclosures for any valid kind of operands. Your interval library needs not to deliver results of maximum accuracy. Of course, it is a matter of fact that the better the enclosures are, the better the algorithms for numerical verification work.

One way around the problem of simulating a dotprecision accumulator is the use of a multi-precision interval arithmetic: You may evaluate a real dot product expression to maximum accuracy by gradually increasing the interval precision as long as the relative diameter of the resulting interval is greater than 1 ulp (unit in the last place, see page 46). Of course, this kind of evaluation is less elegant and, in general, more expensive at runtime than the one provided by C–XSC.

# Chapter 3

# Mathematical Preliminaries

Interval arithmetic is the basic mathematical tool in verified numerical computing. This chapter summarizes the mathematical terms used in this book. After an introduction to real, complex, and extended interval arithmetic, we make some comments on the realization of arithmetics on a digital computer. We also touch on the problem of data conversion. Finally, we point out how to use fixed-point theorems to derive algorithms for verified numerical computing.

Let us start with some remarks on our notation. If $M$ is an arbitrary set, then $M^n$ and $M^{n \times m}$ denote the sets of $n$-dimensional vectors and $n \times m$ matrices over $M$. Vectors are defined to be column vectors. The identity matrix is denoted by $I$. The maximum norm of vectors and matrices is denoted by $\|\cdot\|_\infty$. To specify the index of iterates, we use raised indices enclosed in round brackets: $x^{(k)}$.

## 3.1 Real Interval Arithmetic

Detailed treatments of interval arithmetic are given in the standard textbooks of Moore [67], Alefeld and Herzberger [2], [3], and Neumaier [70]. Here, we give only an introduction and some examples.

A *real interval*, or just an interval, is a non-empty, closed, and bounded subset of the real numbers $I\!R$

$$[x] := [\underline{x}, \overline{x}] := \{x \in I\!R \mid \underline{x} \leq x \leq \overline{x}\},$$

where $\underline{x}$ and $\overline{x}$ denote the *lower* and *upper bounds* of the interval $[x]$, respectively. A real interval covers the range of real numbers between its two bounds. Sometimes, lower and upper bounds are also called *infimum* and *supremum*, respectively. The set of real intervals is denoted by $II\!R$. An interval is called *thin* or a *point interval* if $\underline{x} = \overline{x}$. It is called *thick* if $\underline{x} < \overline{x}$. For a thin interval, we may write $x$ instead of $[x]$. That is, a thin interval is just another notation for a real number.

Since intervals are sets, the terms equality ($=$), membership ($\in$), subset ($\subseteq$), proper subset ($\subset$), superset ($\supseteq$), proper superset ($\supset$), and intersection ($\cap$) are defined in the usual sense of set theory. For instance, the proper subset relation $[x] \subset [y]$ is defined as $[x] \subseteq [y]$ and $[x] \neq [y]$. An interval $[x]$ is said to be *contained in the interior* of $[y]$ if $\underline{y} < \underline{x}$ and $\overline{x} < \overline{y}$. In this case, we write $[x] \overset{\circ}{\subset} [y]$. We also call this relation the *inner inclusion relation.* Sometimes, we need the *hull* of two intervals defined by

$$[x] \,\underline{\cup}\, [y] := [\min\{\underline{x}, \underline{y}\}, \max\{\overline{x}, \overline{y}\}].$$

**Example 3.1** Let $[x] = [1, 3]$, and $[y] = [1, \pi]$. Then the following relations are true: $\frac{4}{3} \in [x]$, $[x] \subseteq [y]$, and $[y] \supset [x]$. But $[x] \not\overset{\circ}{\subset} [y]$ since $\underline{x} = \underline{y}$. Moreover, we have $[x] \underline{\cup} [4, 5] = [1, 5]$, and $[x] \cap [y] = [x]$.

The terms *diameter*, *radius*, and *midpoint* of an interval $[x]$ are defined as

$$d([x]) := \operatorname{diam}([x]) := \overline{x} - \underline{x},$$

$$r([x]) := \operatorname{rad}([x]) := \frac{\overline{x} - \underline{x}}{2}, \text{ and}$$

$$m([x]) := \operatorname{mid}([x]) := \frac{\underline{x} + \overline{x}}{2}.$$

Thus, the membership relation $x \in [x]$ may be expressed as $|x - m([x])| \leq r([x])$ using the radius and midpoint notation. In this sense, the radius of $[x]$ is an upper bound for the absolute error of the midpoint $m([x])$ considered as an approximation of an unknown number $x \in [x]$.

The *smallest* and the *greatest absolute value* of an interval $[x]$ are denoted by

$$\begin{aligned} \langle [x] \rangle &:= \min\{|x| \mid x \in [x]\}, \text{ and} \\ |[x]| &:= \max\{|x| \mid x \in [x]\} = \max\{|\underline{x}|, |\overline{x}|\}. \end{aligned} \tag{3.1}$$

Thus, $\langle [x] \rangle = 0$ if $0 \in [x]$. We note that $\langle [x] \rangle$ and $|[x]|$ are real numbers. However, the absolute value of an interval is an interval and is denoted by $\operatorname{abs}([x])$. Hence, $\operatorname{abs}([x])$ is an elementary interval function, as discussed below. The attributes of an interval are shown graphically in Figure 3.1.

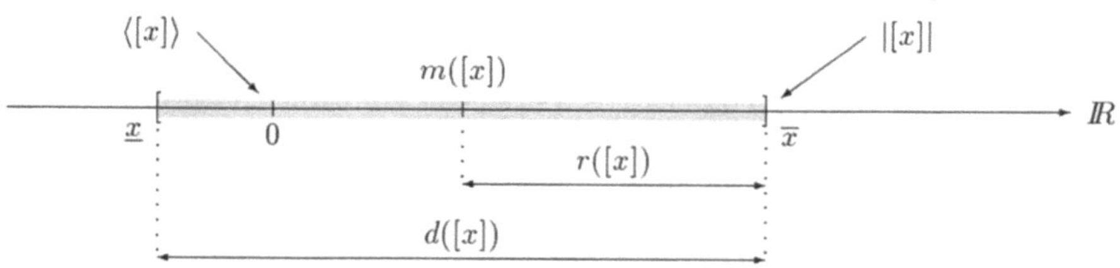


Figure 3.1: Attributes of an interval $[x]$

Suppose $[x]$ is an enclosure of a real number $x$, the exact value of which is unknown. To get a measure for the quality of the enclosure $[x]$, we define the *relative diameter* of an interval $[x]$ by

$$d_{\text{rel}}([x]) := \begin{cases} \dfrac{d([x])}{\langle [x] \rangle} & \text{if } 0 \notin [x] \\ d([x]) & \text{otherwise.} \end{cases} \tag{3.2}$$

The relative diameter is an upper bound for the relative error of $x$ with respect to an arbitrary element of $[x]$.

The *distance* $q([x],[y])$ between two intervals $[x]$ and $[y]$ is defined by

$$q([x],[y]) := \max\{|\underline{x}-\underline{y}|, |\overline{x}-\overline{y}|\}. \tag{3.3}$$

The distance is nonnegative and vanishes if and only if $[x]=[y]$. It does not depend on the order of its arguments, and the triangle inequality holds. Thus, $q$ is a metric, and the set $I\!R$ provided with the metric $q$ is a *metric space*. Moreover, $(I\!R, q)$ is a complete metric space. See Figure 3.2 for a graphical interpretation of the distance between two intervals.

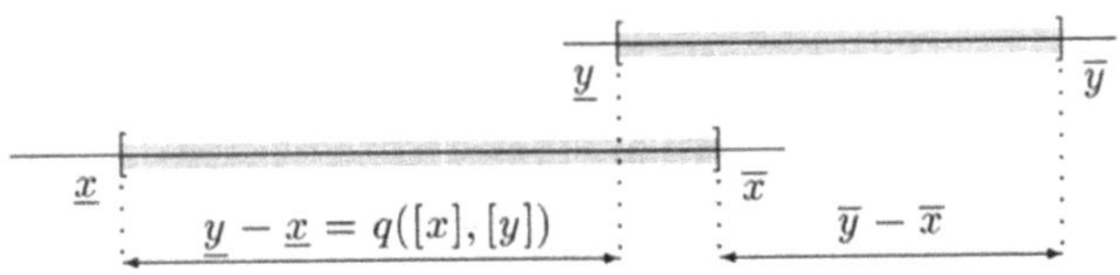


Figure 3.2: Distance between two intervals $[x]$ and $[y]$

Since $(I\!R, q)$ is a metric space, the concepts of convergence and continuity may be introduced in the usual manner. A sequence $\{[x]^{(k)}\}$ of intervals converges to an interval $[x]$ if and only if the sequences of their bounds converge to the bounds of $[x]$. More precisely, we have

$$\begin{aligned}\lim_{k\to\infty}[x]^{(k)}=[x] \quad &\Leftrightarrow \quad \lim_{k\to\infty} q([x]^{(k)},[x])=0\\ &\Leftrightarrow \quad \left(\lim_{k\to\infty}\underline{x}^{(k)}=\underline{x} \ \wedge \ \lim_{k\to\infty}\overline{x}^{(k)}=\overline{x}\right).\end{aligned}$$

The elementary real operations $\circ \in \{+,-,\cdot,/\}$ are extended to interval arguments $[x],[y]$ by defining the result of an *elementary interval operation* to be the set of real numbers which results from combining any two numbers contained in $[x]$ and in $[y]$. That is,

$$[x]\circ[y] := \{x\circ y \mid x\in[x], y\in[y]\}. \tag{3.4}$$

Of course, the definition of $[x]/[y]$ is restricted to intervals $[y]$ with $0\notin[y]$. The operator $\cdot$ is used for clarity only and will be dropped, in general. The right-hand side of (3.4) is an interval since the corresponding real operations are continuous. By using monotonicity properties, we get the more convenient formulae

$$\begin{aligned}[x]+[y] &= [\underline{x}+\underline{y}, \overline{x}+\overline{y}],\\ [x]-[y] &= [\underline{x}-\overline{y}, \overline{x}-\underline{y}],\\ [x]\cdot[y] &= [\min\{\underline{x}\underline{y}, \underline{x}\overline{y}, \overline{x}\underline{y}, \overline{x}\overline{y}\}, \max\{\underline{x}\underline{y}, \underline{x}\overline{y}, \overline{x}\underline{y}, \overline{x}\overline{y}\}], \text{ and}\\ [x]/[y] &= [x]\cdot[1/\overline{y}, 1/\underline{y}], \quad 0\notin[y].\end{aligned} \tag{3.5}$$

For each elementary operation, the bounds of the resulting interval may be expressed in terms of the bounds of its left and right operands. The rules given here for

Table 3.1: Multiplication of intervals $[x] \cdot [y]$

| $[x] \cdot [y]$ | $0 \le \underline{y}$ | $\underline{y} < 0 < \overline{y}$ | $\overline{y} \le 0$ |
|---|---|---|---|
| $0 \le \underline{x}$ | $[\underline{x}\underline{y}, \overline{x}\overline{y}]$ | $[\overline{x}\underline{y}, \overline{x}\overline{y}]$ | $[\overline{x}\underline{y}, \underline{x}\overline{y}]$ |
| $\underline{x} < 0 < \overline{x}$ | $[\underline{x}\overline{y}, \overline{x}\overline{y}]$ | $[\min\{\underline{x}\overline{y}, \overline{x}\underline{y}\}, \max\{\underline{x}\underline{y}, \overline{x}\overline{y}\}]$ | $[\overline{x}\underline{y}, \underline{x}\underline{y}]$ |
| $\overline{x} \le 0$ | $[\underline{x}\overline{y}, \overline{x}\underline{y}]$ | $[\underline{x}\overline{y}, \underline{x}\underline{y}]$ | $[\overline{x}\overline{y}, \underline{x}\underline{y}]$ |

Table 3.2: Division of intervals $[x]/[y]$ with $0 \notin [y]$

| $[x]/[y]$ | $0 < \underline{y}$ | $\overline{y} < 0$ |
|---|---|---|
| $0 \le \underline{x}$ | $[\underline{x}/\overline{y}, \overline{x}/\underline{y}]$ | $[\overline{x}/\overline{y}, \underline{x}/\underline{y}]$ |
| $\underline{x} < 0 < \overline{x}$ | $[\underline{x}/\underline{y}, \overline{x}/\underline{y}]$ | $[\overline{x}/\overline{y}, \underline{x}/\overline{y}]$ |
| $\overline{x} \le 0$ | $[\underline{x}/\underline{y}, \overline{x}/\overline{y}]$ | $[\overline{x}/\underline{y}, \underline{x}/\overline{y}]$ |

elementary interval operations assume exact arithmetic. We describe a floating-point interval arithmetic in Section 3.6. We give more compact rules for multiplication and division in Tables 3.1 and 3.2.

**Example 3.2** Those readers not yet familiar with interval arithmetic should check the following equalities by applying Equations (3.5) and Tables 3.1 and 3.2:

$$
\begin{aligned}
[-1,0]+[0,\pi] &= [-1,\pi], & -1\cdot[2,5] &= [-5,-2],\\
[1,4]-[1,4] &= [-3,3], & [-2,3][-2,3] &= [-6,9],\\
[\tfrac{1}{2},1]-[0,\tfrac{1}{6}] &= [\tfrac{1}{3},1], & [1,\sqrt{2}][-1,1] &= [-\sqrt{2},\sqrt{2}],\\
[2,4]-3 &= [-1,1], & [1,2]/[-2,-1] &= [-2,-\tfrac{1}{2}].
\end{aligned}
$$

It is very important that the elementary operations are *inclusion isotonic.* That means, if $[x]$ is contained in another interval $[x']$, and $[y]$ is contained in $[y']$, then the combination of $[x]$ and $[y]$ is contained in the interval computed by combining the bigger intervals $[x']$ and $[y']$. More precisely,

$$[x] \subseteq [x'],\ [y] \subseteq [y'] \implies [x] \circ [y] \subseteq [x'] \circ [y'], \quad \circ \in \{+,-,\cdot,/\}.$$

By Definition (3.5), the operations of addition and multiplication satisfy the commutative and associative laws

$$
\begin{aligned}
[x] \circ [y] &= [y] \circ [x], \text{ and}\\
[x] \circ ([y] \circ [z]) &= ([x] \circ [y]) \circ [z], \quad \circ \in \{+,\cdot\}.
\end{aligned}
$$

The neutral elements of addition and multiplication are the thin intervals $[0] = 0$ and $[1] = 1$, respectively. Thus, we get $-[x] = 0 - [x] = [-\overline{x}, -\underline{x}]$. In general, there exists neither an additive nor a multiplicative inverse element. On the other hand, we have

$$\begin{aligned} 0 = [0] &\subseteq [x] - [x], \text{ and} \\ 1 = [1] &\subseteq [x] \ / \ [x], \end{aligned}$$

where equality holds if and only if $[x]$ is a thin interval. It is another important property of interval arithmetic that the distributive law is not generally satisfied. However, there is a weaker form called the *subdistributive law*

$$[x] \cdot ([y] + [z]) \subseteq [x] \cdot [y] + [x] \cdot [z]. \tag{3.6}$$

Equality holds only in special cases. For instance, equality holds in (3.6) if both intervals $[y]$ and $[z]$ have the same sign (see Ratschek [76] and Spaniol [87]).

**Example 3.3** With $[x] = [1,2]$, $[y] = [2,3]$, and $[z] = [-4,-3]$, a short computation yields

$$[x] \cdot ([y] + [z]) = [-4, 0] \subset [-6, 3] = [x] \cdot [y] + [x] \cdot [z].$$

Let $\varphi : D \subset I\!R \rightarrow I\!R$ denote a real-valued *elementary function* continuous on every closed interval in its domain $D$. We extend $\varphi$ to interval arguments $[x] \in D$ by

$$\varphi([x]) := \{\varphi(x) \mid x \in [x]\}. \tag{3.7}$$

That is, $\varphi([x])$ denotes the range of the real-valued function $\varphi$ over $[x]$. Since $\varphi$ was assumed to be continuous, $\varphi([x])$ is an interval. By definition, an elementary interval function is inclusion isotonic, i.e. $[x] \subseteq [y] \Rightarrow \varphi([x]) \subseteq \varphi([y])$. We use the set of elementary functions listed in Table 2.3 supplemented by the power function $[x]^n$ defined in Example 3.5. It is possible to express the interval result of most elementary functions in terms of the argument bounds.

**Example 3.4** By using monotonicity properties of the corresponding real-valued function, where $[x]$ is restricted to the domain of $\varphi$, we obtain

$$\begin{aligned} & & \text{abs}([x]) &= [\langle [x] \rangle, |[x]|], \\ & & \varphi([x]) &= [\varphi(\underline{x}), \varphi(\overline{x})], & \varphi \in \{\text{atan}, \text{asinh}, \ln, \sinh\}, \\ & & \varphi([x]) &= [\varphi(\overline{x}), \varphi(\underline{x})], & \varphi \in \{\text{acot}, \text{acoth}\}, \\ [x]^2 &= & \text{sqr}([x]) &= [\langle [x] \rangle^2, |[x]|^2], \\ \sqrt{[x]} &= & \text{sqrt}([x]) &= [\sqrt{\underline{x}}, \sqrt{\overline{x}}], \\ e^{[x]} &= & \exp([x]) &= [e^{\underline{x}}, e^{\overline{x}}]. \end{aligned}$$

We remark that $[-1,2]^2 = [0,4] \neq [-1,2] \cdot [-1,2] = [-2,4]$. If $0 \mathring{\subset} [x]$, $[x]^2 \neq [x] \cdot [x]$. However, $[x]^2 \subset [x] \cdot [x]$ always holds. Thus, it is highly recommended to use the square function instead of multiplication to get a narrow enclosure of $x^2$ for $x \in [x]$.

**Example 3.5** The real power function $x^n$ with positive integer exponent $n \in \mathbb{N}$ is increasing for $x > 0$ and $n$ even, or for all $x$ with $n$ odd, and decreasing otherwise. Thus, we get

$$[x]^n = \begin{cases} [\underline{x}^n, \overline{x}^n] & \text{if } 0 < \underline{x} \text{ or } n \text{ odd}, \\ [0, |[x]|^n] & \text{if } 0 \in [x] \text{ and } n \text{ even}, \\ [\overline{x}^n, \underline{x}^n] & \text{if } \overline{x} < 0 \text{ and } n \text{ even}. \end{cases}$$

For negative exponents with $0 \notin [x]$, a brief computation yields $[x]^n = 1/[x]^{-n}$. For a zero exponent, we set $[x]^0 = 1$.

Perhaps *the* fundamental problem of interval arithmetic is to compute an enclosure with only a slight overestimation of the range of a function $f : D \subset \mathbb{R} \to \mathbb{R}$ defined by an arithmetic expression $f(x)$ including arithmetic operations and elementary functions (see Alefeld and Herzberger [3], Ratschek and Rokne [77], Moore [68], and Mayer [65]). It is easy to get an enclosure of $f([x])$ by simply substituting $[x]$ for $x$ in the defining expression of $f$, and then evaluating $f$ using interval arithmetic. Assuming all interval arithmetic operations are well defined, this kind of evaluation is called an *interval evaluation* or an *interval extension* of $f$ and is denoted by $f_{[]}([x])$. By definition we have

$$f([x]) \subseteq f_{[]}([x]),$$

where equality holds only in rare cases. Unfortunately, an interval extension usually overestimates the range. It is more difficult to get a *tight* enclosure.

An interval extension is inclusion isotonic since elementary interval operations and functions are inclusion isotonic, i.e. $[x] \subseteq [y] \Rightarrow f_{[]}([x]) \subseteq f_{[]}([y])$. A real-valued function may have several interval extensions since it may be defined by several equivalent arithmetic expressions. Mathematically equivalent expressions do not necessarily give rise to equivalent interval extensions, as Example 3.6 shows.

**Example 3.6** Let us consider three different, but mathematically equivalent, expressions of a real-valued function $f$ (see Figure 3.3)

$$f(x) := \underbrace{\frac{1}{2-x} + \frac{1}{2+x}}_{=:f^{(1)}(x)} = \underbrace{\frac{4}{4 - x \cdot x}}_{=:f^{(2)}(x)} = \underbrace{\frac{4}{4 - x^2}}_{=:f^{(3)}(x)} \quad \text{for } |x| < 2.$$

Since $f$ is symmetric, decreasing for $x < 0$, and increasing otherwise, the range of $f$ for $[x] \in (-2, 2)$ is given by $f([x]) = [f(\langle [x] \rangle), f(|[x]|)]$. Let $[x] = [-\frac{1}{2}, \frac{3}{2}]$. If we evaluate the interval extensions of $f^{(i)}$, $i = 1, 2, 3$, we get

$$f_{[]}^{(1)}([x]) = [\tfrac{24}{35}, \tfrac{8}{3}] \supset f_{[]}^{(2)}([x]) = [\tfrac{16}{19}, \tfrac{16}{7}] \supset f_{[]}^{(3)}([x]) = [1, \tfrac{16}{7}].$$

The true range $f([x]) = f_{[]}^{(3)}([x])$. In general, it is difficult to find the best possible interval extension. However, it is an empirical fact that the fewer occurrences of $[x]$ within an expression, the better the result of the corresponding interval evaluation. If $[x]$ appears only once in the expression, and there are no interval-valued parameters (as in $f_{[]}^{(3)}([x])$), then Moore [68] showed that the naive interval extension yields a tight enclosure.

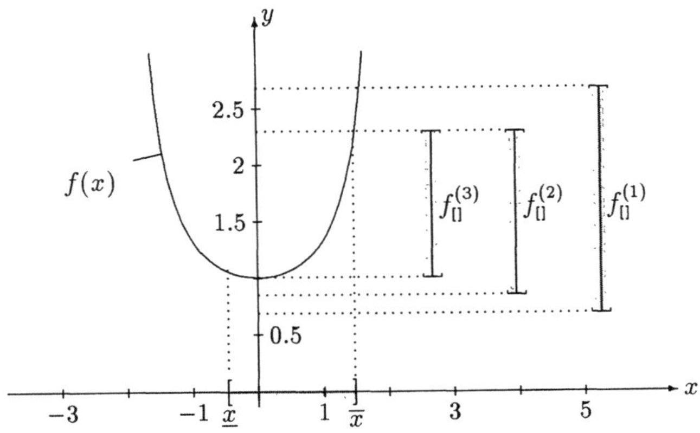


Figure 3.3: Different interval extensions of $f(x)$ as defined in Example 3.6

We have seen that, in general, the range of a function $f$ over some interval $[x]$ is overestimated by evaluating its interval extension. There are different approaches to reduce this overestimation. *Centered forms* are often used to get a narrower enclosure for sufficiently small $[x]$ (see Alefeld and Herzberger [3], Neumaier [70] or Ratschek and Rokne [77]). A centered form is derived from the mean-value theorem. Suppose $f$ is differentiable on its domain $D$. Then $f(x) = f(c) + f'(\xi)(x - c)$ with some fixed $c \in D$ and $\xi$ between $x$ and $c$. Let $c, x \in [x]$, so $\xi \in [x]$. Therefore

$$\begin{aligned} f(x) = f(c) + f'(\xi)(x-c) &\in f(c) + f'([x])(x-c) \\ &\subseteq \underbrace{f(c) + f'([x])([x]-c)}_{=:f_{[],c}([x])}, \quad c, x \in [x]. \end{aligned} \tag{3.8}$$

Here, $f_{[],c}([x])$ is called a *standard centered form* of $f$ on $[x]$ with *center* $c$. Since (3.8) does not depend on $x \in [x]$, the centered form is an enclosure for the range of $f$ on $[x]$. If $c = m(x)$, we write $f_{[],m}([x])$ and call $f_{[],m}([x])$ the *mean-value form* of $f$. A centered form is not inclusion isotonic unless the midpoint is used as center. We only use the mean-value form in this book. A centered form does not depend on the expression of $f$, but it depends on the interval evaluation of $f'$. Thus, different expressions for $f'$ lead to different values of the centered form.

**Example 3.7** Let $f(x) = x + \sin(x)$, and $[x] = [2.7, 3]$. Since $f$ is monotonically increasing, its range on $[x]$ is given by $f([x]) = [f(\underline{x}), f(\overline{x})]$. Hence, we have $f([x]) = [2.7+\sin(2.7), 3+\sin(3)] \subset [3.127, 3.142]$. Table 3.3 gives some enclosures computed for different interval arguments $[y]$ containing $[x]$ using interval evaluation of both $f$ and its mean-value form with $f'(x) = 1 + \cos(x)$. The results are rounded to 4 decimal digits.

Example 3.7 shows that for rough interval arguments, the mean-value form may yield overestimated enclosures, too. For the arguments in the first row of Table 3.3,

Table 3.3: Interval evaluation compared to the mean-value form

| $[y]$ | $f_{[]}([y])$ | $f_{[],m}([y])$ |
|---|---|---|
| $[0,\pi] \subset [0.000, 3.142]$ | $[0, \pi+1] \subset [0.000, 4.142]$ | $1 + [-\frac{\pi}{2}, \frac{3}{2}\pi] \subset [-1.571, 4.713]$ |
| $[\frac{\pi}{2}, \pi] \subset [1.570, 3.142]$ | $[\frac{\pi}{2}, \pi+1] \subset [1.570, 4.142]$ | $\frac{\sqrt{2}}{2} + [\frac{\pi}{2}, \pi] \subset \; [2.277, 3.849]$ |
| $[\frac{3}{4}\pi, \pi] \subset [2.356, 3.142]$ | $[\frac{3}{4}\pi, \pi + \frac{\sqrt{2}}{2}] \subset [2.356, 3.849]$ | $f(\frac{7}{8}\pi) + \frac{\pi}{8}(1+\cos(\frac{3}{4}\pi))[-1,1]$ $\subset \; [3.016, 3.247]$ |

standard interval evaluation delivers significantly tighter results. The effect changes as the diameter of the argument gets smaller. Thus, the mean-value form is recommended for narrow interval arguments. More precisely, the following inequalities relate the quality of an interval evaluation to the quality of a centered form:

$$q(f([x]), f_{[]}([x])) \;\leq\; \alpha\, d([x]), \text{ and} \tag{3.9}$$

$$q(f([x]), f_{[],m}([x])) \;\leq\; \beta\, d^2([x]), \tag{3.10}$$

where $\alpha$ and $\beta$ are nonnegative constants independent of $[x]$. In this sense, an interval extension approximates the range of $f$ linearly, whereas a centered form approximates it quadratically as the width of the argument tends to zero.

## 3.2 Complex Interval Arithmetic

In this section, we will introduce complex intervals, i.e. intervals in the complex plane. We only summarize those terms which are required for the understanding of the following chapters of this book. See Alefeld and Herzberger [3] for more details. Let $[x_{\text{re}}]$, $[x_{\text{im}}] \in I\mathbb{R}$. Then the set

$$[x] := [x_{\text{re}}] + i[x_{\text{im}}] := \{x = x_{\text{re}} + ix_{\text{im}} \mid x_{\text{re}} \in [x_{\text{re}}], x_{\text{im}} \in [x_{\text{im}}]\}$$

is called a *complex interval*, where $i$ denotes the imaginary unit. The set of complex intervals is denoted by $I\mathbb{C}$. A complex interval $[x]$ is said to be *thin* or a *point interval* if both its *real part* $[x_{\text{re}}]$ and its *imaginary part* $[x_{\text{im}}]$ are thin. It is called *thick* otherwise. A complex interval may also be written as an ordered pair $([x_{\text{re}}], [x_{\text{im}}])$ of real intervals. We use rectangular intervals with sides parallel to the coordinate axes, but a complex interval could also be defined as a disk in the complex plane given by its midpoint and its radius. See Alefeld and Herzberger [2], [3] for more details concerning circular interval arithmetic.

Let $[x]$, $[y] \in I\mathbb{C}$. Relations like the equality relation or the inner inclusion relation are valid if and only if they are valid for both the real and the imaginary parts of their operands, i.e.

$$[x] \circ [y] :\Leftrightarrow ([x_{\text{re}}] \circ [y_{\text{re}}] \wedge [x_{\text{im}}] \circ [y_{\text{im}}]), \quad \circ \in \{=, \overset{\circ}{\subset}, \subseteq\}. \tag{3.11}$$

However, the proper subset relation is defined by $[x] \subset [y] :\Leftrightarrow ([x] \subseteq [y] \wedge [x] \neq [y])$. Figure 3.4 illustrates some subset relations.

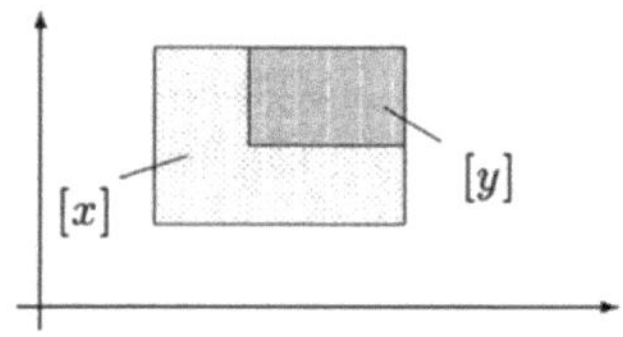


Figure 3.4: Some subset relations: $[y] \subseteq [x]$ and $[y] \subset [x]$, but $[y] \mathring{\not\subset} [x]$

The lattice operators for the intersection and the interval hull of two complex intervals may also be defined by reduction to the corresponding operators for the real and the imaginary parts, i.e.

$$[x] \circ [y] := ([x_{\text{re}}] \circ [y_{\text{re}}]) + i([x_{\text{im}}] \circ [y_{\text{im}}]), \quad \circ \in \{\cap, \underline{\cup}\}. \tag{3.12}$$

See Figure 3.5 for a graphical illustration of an intersection of complex intervals.

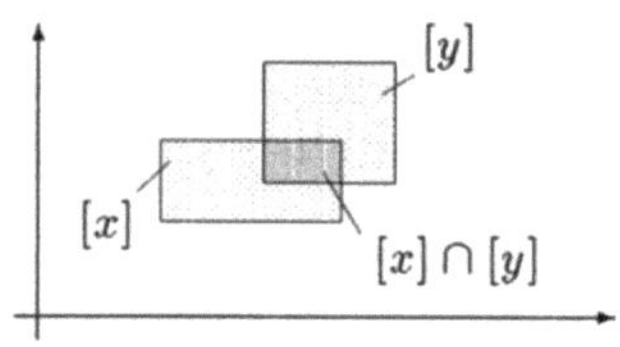


Figure 3.5: Intersection of complex intervals

As for real intervals, we may define a distance in $I\mathbb{C}$ (cf. Definition 3.3). It is easy to prove that $I\mathbb{C}$ provided with an appropriate distance is a complete metric space, so the concepts of convergence and continuity may be transferred to the space of complex intervals, too. For more details, we again refer the reader to Alefeld and Herzberger [2], [3].

Let us now define the *elementary operations* $+$, $-$, $\cdot$, and $/$ for complex intervals. Let $[x], [y] \in I\mathbb{C}$. According to the standard rules of complex arithmetic, we obtain

$$\begin{aligned}
[x] + [y] &= [x_{\text{re}}] + [y_{\text{re}}] + i([x_{\text{im}}] + [y_{\text{im}}]), \\
[x] - [y] &= [x_{\text{re}}] - [y_{\text{re}}] + i([x_{\text{im}}] - [y_{\text{im}}]), \\
[x] \cdot [y] &= [x_{\text{re}}][y_{\text{re}}] - [x_{\text{im}}][y_{\text{im}}] + i([x_{\text{re}}][y_{\text{im}}] + [x_{\text{im}}][y_{\text{re}}]), \text{ and} \\
[x] \,/\, [y] &= \frac{[x_{\text{re}}][y_{\text{re}}] + [x_{\text{im}}][y_{\text{im}}]}{[y_{\text{re}}]^2 + [y_{\text{im}}]^2} + i\frac{[x_{\text{im}}][y_{\text{re}}] - [x_{\text{re}}][y_{\text{im}}]}{[y_{\text{re}}]^2 + [y_{\text{im}}]^2}.
\end{aligned} \tag{3.13}$$

Of course, the definition of $[x]/[y]$ is restricted to intervals with $0 \notin [y_{\text{re}}]^2 + [y_{\text{im}}]^2$. We point out that $[x]/[y]$ is evaluated using the elementary interval square function to guarantee $0 \notin [y_{\text{re}}]^2 + [y_{\text{im}}]^2$ for $[y]$ with $0 \notin [y]$.

**Example 3.8** Let $[y] = [-2,1] + i[1,2]$. Then $0 = (0,0) \notin [y]$, and thus $0 \notin [y_{\text{re}}]^2 + [y_{\text{im}}]^2 = [1,8]$. Using multiplications instead of elementary square functions yields $0 \in [y_{\text{re}}][y_{\text{re}}] + [y_{\text{im}}][y_{\text{im}}] = [-1,8]$. Thus, the division would fail.

There is an important difference between the definitions of elementary operations for real and for complex intervals. The continuous image of a complex interval is not necessarily another complex interval. In Equation (3.4), the result of an elementary real interval operation was defined as the set $\{x \circ y \mid x \in [x], y \in [y]\}$, with $[x]$, $[y] \in I\mathbb{R}$. By definition, this set is a real interval. However, for $[x]$, $[y] \in I\mathbb{C}$ the set $S := \{x \circ y \mid x \in [x], y \in [y]\}$ is not necessarily a complex interval. That is, $S$ may not be a rectangle with sides parallel to the coordinate axes (see Example 3.9). To get a result in $I\mathbb{C}$, we take the smallest rectangle enclosing $S$ with sides parallel to the coordinate axes. This is established by Definition (3.13). In this way, we get easily implementable rules for complex interval arithmetic. Unfortunately, we have a loss of information by overestimating the true set $S$. This effect of overestimation is known as the *wrapping effect* (cf. Moore [68]).

**Example 3.9** Let $[x] \in I\mathbb{C}$ be a thick interval, and let $y = \cos\alpha + i\sin\alpha \in \mathbb{C}$ be a thin complex interval. Multiplication of any $x \in [x]$ with $y$ results in a rotation of $x$ by the angle $\alpha$. Thus, unless $\alpha$ is a multiple of $\frac{\pi}{2}$, the set $S = \{x \cdot y \mid x \in [x]\}$ is a rectangle with sides not parallel to the coordinate axes. The complex interval multiplication $[x] \cdot y$ wraps the set $S$ in a rectangle with sides parallel to the axes as shown in Figure 3.6.

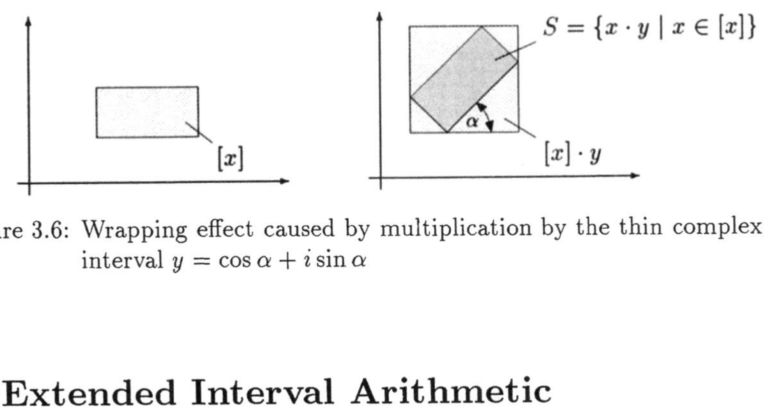


Figure 3.6: Wrapping effect caused by multiplication by the thin complex interval $y = \cos\alpha + i\sin\alpha$

## 3.3 Extended Interval Arithmetic

So far, we have introduced real and complex interval arithmetic and given easy-to-implement rules in Tables 3.1 and 3.2 and Equations (3.13). For these rules, we excluded a division by zero. We will now describe how to remove this restriction by defining a special kind of *extended interval arithmetic*. It is sufficient for our needs in this book to extend the definition of real intervals only. We extend the real number system by adjoining the two ideal points plus and minus infinity $\mathbb{R}^* := \mathbb{R} \cup \{-\infty\} \cup \{+\infty\}$. We now permit the infimum and the supremum of a real interval to be an ideal point. The set of extended real intervals is

$$I\mathbb{R}^* := I\mathbb{R} \cup \{[-\infty, r] \mid r \in \mathbb{R}\} \cup \{[l, +\infty] \mid l \in \mathbb{R}\} \cup \{[-\infty, +\infty]\}. \tag{3.14}$$

That is, $I\mathbb{R}^*$ is the set of real intervals completed by those intervals whose lower and/or upper bound tends to infinity. For instance, $[-\infty, l]$ is another notation for the set $\{x \in \mathbb{R} \mid x \leq l\}$, and $[-\infty, +\infty]$ denotes the entire real axis. We give an example to motivate the rules of extended interval arithmetic. See Ratschek and Rokne [78] for a detailed discussion of extended interval arithmetic.

**Example 3.10** Let $[x], [y] \in I\mathbb{R}$ with $[x] = [4, 5]$ and $[y] = [-1, 2]$. By Definition (3.4), the quotient $[x]/[y]$ is the set $S := \{x/y \mid x \in [x], y \in [y]\}$. Let us now split the denominator at 0 to get $[y] = [y_1] \cup [y_2] = [-1, 0] \cup [0, 2]$. $S$ is divided into two subsets $S = \{x/y \mid x \in [x], y \in [y_1]\} \cup \{x/y \mid x \in [x], y \in [y_2]\}$. Since $x \in [4, 5]$, the quotient $x/y$ tends to $-\infty$ for $y \to 0^-$, and it tends to $+\infty$ for $y \to 0^+$. Thus, the set $S$ may be represented as the union of two extended real intervals, $S = [-\infty, -4] \cup [2, +\infty]$.

Example 3.10 is typical for the common case of an extended interval division. Its result may be represented by one or two extended intervals and may be illustrated by punching out a little gap around the origin of the real axis. For finite intervals $[x], [y] \in I\mathbb{R}$ with $0 \in [y]$, the extended interval division is

$$[x]/[y] := \begin{cases} [-\infty, +\infty] & \text{if } \underline{x} < 0 < \overline{x} \text{ or } [x] = 0 \text{ or } [y] = 0 \\ [\overline{x}/\underline{y}, +\infty] & \text{if } \overline{x} \leq 0 \text{ and } \underline{y} < \overline{y} = 0 \\ [-\infty, \overline{x}/\overline{y}] \cup [\overline{x}/\underline{y}, +\infty] & \text{if } \overline{x} \leq 0 \text{ and } \underline{y} < 0 < \overline{y} \\ [-\infty, \overline{x}/\overline{y}] & \text{if } \overline{x} \leq 0 \text{ and } 0 = \underline{y} < \overline{y} \\ [-\infty, \underline{x}/\underline{y}] & \text{if } 0 \leq \underline{x} \text{ and } \underline{y} < \overline{y} = 0 \\ [-\infty, \underline{x}/\underline{y}] \cup [\underline{x}/\overline{y}, +\infty] & \text{if } 0 \leq \underline{x} \text{ and } \underline{y} < 0 < \overline{y} \\ [\underline{x}/\overline{y}, +\infty] & \text{if } 0 \leq \underline{x} \text{ and } 0 = \underline{y} < \overline{y}. \end{cases} \tag{3.15}$$

In the first case of Definition (3.15), we cannot decide whether $x/y$ with $x \in [x]$ and $y \in [y]$ tends to plus or minus infinity. Thus, we return the entire real axis $[-\infty, +\infty]$ as result. All the other cases are treated as in Example 3.10. In our algorithms, an extended arithmetic division is often followed by an intersection with some finite interval (see Chapter 6). The result after intersection is an empty set or one or two finite intervals. In this sense, extended arithmetic is a tool to generate and deal temporarily with infinite intervals.

**Example 3.11** Consider Example 3.10. We look for $([x]/[y]) \cap [-5, 4]$. The division yields $[-\infty, -4] \cup [2, +\infty]$. Intersection with $[-5, 4]$ yields two finite intervals $[-5, -4] \cup [2, 4]$. Similarly, $([x]/[y]) \cap [-2, 4] = [2, 4]$, and $([x]/[y]) \cap [-3, 1] = \emptyset$.

In addition to the *extended division* operation, we only need one more extended operation in this book. If $x \in \mathbb{R}$, i.e. $x$ may be interpreted as a thin interval, and $[y] \in I\mathbb{R}^*$ has at least one infinite endpoint, we define

$$x - [y] := \begin{cases} [-\infty, +\infty] & \text{if } [y] = [-\infty, +\infty] \\ [-\infty, x - \underline{y}] & \text{if } [y] = [\underline{y}, +\infty] \\ [x - \overline{y}, +\infty] & \text{if } [y] = [-\infty, \overline{y}]. \end{cases} \tag{3.16}$$

An extended arithmetic for complex intervals may be defined using the rules above for the real and imaginary parts.

## 3.4 Interval Vectors and Matrices

An *interval vector* is a vector whose elements are intervals. An *interval matrix* is a matrix whose elements are intervals. The sets of all $n$-dimensional real or complex interval vectors are denoted by $I\mathbb{R}^n$ or $I\mathbb{C}^n$, respectively. In the same manner, $I\mathbb{R}^{n\times m}$ and $I\mathbb{C}^{n\times m}$ denote the sets of all real and complex $n \times m$ interval matrices, respectively. We use the notations

$$[x] := ([x]_i)_{i=1,\ldots,n} := ([x]_1, \ldots, [x]_n)^T \quad \text{for } [x] \in I\mathbb{R}^n \text{ or } I\mathbb{C}^n$$

and

$$[A] := ([a]_{ij})_{\substack{i=1,\ldots,n \\ j=1,\ldots,m}} := \begin{pmatrix} [a]_{11} & \ldots & [a]_{1m} \\ \vdots & & \vdots \\ [a]_{n1} & \ldots & [a]_{nm} \end{pmatrix} \quad \text{for } [A] \in I\mathbb{R}^{n\times m} \text{ or } I\mathbb{C}^{n\times m}.$$

A real interval vector may be interpreted as the set of points in $n$-dimensional space bounded by a parallelepiped with sides parallel to the coordinate axes (see Figure 3.7). For this reason, we often speak of a *box* as a synonym for an interval vector.

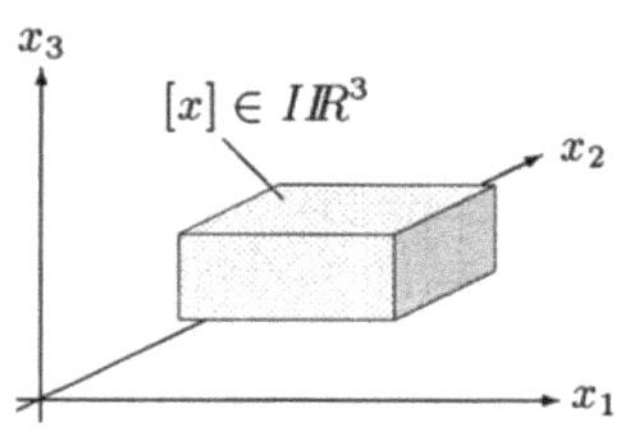


Figure 3.7: A three-dimensional real interval vector or box

The relations $=$, $\overset{\circ}{\subset}$, and $\subseteq$ are defined componentwise. For instance, the inner inclusion relation is defined by $[x] \overset{\circ}{\subset} [y] :\Leftrightarrow [x]_i \overset{\circ}{\subset} [y]_i$, $i = 1, \ldots, n$, for $[x]$, $[y] \in I\mathbb{R}^n$. On the other hand, the proper subset relation is defined by $[x] \subset [y] :\Leftrightarrow ([x] \subseteq [y] \wedge [x] \neq [y])$. The *midpoint* and the *diameter* of an interval vector or matrix are also defined componentwise. For example, $m([x]) := (m([x]_i))$, and $d([A]) := (d([a]_{ij}))$, for $[x] \in I\mathbb{R}^n$, $[A] \in I\mathbb{R}^{n\times m}$. The *maximum norm* is extended to real interval vectors and matrices by

$$||[x]||_\infty := \max_{1\leq i\leq n} |[x]_i| \quad \text{for } [x] \in I\mathbb{R}^n, \text{ and}$$

$$||[A]||_\infty := \max_{1\leq i\leq n} \sum_{j=1}^{m} |[a]_{ij}| \quad \text{for } [A] \in I\mathbb{R}^{n\times m}.$$

Finally, we introduce notations for that component of a real interval vector with maximum absolute or relative diameter

$$\begin{aligned} d_\infty([x]) &:= \max_{1\le i\le n} d([x]_i), \text{ and} \\ d_{\text{rel},\infty}([x]) &:= \max_{1\le i\le n} d_{\text{rel}}([x]_i) \quad \text{for } [x] \in I\mathbb{R}^n. \end{aligned}$$

## 3.5 Floating-Point Arithmetic

Computers support only finite sets of numbers. In general, these numbers are represented in a semilogarithmic manner as *floating-point* numbers. A floating-point or *machine* number is of the form

$$x = \pm m \cdot b^e = \pm 0.m_1 m_2 \dots m_l \cdot b^e,$$

where $m$ is a signed mantissa of fixed length $l$, $b$ is the base, and $e$ is the exponent. The digits of the mantissa are restricted to $1 \le m_1 \le b-1$, and $0 \le m_i \le b-1$, $i = 2, \dots, l$. Because $\frac{1}{b} \le m < 1$, $x$ is called a normalized floating-point number. Its exponent is bounded by $e_{\min} \le e \le e_{\max}$. Floating-point numbers are usually represented in binary format, i.e. with base $b = 2$. The set of numbers is characterized by the above conditions, with $+0.0\dots0 \cdot b^{e_{\min}}$ as the unique representation of zero, and forms a *floating-point system* $R = R(b, l, e_{\min}, e_{\max})$. The elements of smallest and largest absolute value in $R$ are $x_{\min} = 0.10\dots0 \cdot b^{e_{\min}}$ and $x_{\max} = 0.(b-1)(b-1)\dots(b-1) \cdot b^{e_{\max}}$. A floating-point system $R$ is also called a *screen* for all real numbers lying in the interval $[-x_{\max}, +x_{\max}]$. Obviously, the elements of $R$ are not uniformly distributed in the screen, but they are symmetrically arranged with respect to zero. The IEEE standards 754 and 854 provide detailed descriptions of binary and radix-independent floating-point systems [4], [5].

Since it is our aim to carry out arithmetic calculations on a digital computer, we have to approximate real numbers by floating-point numbers. This is done by a special mapping, called a *rounding* $\bigcirc : \mathbb{R} \to R$, defined by the two conditions

$$\bigcirc x = x \qquad \text{for all } x \in R, \text{ and} \tag{3.17}$$

$$x \le y \quad \Rightarrow \quad \bigcirc x \le \bigcirc y \qquad \text{for all } x, y \in \mathbb{R}. \tag{3.18}$$

The first condition guarantees that elements of the screen are not changed by a rounding. The second condition means that a rounding is *monotone*, i.e. the order of elements is maintained if they are rounded. We distinguish the roundings

- $\square$: Rounding *to the nearest* element of the screen
- $\nabla$: Rounding *toward* $-\infty$ or *downwardly* directed
- $\Delta$: Rounding *toward* $+\infty$ or *upwardly* directed

Figure 3.8 illustrates how the different roundings map an element of $\mathbb{R}$ to an element of the screen $R$.

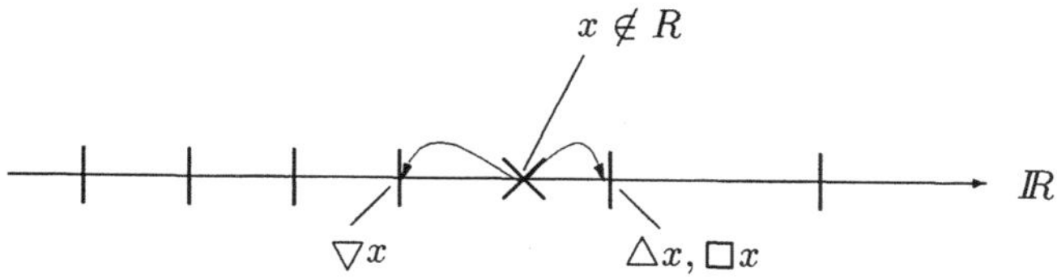


Figure 3.8: The different roundings: To the nearest number (□), downwardly (∇), and upwardly (Δ) directed. The elements of the screen $R$ are marked by vertical lines.

A rounding is said to be *antisymmetric* if it has the property

$$\bigcirc(-x) = -\bigcirc x \qquad \text{for all } x \in \mathbb{R}. \tag{3.19}$$

Thus, the rounding to the nearest number is antisymmetric, but the directed roundings $\nabla$ and $\Delta$ are not antisymmetric. Instead, we have $\nabla(-x) = -\Delta x$, and $\Delta(-x) = -\nabla x$. The directed roundings satisfy the additional conditions $\nabla x \leq x$, and $x \leq \Delta x$ for all $x \in \mathbb{R}$.

One of the essential assumptions in verified numerical computing is that the elementary floating-point operations $\circledcirc$ with $\circ \in \{+,-,\cdot,/\}$ and $\bigcirc \in \{\nabla, \Delta, \square\}$ satisfy the condition

$$x \circledcirc y = \bigcirc(x \circ y) \qquad \text{for all } x, y \in R. \tag{3.20}$$

Formally, Equation (3.20) means that the result of a floating-point operation is defined to be the rounded result of the exactly computed real operation. In general, the exact result is not representable in the number screen of the computer. Thus, for an actual implementation, an auxiliary result $x \,\tilde{\circ}\, y$ is computed which satisfies $x \circledcirc y = \bigcirc(x \circ y) = \bigcirc(x \,\tilde{\circ}\, y)$. That is, we compute an approximate result that is rounded to the same floating-point number as the exact result.

A mapping satisfying Properties (3.17)–(3.20) is called a *semimorphism.* The principle of semimorphism may also be applied to define arithmetic operations for complex floating-point numbers as well as for real and complex floating-point vectors and matrices. For this purpose, the set of arithmetic operations $+$, $-$, $\cdot$, and $/$ is augmented by a fifth operation, the exact dot product, defined according to (3.20) by

$$x \odot y = \bigcirc(x \cdot y) \qquad \text{for all } x, y \in R^n. \tag{3.21}$$

A *complex floating-point number* is a complex number whose real and imaginary parts are elements of a floating-point system $R$. We denote the set of complex floating-point numbers by

$$C = \{x \in \mathbb{C} \mid x_{\mathrm{re}}, x_{\mathrm{im}} \in R\}.$$

Any operation defined according to the principle of semimorphism delivers a result of maximum accuracy. That is, there is no element of the screen lying between

$x \circledcirc y$ and $x \circ y$. In case of operations for complex or higher dimensional operands, this is to be understood componentwise. For a detailed discussion of a semimorphic definition of computer arithmetics and their implementation on digital computers, refer to Kulisch [58] and Kulisch and Miranker [60]. We stress that all arithmetic operations provided by C–XSC are implemented semimorphically. Moreover, C–XSC comes with a more sophisticated tool, the *accurate expressions* (see Chapter 2.3), which may be used to compute certain algebraic expressions with maximum accuracy. For instance, accurate expressions are used to realize a semimorphic implementation of the operators for vectors and matrices. In our algorithms, we use the notation $\square(\ldots)$ to indicate that an expression is evaluated with maximum accuracy.

According to Property (3.20), the floating-point implementation $\varphi_\square$ of an elementary real function $\varphi : D \subset I\!R \to I\!R$ should satisfy the condition

$$\varphi_\square(x) = \square(\varphi(x)) \qquad \text{for all } x \in D \cap R.$$

Thus, $\varphi_\square$ delivers a result of maximum accuracy, too. An equivalent condition should hold for the floating-point implementations of complex elementary functions. Again, we emphasize that the elementary functions provided by C–XSC (see Table 2.3) satisfy this condition.

To point out that an arithmetic expression $f$ composed of elementary operators and functions is to be evaluated using floating-point arithmetic, we provide it with a subscript $\square$-symbol.

**Example 3.12** Let $f(x) = \sin(x) + \sqrt{x} - x$ be a real-valued function. By specifying $f_\square(x) = \sin(x) + \sqrt{x} - x$, we indicate that the argument of $f_\square$ is a floating-point number, and that the operations on the right-hand side of the equality sign are floating-point operations. For clarity, we even might write $f_\square(x) = \sin_\square(x) \boxplus \mathrm{sqrt}_\square(\mathrm{x}) \boxminus \mathrm{x}$.

## 3.6 Floating-Point Interval Arithmetic

Interval arithmetic as introduced in the earlier sections requires exact arithmetic to compute the endpoints of the resulting intervals. Now, we must build it on an actual machine. A *floating-point* or *machine interval* is a real interval whose endpoints are floating-point numbers. Let $R$ be a floating-point system. Then the set of floating-point intervals over $R$ is denoted by

$$IR = \{[x] \in I\!I\!R \mid \underline{x}, \overline{x} \in R\}.$$

A machine interval $[x] \in IR$ denotes the continuum of numbers lying between its bounds. It is a very important fact that, though $\underline{x}$ and $\overline{x}$ are elements of the basic number screen $R$, $[x]$ contains not only every *floating-point* number between $\underline{x}$ and $\overline{x}$, but also every *real* number within that range (see Figure 3.9). That is, if we have proved a certain property to hold for a floating-point interval, this property holds for any real number lying in that interval.

To compute with a computer representation of intervals, we introduce a rounding $\diamond : I\!I\!R \to IR$ which maps an interval to a machine interval. This *interval rounding*

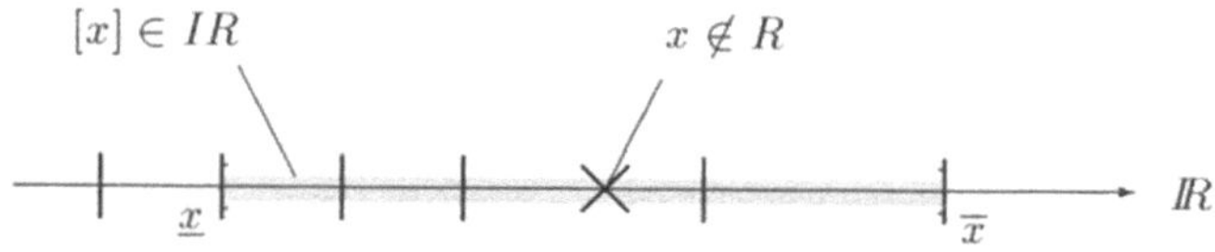


Figure 3.9: A floating-point interval $[x] \in IR$ contains not only elements of $R$ (vertical lines), but also any real number $x$ with $\underline{x} \leq x \leq \overline{x}$.

is assumed to satisfy the Conditions (3.17)–(3.19), where the real spaces $I\!R$ and $R$ and the relation $\leq$ are replaced by $II\!R$, $IR$, and the $\subseteq$ relation, respectively. Additionally, we assume

$$[x] \subseteq \Diamond([x]) \qquad \text{for all } [x] \in II\!R. \tag{3.22}$$

This assumption is quite natural since the rounded image of an interval should always contain its original.

According to Property (3.20), an elementary floating-point interval operation is defined by

$$[x] \mathbin{\Diamond\!\!\!\circ} [y] = \Diamond([x] \circ [y]) \qquad \text{for all } [x], [y] \in IR. \tag{3.23}$$

The principle of semimorphism implies that any operation defined by (3.23) delivers a result of maximum accuracy in the sense that the resulting interval is the smallest machine interval which contains $[x] \circ [y]$. For an actual implementation of (3.23), we may use the rules given in (3.5), using directed-rounding operations to get proper lower and upper bounds.

**Example 3.13** Let $[x], [y] \in IR$. Applying (3.5) for the difference of machine intervals with directed-rounding operations yields $[x] \mathbin{\Diamond\!\!\!-} [y] = \Diamond([x] - [y]) = [\underline{x} \bigtriangledown \overline{y}, \overline{x} \bigtriangleup \underline{y}]$.

A *complex floating-point interval* is an interval whose real and imaginary parts are floating-point intervals. We denote the set of complex floating-point intervals by

$$IC = \{[x] \in I\mathbb{C} \mid [x]_{\text{re}}, [x]_{\text{im}} \in IR\}.$$

See [58] and [60] for a detailed discussion of how to implement a semimorphic floating-point arithmetic for complex intervals and for real and complex interval vectors and matrices. All real and complex interval operations provided by C–XSC are implemented semimorphically. The notation $\Diamond(\ldots)$ used in our algorithms indicates that the specified interval expression is evaluated with maximum accuracy.

According to Definition (3.23), the floating-point implementation $\varphi_\Diamond$ of an elementary interval function $\varphi : D \subset II\!R \to II\!R$ should satisfy the condition

$$\varphi_\Diamond([x]) = \Diamond(\varphi([x])) \qquad \text{for all } [x] \in D \cap IR.$$

Thus, $\varphi_\Diamond$ also delivers a result of maximum accuracy. An equivalent condition should hold for the floating-point implementations of elementary complex interval functions. All elementary interval functions provided by C–XSC (see Table 2.3) satisfy this condition.

As already mentioned, we use $f_{\square}$ to indicate that an expression $f$ is to be evaluated using floating-point arithmetic. We introduce an equivalent notation with a subscript $\diamond$-symbol to indicate that the expression is to be evaluated using interval floating-point arithmetic.

**Example 3.14** The notation $f_{\diamond}([x]) = (e^{[x]} - 1) \cdot [x]$ indicates that the argument of $f_{\diamond}$ is a floating-point interval and that floating-point interval operations are to be used for evaluation. For clarity, we might write $f_{\diamond}([x]) = (\exp_{\diamond}([x]) \,\diamond\!\!\!-\, 1) \,\diamond\!\!\!\cdot\, [x]$.

In Section 3.8, we will see that sometimes it helps to accelerate certain verification steps by slightly enlarging a given interval. Let $\varepsilon \in R$ be some positive machine number. The *epsilon inflation* or *ε-inflation* of a real floating-point interval $[x] \in IR$ is defined by

$$[x] \bowtie \varepsilon := \begin{cases} [x] + [-\varepsilon, +\varepsilon] \cdot d([x]) & \text{if } d([x]) \neq 0 \\ [x] + [-x_{\min}, +x_{\min}] & \text{otherwise,} \end{cases} \tag{3.24}$$

where, $x_{\min}$ denotes the smallest positive element of the floating-point system $R$ as defined on page 42. The $\varepsilon$-inflation is defined componentwise for complex intervals and interval vectors and matrices.

We conclude our discussion of machine interval arithmetics with the definition of a measure for the accuracy of a floating-point interval. A real floating-point interval with non-vanishing diameter is said to be accurate to $n$ *ulp* if its interior contains $n-1$ or fewer elements of the basic number screen $R$. Thus, an accuracy of one ulp is equivalent to maximum accuracy since the infimum and the supremum of the interval are successive elements of the number screen. An accuracy of $n$ ulp indicates that the infimum and the supremum differ by at most $n$ *units* in the *last place* of their mantissae. The ulp-accuracy of complex intervals and higher dimensional interval types is defined to be the maximum of the ulp-accuracies of their components.

**Example 3.15** The shaded interval $[x]$ marked in Figure 3.9 has an accuracy of 4 ulp.

## 3.7 The Problem of Data Conversion

It is *critical* that you understand this section in order to use any of the toolbox routines in this book to achieve the validation they promise. The problem of data conversion is a common problem in numerical computing. It is based on the fact that numerical computations are executed in almost every case on non-decimal floating-point systems. A programmer and, most importantly, a user of numerical software should always be aware of the facts discussed below, yet even the most experienced interval expert gets trapped from time to time by forgetting data conversion questions. Be warned! We emphasize that the problem of data conversion is less a problem of verified numerical computing than of all digital computing (see Auzinger and Stetter [7]).

The floating-point format used internally by almost all modern digital computers is a binary format. For instance, the runtime system of C–XSC is based on the IEEE standard format for binary floating-point numbers [4] with a mantissa of 53 binary digits and an exponent range from $-1021$ to $1024$, $R = R(2, 53, -1021, 1024)$.

Since people are accustomed to thinking in decimal notation, we must convert the decimal numbers of our thoughts to binary numbers the computer can use. Any real constant or input data specified in a user's program has to be converted at runtime. This is done by a rounding $\bigcirc \in \{\square, \triangledown, \triangle\}$. In general, a decimal number has no binary representation of finite length. Thus, the conversion of constants and input data is generally afflicted with a small *conversion error*. It is a common misconception, as the following example demonstrates, that decimal numbers with a short mantissa may be converted without conversion error.

**Example 3.16** The decimal number $x = 0.1$ has no exact representation in any binary floating-point system with a fixed-length mantissa. Since

$$\sum_{k=1}^{\infty} 2^{-4k} + 2^{-(4k+1)} = \frac{3}{2}\sum_{k=1}^{\infty}(2^{-4})^k = \frac{3}{2}\left(\frac{1}{1-2^{-4}} - 1\right) = \frac{1}{10} = x,$$

the unique, but infinite, binary representation of $x$ is $0.000\overline{1100}_2$. Thus, conversion of the mantissa to any finite length results in a conversion error.

It is another common fallacy that integers may always be converted exactly to a binary floating-point format. Large integers also may be afflicted by a conversion error, because the mantissa of the target format is of finite length.

**Example 3.17** $R = R(2, 53, -1021, 1024)$ is the binary floating-point system used by C–XSC. Let

$$x = 9007199254740993 = 2^{53} + 2^0 = \underbrace{100000\ldots000001}_{54 \text{ digits}}{}_2.$$

Error-free conversion of $x$ to the internal binary format requires at least 54 binary digits in the mantissa. Thus, the real value $x$ cannot be mapped exactly to an element of $R$.

If $R(b, l, e_{\min}, e_{\max})$ is the target floating-point system, then *any* integer within the range of $\pm b^l$ is converted without error. There are other integers outside that range which are also exactly representable in $R$, but most integers between $b^l$ and $x_{\max}$ (or $-x_{\max}$ and $-b^l$) are subject to conversion errors. For the floating-point system used by C–XSC, we have $b^l = 9007199254740992$ and $x_{\max} \approx 1.8 \cdot 10^{308}$.

How can we check at runtime if a real input parameter was converted exactly? At first glance, it appears one could enter the real value, write it to a file, and compare these two values. However, this is not the correct way to solve our problem since the method described above results in two successive conversions — one from decimal to binary format and the other from binary to decimal format. Unfortunately, quite often, these two conversions cancel each other out.

To solve the problem correctly, C–XSC provides the facility of converting a real number, given by a string, to the internal floating-point format with respect to the actual rounding mode. So, you can compose an interval that encloses the exact value by rounding its string representation to both directions upwards and downwards (see also Sections 2.4 and 2.5). If the diameter of that interval is zero, the parameter was exactly representable. The following sample program may be used to check if a real number was converted without error.

```
#include <interval.hpp>

int main ( )
{
  static char xString[100];
  real xDown, xUp;

  cout << "Enter a real value: "; cin >> xString;

  // Round 'xString' upwardly into the real variable 'xUp' and
  // round it downwardly into the real variable 'xDown'.
  //------------------------------------------------------------
  xString >> RndUp   >> xUp;
  xString >> RndDown >> xDown;

  if (xDown == xUp)
    cout << "--> " << xString << " is exactly representable" << endl;
  else
    cout << "--> " << xString << " is not exactly representable" << endl;
  return 0;
}
```

This test program produces the results below:

```
Enter a real value: 0.1
--> 0.1 is not exactly representable

Enter a real value: 36452346
--> 36452346 is exactly representable

Enter a real value: 0.50390625
--> 0.50390625 is exactly representable

Enter a real value: 1e50
--> 1e50 is not exactly representable
```

As you use the modules and programs of the following chapters, you should always remember the problem of input conversion. If some of the input data is afflicted with conversion errors, you should be aware that the results of a verification procedure are proved for the converted input data only, but not for the original decimal problem. To avoid confusion about the results, you may wish to check whether the input parameters are exactly representable as demonstrated in the above program. However, this is no restriction in the common case, where the input parameters of a verification procedure result from previous computations.

Some of the problem-solving routines of the following chapters accept interval input parameters. They may be used to solve point problems by entering thin interval parameters. In this context, we want to point out some pitfalls in specifying interval expressions in C–XSC. If you assign real data to an interval variable or if you use explicit type conversion from real to interval type, then the resulting interval is a thin interval.

**Example 3.18** Let $r \in R$ and $[x], [y], [z] \in IR$ be variables

```
real      r;
interval  x, y, z;
```

Then the following statements

```
r = 0.1;
x = _interval(r);
y = r;
z = 0.1;
```

result in identical thin (!) intervals for $[x]$, $[y]$, and $[z]$, none of which contains the real number 0.1! The literal constant 0.1 is rounded to the internally used binary format before being used for an assignment. Thus, actually, we get $[x] = [y] = [z] = [\square(0.1)] = [\square(0.1), \square(0.1)]$.

It is not our intention to cause confusion, but we have to remark that any of the intervals of the previous example, printed by using the standard output operator `<<` and 16 significant decimal digits, results in

```
[ 1.000000000000000E-001, 1.000000000000001E-001]
```

Hence, it *appears* that 0.1 is contained in $[x]$, for instance. But remember, to display $[x]$ which actually is a thin interval stored in binary format, its lower and upper bounds are converted to a decimal format using the downwardly and upwardly directed roundings, respectively. Thus, the conversion error from decimal to binary format was canceled out by the subsequent rounding from binary to decimal format. This demonstrates that the effects of conversion are small, but they must not be neglected in a program claiming to produce validated results.

To get a proper interval enclosure for real literal constants, we again can use the C–XSC feature of rounding controlled string-to-real conversion. For instance, the following statements

```
static char *ConstString = "0.1";

real      ConstDown, ConstUp;
interval x;

ConstString >> RndUp   >> ConstUp;
ConstString >> RndDown >> ConstDown;
x = _interval(ConstDown,ConstUp);
```

result in an interval $[x] = [\bigtriangledown(0.1), \bigtriangleup(0.1)]$. That is, the binary representation of $[x]$ is a proper enclosure of the decimal literal constant 0.1. In practice, the most natural notation to achieve an enclosure of 0.1 is probably

```
x = 1.0 / _interval(10.0);
```

This notation emphasizes the role of type recognition for overloaded operators.

Our final remark on the problem of conversion concerns the evaluation of arithmetic expressions. Since C++ supports operator and function overloading, one should always bear in mind that the type of an operator or function is defined by the type of its operands. Thus, we must be particularly careful when specifying expressions which include both real and interval operands if we want to compute enclosures for these expressions.

**Example 3.19** Let $[x]$, $[y]$, and $[z]$ be defined as in Example 3.18. The C–XSC statements

```
y = x + exp(1.0/3.0);
z = x + exp(_interval(1.0)/3.0);
```

are not equivalent. Since the operands of the division operation in the first expression are both of type *real*, it corresponds to $[x] = [y] \diamond\!\!\!+ [\exp_\square(1 \boxslash 3)]$. In particular, the second summand is not an enclosure of exp(1/3), but a thin interval. On the other hand, the second statement corresponds to $[z] = [y] \diamond\!\!\!+ \exp_\diamond(\diamond(1) \diamond\!\!\!/ 3)$. Here, the second summand is a proper enclosure of exp(1/3).

The example demonstrates that it is recommended to convert any integer or real operands to intervals before they are used within an expression of type interval.

## 3.8 Principles of Numerical Verification

The automatic verification of numerical results is based on two major prerequisites:

1. the theory of interval arithmetic,
2. appropriate algorithms.

The easiest technique for computing verified numerical results is to replace any real or complex operation by its interval equivalent and then to perform the computations using interval arithmetic. This procedure leads to reliable, verified results. However, the diameter of the computed enclosures may be so wide as to be practically useless. We need more sophisticated methods which combine the benefits of interval arithmetic with a mechanism for refining already computed, but rough, enclosures.

In developing algorithms for validated enclosures, we must be careful what we compute an enclosure *of*. For example, if we take a program for the approximate solution of initial value problems in ordinary differential equations by a Runge-Kutta method and replace all floating-point computations by computations with floating-point intervals, we get an enclosure, as suggested in the preceding paragraph. However, we get an enclosure not of the solution to the differential equation, but of the Runge-Kutta approximation. We have enclosed the roundoff errors, but not the truncation errors. An algorithm must enclose all sources of error that come into the computation, including conversion errors discussed in Section 3.7, in order to achieve validated enclosures of the desired answer.

A simple mechanism for the verified solution of point problems, i.e. problems with non-interval input data, is the principle of *iterative refinement.* After computing a first approximation, its error is enclosed using machine interval arithmetic. If the diameter of the error interval is less than a desired accuracy, then a verified enclosure of the solution is given by the sum of the approximation and the enclosure of its error. Otherwise, the approximation may be refined by adding the midpoint of the error interval and repeating the process. See also Stetter [88]. We use iterative refinement methods in this book for the accurate evaluation of polynomials (Chapter 4) and arithmetic expressions (Chapter 8).

Many algorithms for numerical verification are based on the application of well-known fixed-point theorems to interval sets. As an example, we cite the following theorem (cf. [70]).

**Theorem 3.1 (Brouwer's Fixed-Point Theorem)** *Let $f : \mathbb{R}^n \to \mathbb{R}^n$ be a continuous mapping and $X \subseteq \mathbb{R}^n$ a non-empty, closed, convex, and bounded set. If $f(X) \subseteq X$, then $f$ has at least one fixed-point $x^*$ in $X$.*

Let $X = [x] \in I\mathbb{R}^n$ be a machine interval vector. As a box in $n$-dimensional space, $[x]$ satisfies the conditions of Brouwer's Fixed-Point Theorem. Suppose we can find a box with $f([x]) \subseteq [x]$. Then $[x]$ is proved to be an enclosure of at least one fixed-point $x^*$ of $f$. The assertion remains valid if we replace $f$ by its floating-point interval evaluation $f_\diamond$ because $f_\diamond([x]) \subseteq [x]$ implies $f([x]) \subseteq [x]$ since $f_\diamond([x])$ is a superset of $f([x])$.

Viewing Brouwer's Fixed-Point Theorem in an interval context motivates a template for the design of algorithms that compute a verified solution of numerical problems. First, find a fixed-point form $x = f(x)$ equivalent to the original problem. As an example, refer to Newton's method for problems of finding a zero. Replace the generating function of the right-hand side by its floating-point interval extension $f_\diamond$. Start the following iteration scheme with some approximate solution $[x]^{(0)}$

$$[x]^{(k+1)} = f_\diamond([x]^{(k)}) \qquad \text{for } k = 0, 1, 2, \ldots \tag{3.25}$$

Stop the iteration if $[x]^{(k+1)} \subseteq [x]^{(k)}$ for some $k \geq 0$. If the iteration succeeded, then we have proven in the mathematical sense that the original problem has at least one solution $x^*$ contained in $[x]^{(k)}$.

We distinguish *a priori* and *a posteriori* methods for the starting approximation of (3.25). For an *a priori* method, the starting approximation already includes the fixed-point desired. For this case, the iteration scheme (3.25) may be modified by intersecting successive iterates, that is,

$$[x]^{(k+1)} = f_\diamond([x]^{(k)}) \cap [x]^{(k)} \qquad \text{for } k = 0, 1, 2, \ldots$$

The iteration is halted if there are two successive iterates with the same value or if a maximum number of iterations is exceeded. Figure 3.10 gives an illustration of how an *a priori* interval iteration works. Examples for *a priori* methods are found in Chapters 6 and 7.

The starting approximation for an *a posteriori* method does not necessarily have to contain a desired fixed-point. Here, the hope is that successive iterates come

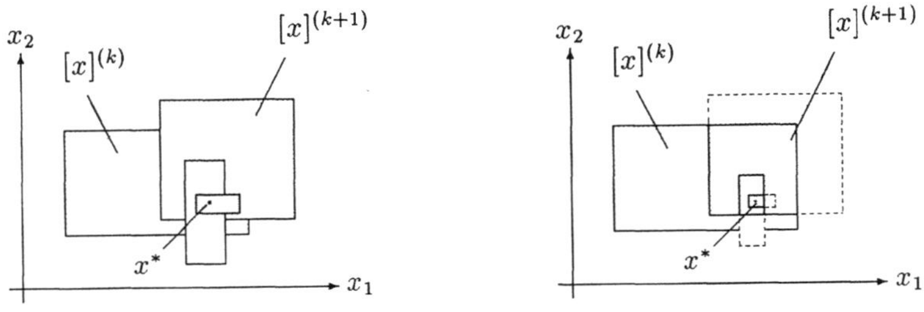


Figure 3.10: *A priori* method without (left picture) and with intersection

closer and closer to a fixed-point and finally will enclose it. Of course, the better the starting approximation is, the faster the iteration will converge. However, in practice the iterates come closer and closer to the fixed-point, but they rarely catch it. A simple trick saves the method. Before starting a new iteration step, the actual iterate is slightly enlarged by means of an $\varepsilon$-inflation, as defined in Section 3.6. Thus, for an *a posteriori* method, the iteration scheme (3.25) is modified to

$$\left.\begin{array}{rcl} [x]^{(k)} & = & [x]^{(k)} \bowtie \varepsilon \\ [x]^{(k+1)} & = & f_\diamond([x]^{(k)}) \end{array}\right\} \quad \text{for } k = 0, 1, 2, \ldots$$

Figure 3.11 illustrates the effects of $\varepsilon$-inflation on an *a posteriori* method. Examples of *a posteriori* methods appear in Chapters 10 and 11.

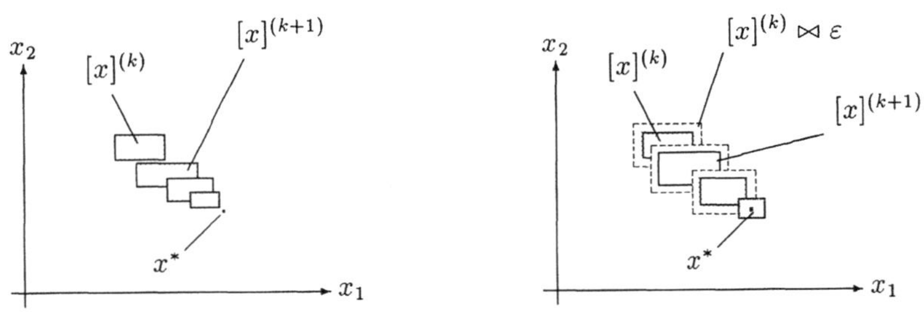


Figure 3.11: *A posteriori* method without (left picture) and with $\varepsilon$-inflation

Fixed-point methods may even be modified for some problems to prove the uniqueness of a fixed-point. For example, in Chapter 10 the uniqueness of the solution of a system of linear equations is proved by simply changing the stopping criterion of the fixed-point iteration to $[x]^{(k+1)} \overset{\circ}{\subset} [x]^{(k)}$.

So far, we have introduced the mathematical basics for verified numerical computing. We have defined the basic terms of real, complex, and extended interval arithmetics and have discussed aspects of an implementation on a computer. We have also discussed the problem of data conversion. Finally, we have introduced the principles of numerical verification by iterative refinement, *a priori*, and *a posteriori* methods. All these terms are used in the following chapters to describe algorithms that compute verified solutions of problems from numerical analysis.

# Part II

# One-Dimensional Problems

# Chapter 4

# Evaluation of Polynomials

In this chapter, we consider the evaluation of a polynomial function of a single variable. We usually compute the value of an arithmetic function by replacing each arithmetic operation by its corresponding floating-point machine operation (see Section 3.5). Roundoff errors and cancellations sometimes cause the calculated result to be drastically wrong. For similar reasons, a naive interval evaluation of a polynomial may lead to intervals so large as to be practically useless. Roundoff and cancellation errors are especially dangerous if we are evaluating a function close to a root, as we will see in Chapter 9 when we compute verified enclosures of zeros of polynomials.

We present an algorithm to evaluate a real polynomial $p : I\!R \rightarrow I\!R$ defined as

$$p(t) = \sum_{i=0}^{n} p_i t^i, \qquad p_i, t \in I\!R, \quad i = 0, .., n, \qquad p_n \neq 0. \tag{4.1}$$

We assume the coefficients $p_i$ to be representable in the floating-point number system of the host computer. The algorithm achieves maximum accuracy, even in the neighborhood of a root where cancellation dooms an ordinary floating-point evaluation.

## 4.1 Theoretical Background

### 4.1.1 Description of the Problem

The basic idea of the algorithm is to transform the evaluation of a polynomial into a linear system of equations. The linear system can be solved efficiently and with maximum accuracy using vector and matrix operations based on the accurate dot product as described in Section 3.5.

We rewrite the polynomial $p(t)$ in Horner's nested multiplication form

$$p(t) = (\cdots (p_n t + p_{n-1})t + \cdots + p_1)t + p_0. \tag{4.2}$$

Labeling the intermediate results in Equation (4.2) as

$$\begin{aligned} x_n &= p_n \\ x_i &= x_{i+1} t + p_i, \quad i = n-1, \ldots, 0, \end{aligned} \tag{4.3}$$

we get the linear system

$$Ax = p \tag{4.4}$$

with $A \in \mathbb{R}^{(n+1)\times(n+1)}$, $x$ and $p \in \mathbb{R}^{n+1}$ where

$$A = \begin{pmatrix} 1 & & & \\ -t & 1 & & \\ & \ddots & \ddots & \\ & & -t & 1 \end{pmatrix}, \quad x = \begin{pmatrix} x_n \\ \vdots \\ x_0 \end{pmatrix}, \quad \text{and} \quad p = \begin{pmatrix} p_n \\ \vdots \\ p_0 \end{pmatrix}. \tag{4.5}$$

The last unknown $x_0$ is the desired value of the polynomial $p(t)$. There are well-known techniques and algorithms to solve a linear system of equations with maximum accuracy and verification of the result (see Chapter 10), but a special method for this very special bidiagonal Toeplitz form of $A$ is much faster than the general methods.

### 4.1.2 Iterative Solution

First, we compute an approximation $\widetilde{x} \in \mathbb{R}^{n+1}$ for the solution of (4.4) by direct forward substitution in floating-point arithmetic

$$\begin{aligned} \widetilde{x}_n &= p_n \\ \widetilde{x}_i &= \widetilde{x}_{i+1} \cdot t + p_i, \quad i = n-1, \ldots, 0, \end{aligned}$$

such that $\widetilde{x}_0 \approx p(t)$. We will use a residual iteration to improve this approximation, to achieve the desired maximum accuracy, and to get verified bounds for the result. To determine the accuracy of the first approximation $\widetilde{x}$, we have to calculate an enclosure of the residual with sharp bounds (in floating-point interval arithmetic)

$$[r]^{(1)} := \Diamond(p - A\widetilde{x}).$$

The dot product operation is essential to avoid both catastrophic cancellation and gross overestimation of the residual.

Let $x^*$ denote the true solution to the system $Ax = p$. Let the error in the approximate solution be $y := x^* - \widetilde{x}$. Then $p = Ax^* = A(\widetilde{x} + y) = A\widetilde{x} + Ay$, and the residual $r := p - A\widetilde{x} = Ay$. Now the true residual $r \in [r]^{(1)} := \Diamond(p - A\widetilde{x})$ (evaluated in floating-point interval arithmetic), so the true error $y$ is contained in the solution (using floating-point interval arithmetic) to the linear system

$$A[y]^{(1)} = [r]^{(1)} \tag{4.6}$$

Because of the special form of $A$, Equation (4.6) can be solved using the same forward substitution as used to determine $\widetilde{x}$, except that we use floating-point interval arithmetic.

$$\begin{aligned} [y]_n^{(1)} &= [r]_n^{(1)} \\ [y]_i^{(1)} &= [y]_{i+1}^{(1)} \cdot t + [r]_i^{(1)}, \quad i = n-1, \ldots, 0. \end{aligned}$$

It is guaranteed that the exact value of $p(t)$ lies in the interval $[z] = \widetilde{x}_0 + [y]_0^{(1)}$. If the diameter of this interval is not sufficiently small, the residual iteration is continued. Let $y^{(1)} = m([y]^{(1)})$ denote the midpoint of the vector $[y]^{(1)}$, and use $\widetilde{x} + y^{(1)}$ as the new approximation for the solution of the linear system.

Let $y^{(0)} := \widetilde{x}$. At the beginning of the $(k+1)^{st}$ iteration step, we have $k+1$ vectors $y^{(0)}, \ldots, y^{(k)}$ determining the approximate solution $\sum_{j=0}^{k} y^{(j)}$ for $Ax = p$. The residual of this approximation is enclosed in the interval vector

$$[r]^{(k+1)} = \Diamond \left( p - A \cdot \sum_{j=0}^{k} y^{(j)} \right) \in I\!R^{n+1}.$$

The solution of the linear system $Ay^{(k+1)} = r^{(k+1)}$ is enclosed by the solution of the linear system $A[y]^{(k+1)} = [r]^{(k+1)}$. Hence, the enclosure

$$p(t) = x_0^* \in [z] = \sum_{j=0}^{k} y_0^{(j)} + [y]_0^{(k+1)}$$

is always guaranteed. It is easy to prove that the sequence of interval vectors $[y]^{(k)}$ converges towards the zero vector for $k \to \infty$ as $\sum_{j=0}^{k} y^{(j)}$ approximates $x^*$ (see [10]). With this we have

$$\lim_{k \to \infty} \left( \sum_{j=0}^{k} y_0^{(j)} \right) = x_0^* = p(t)$$

as an exact evaluation of polynomial (4.1) at the point $t$. The rate of convergence of the sequence $\left( [y]_0^{(k)} \right)_{k=0}^{\infty}$ to zero, or of $\left( \sum_{j=0}^{k} y_0^{(j)} \right)_{k=0}^{\infty}$ to $p(t)$ is linear and is proportional to the condition number of matrix $A$. In fact, the number of residual iterations necessary to achieve maximum accuracy is a rough indicator for the condition number of $A$.

The complexity of the algorithm is only linear in the degree of the polynomial because of the special form of matrix $A$. However, execution time tends to grow faster than linearly because the condition number of $A$ tends to grow with the degree, forcing more iterations to achieve maximum accuracy.

## 4.2 Algorithmic Description

We present the algorithm RPolyEval for the evaluation of a real polynomial $p(t) = \sum_{i=0}^{n} p_i t^i$ with maximum accuracy. Except for the special cases $n = 0$ and $n = 1$, which can be calculated directly, an iterative solution method is used. As described in the preceding section, we first compute a floating-point approximation of $p(t)$. We then carry out a residual iteration by solving a linear system of equations. Because of the shape (4.5) of the matrix $A$, this can be done by a direct forward substitution. The new solution interval determined in the next step is checked for being of maximum accuracy, i.e. for being exact to one unit in the last place of the mantissa (1 ulp) (see Section 3.6).

**Algorithm 4.1:** RPolyEval $(p, t, z, [z], k, Err)$

1. {Initialization and treatment of special cases}
   $Err$ := "No Error"; $k := 0$;
   **if** $(n = 0)$ **then** $z := p_0$; $[z] := p_0$;
   **if** $(n = 1)$ **then** $z := t \cdot p_1 + p_0$; $[z] := \Diamond(t \cdot p_1 + p_0)$;

2. {Usual case}
   **if** $(n > 1)$ **then**
   (a) {Computation of a first approximation of $A\tilde{x} = p$}
   $$\tilde{x}_n := p_n;$$
   $$\tilde{x}_i := \tilde{x}_{i+1} \cdot t + p_i; \quad (i = n-1, \dots, 0)$$
   $$z := \tilde{x}_0;$$
   (b) {Iterative refinement of the solution of $A\tilde{x} = p$}
   $y^{(0)} := \tilde{x}$;
   **repeat**
   i. **if** $(k > 0)$ **then** $y^{(k)} := m([y]^{(k)})$;
   ii. {Computation of the residual $[r]^{(k+1)}$ and evaluation}
   {of the interval system $A \cdot [y]^{(k+1)} = [r]^{(k+1)}$ (see 4.6)}
   $$[y]_n^{(k+1)} := 0;$$
   $$\left.\begin{aligned} [r]_i^{(k+1)} &:= \Diamond\Big(p_i - \sum_{j=0}^{k} y_i^{(j)} + \sum_{j=0}^{k} y_{i+1}^{(j)} \cdot t\Big); \\ [y]_i^{(k+1)} &:= [y]_{i+1}^{(k+1)} \cdot t + [r]_i^{(k+1)}; \end{aligned}\right\} (i = n-1, \dots, 0)$$
   iii. {Determination of a new enclosure $[z]$ of $p(t)$}
   $$[z] := \Diamond\Big(\sum_{j=0}^{k} y_0^{(j)} + [y]_0^{(k+1)}\Big);$$
   iv. $k := k + 1$;
   **until** (UlpAcc $([z])$ **or** $(k > k_{\max})$)

3. **if not** UlpAcc $([z])$ **then** $Err$ := "Iteration failed";

4. **return** $z$, $[z]$, $k$, $Err$;

## Applicability of the Algorithm

The theoretical statement about the convergence of the sequence of iterates $[y]^{(k)}$ in Section 4.1.2 is also true when the calculations are done on a computer, if a precisely defined interval arithmetic as the one of C–XSC is used. Since all components of the residual interval vectors $[r]^{(k)}$ are results of scalar products, they can be calculated with maximum accuracy. That is, each component of $[r]^{(k)}$ is calculated with just one rounding. This guarantees that the upper and lower bounds of each interval component $[r]_i^{(k)}$ are identical or adjacent machine numbers. Even if the width of

$[y]_0^{(k+1)}$ vanishes, the C–XSC runtime system returns an enclosure of the floating-point element of smallest absolute value (see Section 3.5). With this we always obtain a true enclosure of the defect.

If the condition number of the matrix $A$ is extremely large, then the convergence of the residual iteration is slow. To avoid the possibility of an unbounded number of iterations at Step 2(b), we halt after $k_{\max}$ iterations. Our implementation uses $k_{\max} = 10$. This is large enough to achieve maximum accuracy unless the condition number of $A$ is larger than approximately $10^{10} \cdot \|p(t)\|$.

# 4.3 Implementation and Examples

## 4.3.1 C++ Program Code

We list the C++ program code for the evaluation of a real polynomial with maximum accuracy. Interval variables are named with double characters, e.g. `rr[i]` denotes the interval $[r]_i$.

### 4.3.1.1 Module rpoly

The header file of module *rpoly* supplies the definition of the class *RPolynomial* representing a real polynomial $p(t) = \sum_{i=0}^{n} p_i t^i$, $p_i, t \in R$. Since arithmetic operations for real polynomials are not required by Algorithm 4.1, they are not provided by module *rpoly*.

```
//----------------------------------------------------------------------------
// File: rpoly (header)
// Purpose: Definition of the class for the representation of a real
//    polynomial by its coefficients.
// Class RPolynomial:
//    Deg()          : to get the degree of the polynomial
//    RPolynomial(): constructors
//    operator []   : component access
//    operator >>   : input operator for data of type RPolynomial
//    operator <<   : output operator for data of type RPolynomial
//----------------------------------------------------------------------------
#ifndef __RPOLY_HPP
#define __RPOLY_HPP

#include <rvector.hpp>      // Real vector arithmetic

class RPolynomial {
  private:
    rvector coeff;
  public:
    RPolynomial ( int );
    RPolynomial ( RPolynomial& );
    real& operator[] ( int i ) { return coeff[i]; }

    friend int Deg ( RPolynomial& );
    friend istream& operator>> ( istream&, RPolynomial& );
    friend ostream& operator<< ( ostream&, RPolynomial );
};
#endif
```

The input operator >> echos x↑0 *, + x↑1 *, + x↑2 *, ... on the standard output device to give an information on the index of the coefficient actually to be entered.

```
//----------------------------------------------------------------------------
// File: rpoly (implementation)
// Purpose: Definition of the class for the representation of a real
//     polynomial by its coefficients.
// Class RPolynomial:
//     Deg()         : to get the degree of the polynomial
//     RPolynomial(): constructors
//     operator []  : component access
//     operator >>  : input operator for data of type RPolynomial
//     operator <<  : output operator for data of type RPolynomial
//----------------------------------------------------------------------------
#include <rpoly.hpp>

int Deg ( RPolynomial& p ) { return Ub(p.coeff); }

RPolynomial::RPolynomial( int n )
{
  Resize(coeff,0,n);
  coeff = 0.0;
}

RPolynomial::RPolynomial( RPolynomial& p )
{
  Resize(coeff,0,Deg(p));
  coeff = p.coeff;
}

istream& operator>> ( istream& in, RPolynomial& p )
{
  cout << "  x↑0 * ";  in >> p[0];
  for (int i = 1; i <= Deg(p); i++)
    { cout << "+ x↑" << i << " * ";  in >> p[i]; }
  return in;
}

ostream& operator<< ( ostream& out, RPolynomial p )
{
  int PolyIsZero, n = Deg(p);

  PolyIsZero = 1;
  for (int i = 0; i <= n; i++) {
    if (p[i] == 0.0) continue;
    if (PolyIsZero)
      out << "  ";
    else
      out << "+ ";
    out << p[i] << " * x↑" << i << endl;
    PolyIsZero = 0;
  }
  if (PolyIsZero) out << "  0 (= zero polynomial)" << endl;
  return out;
}
```

#### 4.3.1.2 Module rpeval

The header file of module *rpeval* supplies the interface of the function *RPolyEval()*, for evaluating a real polynomial with maximum accuracy and the interface of the function *RPolyEvalErrMsg()* which can be used to get an error message for the error code returned by *RPolyEval()*.

```
//----------------------------------------------------------------------------
// File: rpeval (header)
// Purpose: Evaluation of a real polynomial p with maximum accuracy.
// Global functions:
//    RPolyEval()      : computes an enclosure for the exact value p(t) with
//                       maximum accuracy
//    RPolyEvalErrMsg(): delivers an error message text
//----------------------------------------------------------------------------
#ifndef __RPEVAL_HPP
#define __RPEVAL_HPP

#include <interval.hpp>      // Interval arithmetic
#include <rpoly.hpp>         // Real polynomial type

extern char* RPolyEvalErrMsg ( int );
extern void  RPolyEval ( RPolynomial, real, real&, interval&, int&, int& );
#endif
```

The function *RPolyEval()* is an implementation of Algorithm 4.1. It computes an approximation and an enclosure of the value of the real polynomial $p(t) = \sum_{i=0}^{n} p_i t^i$ with $p_i \in R$ at a point $t \in R$.

```
//----------------------------------------------------------------------------
// File: rpeval (implementation)
// Purpose: Evaluation of a real polynomial p with maximum accuracy.
// Method: Transformation of Horner's scheme for evaluating p(t) to a linear
//    system of equations A(t)*x = p, where A(t) is a bidiagonal, Toeplitz
//    matrix and x = (x_n,...,x_0). By iterative refinement the floating-
//    point approximation is improved and the exact solution x is enclosed in
//    an interval vector [x].
// Global functions:
//    RPolyEval()      : computes an enclosure for the exact value p(t) with
//                       maximum accuracy
//    RPolyEvalErrMsg(): delivers an error message text
//----------------------------------------------------------------------------
#include <string.h>          // String handling
#include <imatrix.hpp>       // Interval matrix/vector arithmetic
#include <mvi_util.hpp>      // Interval matrix/vector utility functions
#include <rpeval.hpp>

const int kmax = 10;         // Maximum number of iteration steps

//----------------------------------------------------------------------------
// Error codes used in this module. In the comments below p[0],..., p[n] are
// the coefficients of a polynomial p.
//----------------------------------------------------------------------------
const int
  NoError  = 0,     // No error occurred
  ItFailed = 1;     // Maximum number of iterations exceeded

//----------------------------------------------------------------------------
// Error messages depending on the error code.
//----------------------------------------------------------------------------
char* RPolyEvalErrMsg ( int Err )
```

```
{
  static char Msg[80] = "";

  if (Err != NoError) {
    char Hlp[60];

    if (Err == ItFailed)
      sprintf(Hlp,"Maximum number of iterations (=%d) exceeded",kmax);
    else
      strcpy(Hlp,"Code not defined");
    sprintf(Msg,"Error: %s!",Hlp);
  }
  return(Msg);
}

//----------------------------------------------------------------------------
// Purpose: Determination of p(t) (a polynomial p with argument t) with
//    maximum accuracy.
// Parameters:
//    In : 'p'   : represents a real polynomial by its coefficients
//         't'   : specifies the point of evaluation
//    Out: 'z'   : floating-point approximation computed by Horner's scheme
//         'zz'  : enclosure of p(t)
//         'k'   : number of iterations needed
//         'Err' : error flag
// Description: The polynomial 'p' is defined by its coefficients and, 't'
//    denotes the evaluation point. Horner's scheme for evaluating p is equi-
//    valent to computing the solution of a linear system of equations
//    A(t)*x = p, where A(t) is a bidiagonal, Toeplitz matrix and
//    x =(x_n,...,x_0). The component x_0 of x is equal to p(t). The solution
//    x is enclosed in an interval vector [x] by iterative refinement. The
//    first element of [x] is an enclosure of p(t). It is returned in the
//    variable 'zz'. A floating-point approximation computed by Horner's
//    scheme is returned in 'z'. The number of iterations is returned in 'k'
//    and the state of success is returned in 'Err'.
// Remark: The polynomial's coefficients and the argument are assumed to be
//    exact floating-point numbers!
//----------------------------------------------------------------------------
void RPolyEval ( RPolynomial p,  real t, real& z,
                 interval&  zz, int& k, int&  Err )
{
  int           i, j, n = Deg(p);         // The j-th column of the matrix
  ivector       rr(0,n), yy(0,n);         // 'y' is used to store the j-th
  rvector       x(0,n);                   // correction for the solution of
  rmatrix       y(0,n,0,kmax);            // A(t)*y = p.
  dotprecision  Accu;                     //--------------------------------
  idotprecision IAccu;

  k = 0;                                  // Initialization
  Err = NoError;                          //---------------

  if (n == 0)                             // Constant polynomial
    { z = p[0]; zz = p[0]; }              //--------------------

  else if (n == 1) {                      // Linear polynomial
    Accu = p[0];                          //------------------
    accumulate(Accu,t,p[1]);
    rnd(Accu,z);
    rnd(Accu,zz);
  }
  else { // n > 1
    x = 0.0; y = 0.0;                     // Initialization x = 0 and y = 0
                                          //--------------------------------
```

```
    x[n] = p[n];                                // Computation of a first approxi-
    for (i = n-1; i >= 0; i--)                  // mation using Horner's scheme
      x[i] = x[i+1]*t + p[i];                   //--------------------------------
    z = x[0];

    y[Col(0)] = x;                              // Iterative refinement for the
    do {                                        // solution of A*x = p
                                                //--------------------------------
      if (k > 0)                        // If a residual was computed, its middle
        y[Col(k)] = mid(yy);            // is stored as the next correction of 'y'

      yy[n] = 0.0;                              // Computation of the residual [r]
      for (i = n-1; i >= 0; i--) {              // and evaluation of the interval
        Accu = p[i];                            // system A*[y] = [r]
        for(j = 0; j <= k; j++) Accu -= y[i][j];
        for(j = 0; j <= k; j++) accumulate(Accu,t,y[i+1][j]);
        rnd(Accu,rr[i]);
        yy[i] = yy[i+1]*t + rr[i];
      }

      IAccu = yy[0];                                    // Determination of a new
      for (j = 0; j <= k; j++) IAccu += y[0][j];        // enclosure [z] of p(t)
      rnd(IAccu,zz);

      k++;
    } while ( !(UlpAcc(zz,1) || (k > kmax)) );

    if ( !UlpAcc(zz,1) ) Err = ItFailed;
  }
}
```

## 4.3.2 Examples

We consider the polynomial $p(t) = t^4 - 8t^3 + 24t^2 - 32t + 16$ to be evaluated in the neighborhood of the real value $t = 2.0001$ and the polynomial $q(t) = -t^3+3t^2-3t+1$ to be evaluated in the neighborhood of the real value $t = 1.000005$. To make sure that the arguments are representable on the computer, we use the machine numbers $t = \square(2.0001)$ and $t = \square(1.000005)$, respectively.

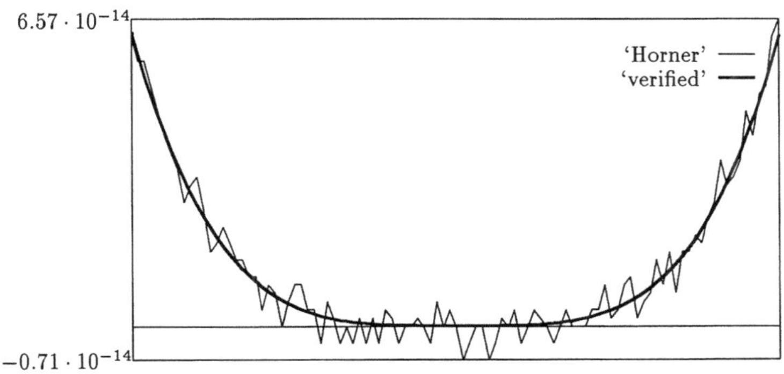


Figure 4.1: Polynomial $p(t)$ for $t \in [1.9995, 2.0005]$

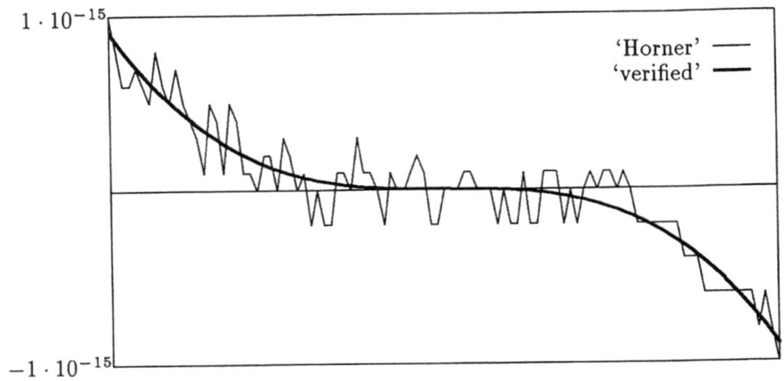


Figure 4.2: Polynomial $q(t)$ for $t \in [0.99999, 1.00001]$

To illustrate the difficulties that may occur with the calculation of a polynomial value when using floating-point arithmetic, these two polynomials have been evaluated in floating-point arithmetic for 100 values in the neighborhood of their roots. The corresponding plots are given in Figures 4.1 and 4.2. These two examples are solved reliably using the following sample program to call *RPolyEval()*.

```
//----------------------------------------------------------------------------
// Example: Evaluation of a real polynomial with maximum accuracy
//----------------------------------------------------------------------------
#include <rpeval.hpp>      // Polynomial evaluation

int main ( )
{
  int      Err, No, n;
  real     t, y;
  interval yy;

  do {
    cout << "Enter the degree of the polynomial (>=0): ";
    cin  >> n; cout << endl;
  } while (n < 0);

  RPolynomial p(n);

  cout << SetPrecision(23,15) << Scientific;   // Output format

  cout << "Enter the coefficients of p in increasing order:" << endl;
  cin  >> p; cout << endl;
  cout << "Enter the argument t for evaluation:" << endl;
  cin  >> t; cout << endl;

  RPolyEval(p,t,y,yy,No,Err);

  if (!Err) {
    cout << "Polynomial:" << endl << p <<endl;
    cout << "Floating-point evaluation of p(t) using Horner's scheme:"
         << endl << " " << y << endl << endl;
    cout << "Verified inclusion of p(t):"
         << endl << yy << endl << endl;
    cout << "Number of iterations needed: " << No << endl;
  }
  else
    cout << RPolyEvalErrMsg(Err) << endl;
```

```
  return 0;
}
```

Our implementation of Algorithm 4.1 produces the output listed below. In both cases the floating-point approximation, naively using Horner's nested multiplication form, does *not* lie within the verified enclosure of $p(\square(t))$ and $q(\square(t))$. An even worse side effect of rounding errors occurring during the evaluation using Horner's scheme is that the standard algorithm returns not only wrong digits but also an incorrect sign for the values of $p(\square(2.0001))$ and $q(\square(1.000005))$. To avoid an incorrect interpretation of the resulting interval, we stress that this interval is numerically proved to be an enclosure of the exact value of the given polynomials for the machine numbers $t = \square(2.0001)$ and $t = \square(1.000005)$.

Output for polynomial $p(t) = t^4 - 8t^3 + 24t^2 - 32t + 16$:

```
Enter the degree of the polynomial (>=0): 4

Enter the coefficients of p in increasing order:
  x^0 * 16
+ x^1 * -32
+ x^2 * 24
+ x^3 * -8
+ x^4 * 1

Enter the argument t for evaluation: 2.0001

Polynomial:
   1.600000000000000E+001 * x^0
+ -3.200000000000000E+001 * x^1
+  2.400000000000000E+001 * x^2
+ -8.000000000000000E+000 * x^3
+  1.000000000000000E+000 * x^4

Floating-point evaluation of p(t) using Horner's scheme:
 -3.552713678800501E-015

Verified inclusion of p(t):
[ 1.000000000008441E-016, 1.000000000008442E-016]

Number of iterations needed: 2
```

Output for polynomial $q(t) = -t^3 + 3t^2 - 3t + 1$:

```
Enter the degree of the polynomial (>=0): 3

Enter the coefficients of p in increasing order:
  x^0 * 1
+ x^1 * -3
+ x^2 * 3
+ x^3 * -1

Enter the argument t for evaluation: 1.000005

Polynomial:
   1.000000000000000E+000 * x^0
+ -3.000000000000000E+000 * x^1
+  3.000000000000000E+000 * x^2
+ -1.000000000000000E+000 * x^3
```

```
Floating-point evaluation of p(t) using Horner's scheme:
  1.110223024625157E-016

Verified inclusion of p(t):
[-1.250000000024568E-016,-1.250000000024566E-016]

Number of iterations needed: 2
```

### 4.3.3 Restrictions and Hints

Ours is a method for evaluating *real* polynomials, so we must be careful that the coefficients of $p$ used for evaluation are exactly representable on the computer. To get an exact representation of real coefficients $p_i$ such as 0.01, the entire polynomial might be scaled by a factor of 100. For more details about conversion of input data, see the remarks in Section 3.7.

## 4.4 Exercises

**Exercise 4.1** Let $p(t) = 543339720t^3 - 768398401t^2 - 1086679440t + 1536796802$ and $q(t) = 67872320568t^3 - 95985956275t^2 - 135744641136t + 191971912515$ be real polynomials. Evaluate them at the machine number nearest to $t = \sqrt{2}$, i.e. $t = \Box(\sqrt{2})$. Compare the results of floating-point evaluation, high accuracy evaluation by residual correction, and the interval enclosure of the exact value.

**Exercise 4.2** Let $p_n(t) = \sum_{i=0}^{n} t^i/i!$ be the truncated Taylor series for $e^t$. Use the example program to evaluate $p_n(-100)$ for various $n$. Graph the execution time as a function of the degree $n$. Can you ever elicit the error message "Maximum number of iterations exceeded"? Note that the coefficients $t^i/i!$ are not exactly representable. Therefore, you may either evaluate an approximation to $p_n(t)$, or else follow the suggestion of Section 4.3.3 and evaluate $n! \cdot p_n(t)$. Is the enclosure for $p_n(-100)$ also an enclosure for $e^{-100}$?

**Exercise 4.3** This exercise requires a C++ compiler that supports class templates: Implement a module *poly* that defines a template class *Polynomial* for polynomials of arbitrary component type. Use this module instead of module *rpoly* in our programs. Replace all declarations of type *RPolynomial* by declarations of type *Polynomial<real>*. See Exercise 9.4 for another application of that template class.

## 4.5 References and Further Reading

A first algorithm for the evaluation of polynomials with maximum accuracy was given by Böhm in [10] and [63] and later by Krämer [55]. This algorithm is easily modified for rational expressions. A more general approach which also includes the evaluation of certain arithmetic expressions is described in Chapter 8. Finally, let us remark that there are similar commercial implementations of Algorithm 4.1 in the subroutine libraries ACRITH of IBM [36] and ARITHMOS of SIEMENS [86].

# Chapter 5
# Automatic Differentiation

In many applications of numerical and scientific computing, it is necessary to compute derivatives of functions. Simple examples are methods for finding the zeros, maxima, or minima of nonlinear functions. There are three different methods to get the values of the derivatives: *numerical differentiation*, *symbolic differentiation*, and *automatic differentiation.*

Numerical differentiation uses difference approximations to compute approximations of the derivative values. Symbolic differentiation computes explicit formulas for the derivative functions by applying differentiation rules. Those formulas must then be evaluated to get numerical values. Automatic differentiation also uses the well-known differentiation rules, but it propagates numerical values for the derivatives. This combines the advantages of symbolic and numerical differentiation. Numbers instead of symbolic formulas must be handled, and the computation of the derivative values is done automatically together with the computation of the function values. The main advantage of this process is that only the algorithm or formula for the function is needed. No explicit formulas for the derivatives are required.

In this chapter, we deal with automatic differentiation based on interval operations to get guaranteed enclosures for the function value and the derivative values. Automatic differentiation is a fundamental enabling technology for validated computation since so many interval algorithms require an enclosure for high-order derivative terms to capture truncation errors in the algorithms.

## 5.1 Theoretical Background

Automatic differentiation evaluates functions specified by algorithms or formulas where all operations are performed according to the rules of a *differentiation arithmetic* given below (see also [72], [74]).

### First Derivatives

In the one-dimensional case, first order differentiation arithmetic is an arithmetic for ordered pairs of the form

$$U = (u, u') \quad \text{with} \quad u, u' \in I\!R.$$

The first component of $U$ contains the value $u(x)$ of the function $u : \mathbb{R} \to \mathbb{R}$ at the point $x \in \mathbb{R}$. The second component contains the value of the derivative $u'(x)$. The rules for the arithmetic are

$$\begin{array}{ccccccccc} U & + & V & = & (u,u') & + & (v,v') & = & (u+v, u'+v') \\ U & - & V & = & (u,u') & - & (v,v') & = & (u-v, u'-v') \\ U & \cdot & V & = & (u,u') & \cdot & (v,v') & = & (u \cdot v, u \cdot v' + u' \cdot v) \\ U & / & V & = & (u,u') & / & (v,v') & = & (u/v, (u' - u/v \cdot v')/v), \quad v \neq 0, \end{array} \tag{5.1}$$

The familiar rules of calculus are used in the second component. The operations inside the parentheses in these definitions are operations on real numbers. An independent variable $x$ and the arbitrary constant $c$ correspond to the ordered pairs

$$X = (x, 1) \quad \text{and} \quad C = (c, 0),$$

because $\frac{dx}{dx} = 1$, and $\frac{dc}{dx} = 0$.

Let the independent variable $x$ of a formula for a function $f : \mathbb{R} \to \mathbb{R}$ be replaced by $X = (x,1)$, and let all constants $c$ be replaced by their $(c, 0)$ representation. Then evaluation of $f$ using the rules of differentiation arithmetic gives the ordered pair

$$f(X) = f((x,1)) = (f(x), f'(x)).$$

**Example 5.1** We want to compute the function value and the derivative value of the function $f(x) = x \cdot (4+x)/(3-x)$ at the point $x = 1$. Using the differentiation arithmetic defined by (5.1) we get

$$\begin{aligned} f(X) \;=\; (f, f') &= (x,1) \cdot ((4,0) + (x,1))/((3,0) - (x,1)) \\ &= (1,1) \cdot ((4,0) + (1,1))/((3,0) - (1,1)) \\ &= (1,1) \cdot (5,1)/(2,-1) \\ &= (5,6)/(2,-1) \\ &= (2.5, 4.25). \end{aligned}$$

That is, $f(1) = 2.5$, and $f'(1) = 4.25$.

For an elementary function $s : \mathbb{R} \to \mathbb{R}$ (see Section 3.1 for the set of elementary function used within this scope), the rules of differentiation arithmetic must be extended using the chain rule

$$s(U) = s((u,u')) = (s(u), u' \cdot s'(u)). \tag{5.2}$$

For example, the sine function is defined by

$$\sin U = \sin(u, u') = (\sin u, u' \cdot \cos u).$$

### Second Derivatives

For a second order differentiation arithmetic, we use triples

$$U = (u, u', u''), \quad \text{with } u, u', u'' \in \mathbb{R},$$

for the description of the arithmetic rules. Here $u$, $u'$, and $u''$ denote the function value, the value of the first derivative, and the value of the second derivative, respectively, each evaluated at a fixed point $x \in \mathbb{R}$. For the constant function $u(x) = c$, we set $U = (u, u', u'') = (c, 0, 0)$. For the function $u(x) = x$, we define $U = (u, u', u'') = (x, 1, 0)$.

We define the arithmetic operations $W = U \circ V$ with $\circ \in \{+, -, \cdot, /\}$ for two triples $U = (u, u', u'')$ and $V = (v, v', v'')$ using the rules of first order differentiation arithmetic in the first and second components. The third components are defined by

$$\begin{aligned} W &= U + V \Rightarrow w'' = u'' + v'', \\ W &= U - V \Rightarrow w'' = u'' - v'', \\ W &= U \cdot V \Rightarrow w'' = u \cdot v'' + 2 \cdot u' \cdot v' + v \cdot u'', \\ W &= U \,/\, V \Rightarrow w'' = (u'' - 2 \cdot w' \cdot v' - w \cdot v'')/v, \quad v \neq 0. \end{aligned} \tag{5.3}$$

For an elementary function $s : \mathbb{R} \to \mathbb{R}$ and $U = (u, u', u'')$, we define

$$s(U) = (s(u),\ s'(u) \cdot u',\ s'(u) \cdot u'' + s''(u) \cdot (u')^2). \tag{5.4}$$

Here $s' : \mathbb{R} \to \mathbb{R}$ and $s'' : \mathbb{R} \to \mathbb{R}$ are the first and second derivatives of $s$, assuming they exist.

We compute *enclosures* for the true values of the function and its derivatives using a differentiation arithmetic based on interval arithmetic. That is, the components $u$, $u'$, and $u''$ are replaced by the corresponding interval values, and all arithmetic operations and function evaluations performed to compute the components are replaced by the corresponding interval operations and function evaluations. The evaluation of a function $f : \mathbb{R} \to \mathbb{R}$ for an argument $[x] \in I\mathbb{R}$ using interval differentiation arithmetic delivers

$$f(X) = f(([x], 1, 0)) = ([f], [f'], [f''])$$

satisfying

$$f([x]) \subseteq [f], \quad f'([x]) \subseteq [f'], \quad \text{and} \quad f''([x]) \subseteq [f''].$$

## 5.2 Algorithmic Description

We now give the description of the algorithms for the elementary operators $+$, $-$, $\cdot$, and $/$, and for an elementary function $s \in \{$sqr, sqrt, power, exp, ln, sin, cos, tan, cot, asin, acos, atan, acot, sinh, cosh, tanh, coth, asinh, acosh, atanh, acoth$\}$. For the operands and arguments $U$, we use the triple representation $U = ([u_f], [u_{df}], [u_{ddf}])$ with $[u_f], [u_{df}], [u_{ddf}] \in I\mathbb{R}$ to be close to the notation of our implementation.

**Algorithm 5.1:** Operator $+$ $(U, V)$

1. $[w_f] := [u_f] + [v_f]$; {Function value}
2. $[w_{df}] := [u_{df}] + [v_{df}]$; {First derivative}
3. $[w_{ddf}] := [u_{ddf}] + [v_{ddf}]$; {Second derivative}
4. **return** $W := ([w_f], [w_{df}], [w_{ddf}])$;

**Algorithm 5.2:** Operator $-$ $(U, V)$

1. $[w_f] := [u_f] - [v_f]$; {Function value}
2. $[w_{df}] := [u_{df}] - [v_{df}]$; {First derivative}
3. $[w_{ddf}] := [u_{ddf}] - [v_{ddf}]$; {Second derivative}
4. **return** $W := ([w_f], [w_{df}], [w_{ddf}])$;

**Algorithm 5.3:** Operator $\cdot$ $(U, V)$

1. $[w_f] := [u_f] \cdot [v_f]$; {Function value}
2. $[w_{df}] := [u_f] \cdot [v_{df}] + [u_{df}] \cdot [v_f]$; {First derivative}
3. $[w_{ddf}] := [u_f] \cdot [v_{ddf}] + 2 \cdot [u_{df}] \cdot [v_{df}] + [v_f] \cdot [u_{ddf}]$; {Second derivative}
4. **return** $W := ([w_f], [w_{df}], [w_{ddf}])$;

In Algorithm 5.4, we do not take care of the case $0 \in [v_f]$ because it does not make sense to go any further in computations when this case occurs. In an implementation, the standard error handling (runtime error) should be invoked if a division by zero occurs while computing the function value. We chose a special form of the rule for the differentiation of a quotient to be close to our implementation, where this form can save some computation time.

**Algorithm 5.4:** Operator $/$ $(U, V)$

1. $[w_f] := [u_f]/[v_f]$; {Function value}
2. $[w_{df}] := ([u_{df}] - [w_f] \cdot [v_{df}])/[v_f]$; {First derivative}
3. $[w_{ddf}] := ([u_{ddf}] - 2 \cdot [w_{df}] \cdot [v_{df}] - [w_f] \cdot [v_{ddf}])/[v_f]$; {Second derivative}
4. **return** $W := ([w_f], [w_{df}], [w_{ddf}])$;

In Algorithm 5.5, the derivative rules for the elementary functions are applied to compute the temporary values. For a better understanding of our implementation, these functions are listed in Table 5.1. There we use $\tan^2 + 1$ as the derivative of $\tan x$ to avoid a separate evaluation of $\cos x$.

Except for those cases where a division by zero occurs, we do not have to consider non-differentiability because almost all of the elementary functions

Table 5.1: Elementary functions and their derivatives

| $s(x)$ | $s'(x)$ | $s''(x)$ |
|---|---|---|
| $x^k$ | $kx^{k-1}$ | $k(k-1)x^{k-2}$ |
| $\sqrt{x}$ | $1/(2\sqrt{x})$ | $-1/(4\sqrt{x}^3)$ |
| $\sin x$ | $\cos x$ | $-\sin x$ |
| $\cos x$ | $-\sin x$ | $-\cos x$ |
| $\tan x$ | $\tan^2 x + 1$ | $2(\tan^3 x + \tan x)$ |
| $\cot x$ | $-\cot^2 x - 1$ | $2(\cot^3 x + \cot x)$ |
| $\operatorname{asin} x$ | $1/\sqrt{1-x^2}$ | $x/(1-x^2)^{\frac{3}{2}}$ |
| $\operatorname{acos} x$ | $-1/\sqrt{1-x^2}$ | $-x/(1-x^2)^{\frac{3}{2}}$ |
| $\operatorname{atan} x$ | $1/(1+x^2)$ | $-2x/(1+x^2)^2$ |
| $\operatorname{acot} x$ | $-1/(1+x^2)$ | $2x/(1+x^2)^2$ |

| $s(x)$ | $s'(x)$ | $s''(x)$ |
|---|---|---|
| $e^x$ | $e^x$ | $e^x$ |
| $\ln x$ | $1/x$ | $-1/x^2$ |
| $\sinh x$ | $\cosh x$ | $\sinh x$ |
| $\cosh x$ | $\sinh x$ | $\cosh x$ |
| $\tanh x$ | $1 - \tanh^2 x$ | $-2(\tanh^3 x - \tanh x)$ |
| $\coth x$ | $1 - \coth^2 x$ | $-2(\coth^3 x - \coth x)$ |
| $\operatorname{asinh} x$ | $1/\sqrt{x^2+1}$ | $-x/(x^2+1)^{\frac{3}{2}}$ |
| $\operatorname{acosh} x$ | $1/\sqrt{x^2-1}$ | $-x/(x^2-1)^{\frac{3}{2}}$ |
| $\operatorname{atanh} x$ | $1/(1-x^2)$ | $2x/(x^2-1)^2$ |
| $\operatorname{acoth} x$ | $1/(1-x^2)$ | $2x/(x^2-1)^2$ |

are differentiable whenever they are defined. That is, except for functions $s \in E := \{\text{sqrt}, \text{asin}, \text{acos}, \text{acosh}, \text{atanh}, \text{acoth}\}$ the domains for $s$, $s'$, and $s''$ are the same. However, note that many derivative values may overflow, even when the function values are evaluated safely. In an implementation of Algorithm 5.5, the standard error handling of interval arithmetic should be invoked for domain violations in Step 1 (or Step 2 for $s \in E$).

## Algorithm 5.5: s$(U)$

1. $[w_f] := s([u_f])$; {Function value}
2. $[h_1] := s'([u_f])$; $[h_2] := s''([u_f])$; {Temporary values}
3. $[w_{df}] := [h_1] \cdot [u_{df}]$; {First derivative}
4. $[w_{ddf}] := [h_1] \cdot [u_{ddf}] + [h_2] \cdot ([u_{df}])^2$; {Second derivative}
5. **return** s $:= ([w_f], [w_{df}], [w_{ddf}])$;

The computational complexity of this forward mode of differentiation arithmetic depends somewhat on the mix of arithmetic operations and elementary functions. The cost of computing first and second derivatives is at most a small multiple of the cost of evaluating $f$ itself.

# 5.3 Implementation and Examples

## 5.3.1 C++ Program Code

### 5.3.1.1 Module ddf_ari

The header file of module *ddf_ari* supplies the definition of the class *DerivType* including operators and elementary functions for an interval differentiation arithmetic for derivatives up to second order. An object of type *DerivType* is implemented as a triple of intervals representing the function value, the value of the first derivative, and the value of the second derivative, respectively. The type name *ddf_FctPtr* is used for pointers to functions of type *DerivType*.

```
//----------------------------------------------------------------------------
// File: ddf_ari (header)
// Purpose: Definition of an interval differentiation arithmetic which allows
//    function evaluation with automatic differentiation up to second order.
// Type:
//    ddf_FctPtr              : pointer for a function of type 'DerivType'
// Class DerivType:
//    DerivType()             : constructors
//    operators +, -, *, /    : operators of diff. arithmetic
//    operator =              : assignment operator
//    DerivConst()
//    DerivVar()              : to define derivative constants/variables
//    fValue()
//    dfValue()
//    ddfValue()              : to get function and derivative values
//    sqr(), sqrt(), power(),
//    exp(), sin(), cos(), ...: elementary functions of diff. arithmetic
//    fEval()                 : to compute function value only
//    dfEval()                : to compute function and first derivative
//                              value
//    ddfEval()               : to compute function, first, and second
//                              derivative value
//----------------------------------------------------------------------------
#ifndef __DDF_ARI_HPP
#define __DDF_ARI_HPP

#include <interval.hpp>     // Interval arithmetic
#include <i_util.hpp>       // Interval utility functions

class DerivType;

typedef DerivType (*ddf_FctPtr)(DerivType);

class DerivType {
  private:
    interval f, df, ddf;

  public:
    DerivType ( );
    DerivType ( interval&, interval&, interval& );
    DerivType ( DerivType& );

    DerivType& operator= ( DerivType& );

    friend DerivType DerivConst ( const real& );
    friend DerivType DerivConst ( interval& );
    friend DerivType DerivVar   ( const real& );
    friend DerivType DerivVar   ( interval& );

    friend interval fValue   ( DerivType& );
    friend interval dfValue  ( DerivType& );
    friend interval ddfValue ( DerivType& );

    friend DerivType operator+ ( DerivType& );
    friend DerivType operator- ( DerivType& );

    friend DerivType operator+ ( DerivType&, DerivType& );
    friend DerivType operator- ( DerivType&, DerivType& );
    friend DerivType operator* ( DerivType&, DerivType& );
```

```
    friend DerivType operator/ ( DerivType&, DerivType& );

    friend DerivType operator+ ( DerivType&, interval& );
    friend DerivType operator- ( DerivType&, interval& );
    friend DerivType operator/ ( DerivType&, interval& );
    friend DerivType operator* ( DerivType&, interval& );

    friend DerivType operator+ ( interval&, DerivType& );
    friend DerivType operator- ( interval&, DerivType& );
    friend DerivType operator* ( interval&, DerivType& );
    friend DerivType operator/ ( interval&, DerivType& );

    friend DerivType operator+ ( DerivType&, const real& );
    friend DerivType operator- ( DerivType&, const real& );
    friend DerivType operator* ( DerivType&, const real& );
    friend DerivType operator/ ( DerivType&, const real& );

    friend DerivType operator+ ( const real&, DerivType& );
    friend DerivType operator- ( const real&, DerivType& );
    friend DerivType operator* ( const real&, DerivType& );
    friend DerivType operator/ ( const real&, DerivType& );

    friend DerivType sqr   ( DerivType& );
    friend DerivType power ( DerivType&, int );
    friend DerivType sqrt  ( DerivType& );
    friend DerivType exp   ( DerivType& );
    friend DerivType ln    ( DerivType& );

    friend DerivType sin    ( DerivType& );
    friend DerivType cos    ( DerivType& );
    friend DerivType tan    ( DerivType& );
    friend DerivType cot    ( DerivType& );
    friend DerivType asin   ( DerivType& );
    friend DerivType acos   ( DerivType& );
    friend DerivType atan   ( DerivType& );
    friend DerivType acot   ( DerivType& );

    friend DerivType sinh   ( DerivType& );
    friend DerivType cosh   ( DerivType& );
    friend DerivType tanh   ( DerivType& );
    friend DerivType coth   ( DerivType& );
    friend DerivType asinh  ( DerivType& );
    friend DerivType acosh  ( DerivType& );
    friend DerivType atanh  ( DerivType& );
    friend DerivType acoth  ( DerivType& );

    friend void fEval   ( ddf_FctPtr f, interval, interval&);
    friend void dfEval  ( ddf_FctPtr f, interval, interval&, interval& );
    friend void ddfEval( ddf_FctPtr f, interval,  interval&, interval&,
                                                   interval& );
};
#endif
```

The static variable *DerivOrder* is used to select the highest order of derivative which is computed up to second order. This enables the user to save computation time computing only the function value or the first derivative. The default value of *DerivOrder* is 2, so normally the first and the second derivatives are computed. The functions *fEval()*, *dfEval()*, and *ddfEval()* simplify the mechanism of function evaluating and automate the setting and resetting of the *DerivOrder* variable. For a function of type *DerivType*, *fEval()* sets *DerivOrder* to 0 before the evaluation is done, computes, and delivers only the function value. If the first derivative also is desired, *dfEval()* sets *DerivOrder* to 1 before the evaluation is done. The function *ddfEval()* uses the default value of *DerivOrder*, computes, and returns the values of $f(x)$, $f'(x)$, and $f''(x)$.

Module *ddf_ari* can easily be modified to get a *real* version of this second-order differentiation arithmetic by replacing data type *interval* by data type *real* everywhere it occurs.

```
//----------------------------------------------------------------------------
// File: ddf_ari (implementation)
// Purpose: Definition of an interval differentiation arithmetic which allows
//    function evaluation with automatic differentiation up to second order.
// Method: Overloading of operators and elementary functions for operations
//    of data type 'DerivType'.
// Class DerivType:
//    DerivType()              : constructors
//    operators +, -, *, /     : operators of diff. arithmetic
//    operator =               : assignment operator
//    DerivConst()
//    DerivVar()               : to define derivative constants/variables
//    fValue()
//    dfValue()
//    ddfValue()               : to get function and derivative values
//    sqr(), sqrt(), power(),
//    exp(), sin(), cos(), ...: elementary functions of diff. arithmetic
//    fEval()                  : to compute function value only
//    dfEval()                 : to compute function and first derivative
//                               value
//    ddfEval()                : to compute function, first, and second
//                               derivative value
//----------------------------------------------------------------------------
#include <imath.hpp>       // Interval mathematical functions
#include <i_util.hpp>      // Interval utility functions
#include <ddf_ari.hpp>

// The local variable 'DerivOrder' is used to select the highest order of
// derivative which is computed. Its default value is 2, and normally the
// first and the second derivatives are computed.
//----------------------------------------------------------------------------
static int DerivOrder = 2;

//----------------------------------------------------------------------------
// Constructors and assignment operator
//----------------------------------------------------------------------------
DerivType::DerivType ( )
{
  f   = 0.0;
  df  = 0.0;
  ddf = 0.0;
}

DerivType::DerivType ( interval& x_f, interval& x_df, interval& x_ddf )
{
  f   = x_f;
  df  = x_df;
  ddf = x_ddf;
}

DerivType::DerivType ( DerivType& u )
{
  f   = u.f;
```

```
  df  = u.df;
  ddf = u.ddf;
}

DerivType& DerivType::operator= ( DerivType& u )
{
  f   = u.f;
  df  = u.df;
  ddf = u.ddf;
  return *this;
}

//----------------------------------------------------------------------------
// Transfer functions for constants and variables
//----------------------------------------------------------------------------
DerivType DerivConst ( const real& c )                    // Generate constant
{                                                         //------------------
  DerivType u;

  u.f   = c;
  u.df  = 0.0;
  u.ddf = 0.0;
  return u;
}

DerivType DerivConst ( interval& c )                      // Generate constant
{                                                         //------------------
  DerivType u;

  u.f   = c;
  u.df  = 0.0;
  u.ddf = 0.0;
  return u;
}

DerivType DerivVar ( const real& v )                      // Generate variable
{                                                         //------------------
  DerivType u;

  u.f   = v;
  u.df  = 1.0;
  u.ddf = 0.0;
  return u;
}

DerivType DerivVar ( interval& v )                        // Generate variable
{                                                         //------------------
  DerivType u;

  u.f   = v;
  u.df  = 1.0;
  u.ddf = 0.0;
  return u;
}

//----------------------------------------------------------------------------
// Access functions for function and derivative values
//----------------------------------------------------------------------------
inline interval fValue ( DerivType& u )                  // Get function value
  { return u.f; }                                        //-------------------

inline interval dfValue ( DerivType& u )             // Get 1. derivative value
  { return u.df; }                                   //------------------------
```

```
inline interval ddfValue ( DerivType& u )          // Get 2. derivative value
  { return u.ddf; }                                //------------------------

//----------------------------------------------------------------------------
// Unary operators + and - for DerivType operands
//----------------------------------------------------------------------------
inline DerivType operator+ ( DerivType& u )
  { return u; }

DerivType operator- ( DerivType& u )
{
  DerivType res;

  res.f = -u.f;
  if (DerivOrder > 0) {
    res.df = -u.df;
    if (DerivOrder > 1) res.ddf = -u.ddf;
  }
  return res;
}

//----------------------------------------------------------------------------
// Operators +, -, *, and / for two DerivType operands
//----------------------------------------------------------------------------
DerivType operator+ ( DerivType& u, DerivType& v )
{
  DerivType res;

  res.f = u.f + v.f;
  if (DerivOrder > 0) {
    res.df = u.df + v.df;
    if (DerivOrder > 1) res.ddf = u.ddf + v.ddf;
  }
  return res;
}

DerivType operator- ( DerivType& u, DerivType& v )
{
  DerivType res;

  res.f = u.f - v.f;
  if (DerivOrder > 0) {
    res.df = u.df - v.df;
    if (DerivOrder > 1) res.ddf = u.ddf - v.ddf;
  }
  return res;
}

DerivType operator* ( DerivType& u, DerivType& v )
{
  DerivType res;

  res.f = u.f * v.f;
  if (DerivOrder > 0) {
    res.df = u.df*v.f + u.f*v.df;
    if (DerivOrder > 1) res.ddf = u.ddf*v.f + 2.0*u.df*v.df + u.f*v.ddf;
  }
  return res;
}

DerivType operator/ ( DerivType& u, DerivType& v )
{
  DerivType res;
  interval  h1, h2;
```

```
  h1 = u.f / v.f;   // Can propagate 'division by zero' error
  res.f = h1;       //---------------------------------------
  if (DerivOrder > 0) {
    h2 = (u.df - h1*v.df) / v.f;
    res.df = h2;
    if (DerivOrder > 1) res.ddf = (u.ddf - h1*v.ddf - 2.0*h2*v.df)/v.f;
  }
  return res;
}

//----------------------------------------------------------------------------
// Operators +, -, *, and / for one interval and one DerivType operand
//----------------------------------------------------------------------------
DerivType operator+ ( interval& u, DerivType& v )
{
  DerivType res;

  res.f = u + v.f;
  if (DerivOrder > 0) {
    res.df = v.df;
    if (DerivOrder > 1) res.ddf = v.ddf;
  }
  return res;
}

DerivType operator- ( interval& u, DerivType& v )
{
  DerivType res;

  res.f = u - v.f;
  if (DerivOrder > 0) {
    res.df = -v.df;
    if (DerivOrder > 1) res.ddf = - v.ddf;
  }
  return res;
}

DerivType operator* ( interval& u, DerivType& v )
{
  DerivType res;

  res.f = u*v.f;
  if (DerivOrder > 0) {
    res.df = u * v.df;
    if (DerivOrder > 1) res.ddf = u*v.ddf;
  }
  return res;
}

DerivType operator/ ( interval& u, DerivType& v )
{
  DerivType res;
  interval  h1, h2;

  h1  = u / v.f;    // Can propagate 'division by zero' error
  res.f = h1;       //---------------------------------------
  if (DerivOrder > 0) {
    h2   = -h1 * v.df / v.f;
    res.df = h2;
    if (DerivOrder > 1) res.ddf = (-h1*v.ddf - 2.0*h2*v.df)/v.f;
  }
  return res;
}
```

```
DerivType operator+ ( DerivType& u, interval& v )
{
  DerivType res;

  res.f = u.f + v;
  if (DerivOrder > 0) {
    res.df = u.df;
    if (DerivOrder > 1) res.ddf = u.ddf;
  }
  return res;
}

DerivType operator- ( DerivType& u, interval& v )
{
  DerivType res;

  res.f = u.f - v;
  if (DerivOrder > 0) {
    res.df = u.df;
    if (DerivOrder > 1) res.ddf = u.ddf;
  }
  return res;
}

DerivType operator* ( DerivType& u, interval& v )
{
  DerivType res;

  res.f = u.f * v;
  if (DerivOrder > 0) {
    res.df = u.df * v;
    if (DerivOrder > 1) res.ddf = u.ddf * v;
  }
  return res;
}

DerivType operator/ ( DerivType& u, interval& v )
{
  DerivType res;

  res.f = u.f / v;        // Can propagate 'division by zero' error
  if (DerivOrder > 0) {   //--------------------------------------
    res.df = u.df / v;
    if (DerivOrder > 1) res.ddf = u.ddf / v;
  }
  return res;
}

//----------------------------------------------------------------------------
// Operands for +, -, *, and / for one real and one DerivType operand
//----------------------------------------------------------------------------
DerivType operator+ ( const real& u, DerivType& v )
  { return( _interval(u) + v ); }

DerivType operator- ( const real& u, DerivType& v )
  { return( _interval(u) - v ); }

DerivType operator* ( const real& u, DerivType& v )
  { return( _interval(u) * v ); }

DerivType operator/ ( const real& u, DerivType& v )
  { return( _interval(u) / v ); }    // Can propagate 'division by zero' error
                                     //--------------------------------------
```

```
DerivType operator+ ( DerivType& u, const real& v )
  { return( u + _interval(v) ); }

DerivType operator- ( DerivType& u, const real& v )
  { return( u - _interval(v) ); }

DerivType operator* ( DerivType& u, const real& v )
  { return( u * _interval(v) ); }

DerivType operator/ ( DerivType& u, const real& v )
  { return( u / _interval(v) ); }    // Can propagate 'division by zero' error
                                     //--------------------------------------

//----------------------------------------------------------------------------
// Elementary functions for DerivType arguments
//----------------------------------------------------------------------------
DerivType sqr ( DerivType& u )
{
  DerivType res;

  res.f = Power(u.f,2);
  if (DerivOrder > 0) {
    res.df = 2.0*u.f*u.df;
    if (DerivOrder > 1) res.ddf = 2.0 * (Power(u.df,2) + u.f*u.ddf);
  }
  return res;
}

DerivType power ( DerivType& u, int k )
{
  DerivType res;
  interval  h1;

  if (k == 0)
     res = DerivConst(1.0);
  else if (k == 1)
    res = u;
  else {
    res.f = Power(u.f, k);
    if (DerivOrder > 0) {
      h1 = double(k) * Power(u.f, k-1);
      res.df = h1 * u.df;
      if (DerivOrder > 1)
        res.ddf = h1 * u.ddf + double(k*(k-1))*Power(u.f,k-2)*Power(u.df,2);
    }
  }
  return res;
}

DerivType sqrt ( DerivType& u )
{
  DerivType res;
  interval  h1, h2;

  h1 = sqrt(u.f);               // Can propagate domain error
  res.f = h1;                   //----------------------------
  if (DerivOrder > 0) {
    h1 = 0.5/h1;
    h2 = u.df*h1;
    res.df = h2;
    if (DerivOrder > 1) res.ddf = u.ddf*h1 - 0.5*u.df/u.f*h2;
  }
  return res;
}
```

```
DerivType exp ( DerivType& u )
{
  DerivType res;
  interval  h1, h2;

  h1 = exp(u.f);
  res.f = h1;
  if (DerivOrder > 0) {
    h2 = h1*u.df;
    res.df = h2;
    if (DerivOrder > 1) res.ddf = h1*u.ddf + h2*u.df;
  }
  return res;
}

DerivType ln ( DerivType& u )
{
  DerivType res;
  interval  h;

  res.f = ln(u.f);              // Can propagate domain error
  if (DerivOrder > 0) {         //----------------------------
    h = u.df/u.f;
    res.df = h;
    if (DerivOrder > 1) res.ddf = (u.ddf - h*u.df) / u.f;
  }
  return res;
}

DerivType sin ( DerivType& u )
{
  DerivType res;
  interval  h0, h1;

  h0 = sin(u.f);
  res.f = h0;
  if (DerivOrder > 0) {
    h1 = cos(u.f);
    res.df = h1*u.df;
    if (DerivOrder > 1) res.ddf = h1*u.ddf - h0*sqr(u.df);
  }
  return res;
}

DerivType cos ( DerivType& u )
{
  DerivType res;
  interval  h0, h1;

  h0 = cos(u.f);
  res.f = h0;
  if (DerivOrder > 0) {
    h1 = -sin(u.f);
    res.df = h1*u.df;
    if (DerivOrder > 1) res.ddf = h1*u.ddf - h0*sqr(u.df);
  }
  return res;
}

DerivType tan ( DerivType& u )
{
  DerivType res;
  interval  h0, h1, h2;
```

```
  h0 = tan(u.f);                // Can propagate domain error
  res.f = h0;                   //--------------------------
  if (DerivOrder > 0) {                        // The subdistributive law implies
    h1 = sqr(h0)+1.0;                          //   h0 * (h0↑2 + 1) <= h0↑3 + h0.
    h2 = 2.0*h0*h1;                            // So, the first form is used.
    res.df = h1*u.df;                          //--------------------------------
    if (DerivOrder > 1) res.ddf = h1*u.ddf + h2*sqr(u.df);
  }
  return res;
}

DerivType cot ( DerivType& u )
{
  DerivType res;
  interval  h0, h1, h2;

  h0 = cot(u.f);                // Can propagate domain error
  res.f = h0;                   //--------------------------
  if (DerivOrder > 0) {                        // The subdistributive law implies
    h1 = -(sqr(h0)+1.0);                       //   h0 * (h0↑2 + 1) <= h0↑3 + h0.
    h2   = -2.0*h0*h1;                         // So, the first form is used.
    res.df = h1*u.df;                          //--------------------------------
    if (DerivOrder > 1) res.ddf = h1*u.ddf + h2*sqr(u.df);
  }
  return res;
}

DerivType asin ( DerivType& u )
{
  DerivType res;
  interval  h, h1, h2;

  res.f = asin(u.f);            // Can propagate domain error
  if (DerivOrder > 0) {         //--------------------------
    h  = 1.0 - sqr(u.f);
    h1 = 1.0/sqrt(h);
    res.df = h1*u.df;
    h2 = u.f*h1/h;
    if (DerivOrder > 1) res.ddf = h1*u.ddf + h2*sqr(u.df);
  }
  return res;
}

DerivType acos ( DerivType& u )
{
  DerivType res;
  interval  h, h1, h2;

  res.f = acos(u.f);            // Can propagate domain error
  if (DerivOrder > 0) {         //--------------------------
    h  = 1.0 - sqr(u.f);
    h1 = -1.0/sqrt(h);
    res.df = h1*u.df;
    h2 = u.f*h1/h;
    if (DerivOrder > 1) res.ddf = h1*u.ddf + h2*sqr(u.df);
  }
  return res;
}

DerivType atan ( DerivType& u )
{
  DerivType res;
  interval  h1, h2;
```

```
  res.f = atan(u.f);
  if (DerivOrder > 0) {
    h1 = 1.0 / (1.0 + sqr(u.f));
    res.df = h1*u.df;
    h2 = -2.0*u.f*sqr(h1);
    if (DerivOrder > 1) res.ddf = h1*u.ddf + h2*sqr(u.df);
  }
  return res;
}

DerivType acot ( DerivType& u )
{
  DerivType res;
  interval  h1, h2;

  res.f = acot(u.f);            // Can propagate domain error
  if (DerivOrder > 0) {         //----------------------------
    h1 = -1.0 / (1.0 + sqr(u.f));
    res.df = h1*u.df;
    h2 = 2.0*u.f*sqr(h1);
    if (DerivOrder > 1) res.ddf = h1*u.ddf + h2*sqr(u.df);
  }
  return res;
}

DerivType sinh ( DerivType& u )
{
  DerivType res;
  interval  h0, h1;

  h0 = sinh(u.f);
  res.f = h0;
  if (DerivOrder > 0) {
    h1 = cosh(u.f);
    res.df = h1*u.df;
    if (DerivOrder > 1) res.ddf = h1*u.ddf + h0*sqr(u.df);
  }
  return res;
}

DerivType cosh ( DerivType& u )
{
  DerivType res;
  interval  h0, h1;

  h0 = cosh(u.f);
  res.f = h0;
  if (DerivOrder > 0) {
    h1 = sinh(u.f);
    res.df = h1*u.df;
    if (DerivOrder > 1) res.ddf = h1*u.ddf + h0*sqr(u.df);
  }
  return res;
}

DerivType tanh ( DerivType& u )
{
  DerivType res;
  interval  h0, h1, h2;

  h0 = tanh(u.f);
  res.f = h0;
  if (DerivOrder > 0) {                        // The subdistributive law implies
```

```
    h1 = 1.0 - sqr(h0);                        //   h0 * (h0↑2 - 1) <= h0↑3 - h0.
    h2 = -2.0*h0*h1;                           // So, the first form is used.
    res.df = h1*u.df;                          //--------------------------------
    if (DerivOrder > 1) res.ddf = h1*u.ddf + h2*sqr(u.df);
  }
  return res;
}

DerivType coth ( DerivType& u )
{
  DerivType res;
  interval  h0, h1, h2;

  h0 = coth(u.f);                 // Can propagate domain error
  res.f = h0;                     //--------------------------
  if (DerivOrder > 0) {                        // The subdistributive law implies
    h1 = 1.0 - sqr(h0);                        //   h0 * (h0↑2 - 1) <= h0↑3 - h0.
    h2 = -2.0*h0*h1;                           // So, the first form is used.
    res.df = h1*u.df;                          //--------------------------------
    if (DerivOrder > 1) res.ddf = h1*u.ddf + h2*sqr(u.df);
  }
  return res;
}

DerivType asinh ( DerivType& u )
{
  DerivType res;
  interval  h, h1, h2;

  res.f = asinh(u.f);             // Can propagate domain error
  if (DerivOrder > 0) {           //--------------------------
    h  = 1.0 + sqr(u.f);
    h1 = 1.0/sqrt(h);
    res.df = h1*u.df;
    h2 = -u.f*h1/h;
    if (DerivOrder > 1) res.ddf = h1*u.ddf + h2*sqr(u.df);
  }
  return res;
}

DerivType acosh ( DerivType& u )
{
  DerivType res;
  interval  h, h1, h2;

  res.f = acosh(u.f);             // Can propagate domain error
  if (DerivOrder > 0) {           //--------------------------
    h  = sqr(u.f) - 1.0;
    h1 = 1.0/sqrt(h);
    res.df = h1*u.df;
    h2 = -u.f*h1/h;
    if (DerivOrder > 1) res.ddf = h1*u.ddf + h2*sqr(u.df);
  }
  return res;
}

DerivType atanh ( DerivType& u )
{
  DerivType res;
  interval  h1, h2;

  res.f = atanh(u.f);             // Can propagate domain error
  if (DerivOrder > 0) {           //--------------------------
    h1 = 1.0 / (1.0 - sqr(u.f));
```

```
    res.df = h1*u.df;
    h2 = 2.0*u.f*sqr(h1);
    if (DerivOrder > 1) res.ddf = h1*u.ddf + h2*sqr(u.df);
  }
  return res;
}

DerivType acoth ( DerivType& u )
{
  DerivType res;
  interval  h1, h2;

  res.f = acoth(u.f);          // Can propagate domain error
  if (DerivOrder > 0) {        //---------------------------
    h1 = 1.0 / (1.0 - sqr(u.f));
    res.df = h1*u.df;
    h2 = 2.0*u.f*sqr(h1);
    if (DerivOrder > 1) res.ddf = h1*u.ddf + h2*sqr(u.df);
  }
  return res;
}

//----------------------------------------------------------------------------
// Predefined routines for evaluation of DerivType-functions
//----------------------------------------------------------------------------

//----------------------------------------------------------------------------
// Purpose: Evaluation of function 'f' for argument 'x' in differentiation
//    arithmetic computing only the function value.
// Parameters:
//    In : 'f' : function of 'DerivType'
//         'x' : argument for evaluation of 'f'
//    Out: 'fx': returns the function value 'f(x)'
// Description: This function sets 'DerivOrder' to 0, evaluates 'f(x)' in
//    differentiation arithmetic, and returns the function value only.
//----------------------------------------------------------------------------
void fEval ( ddf_FctPtr f, interval x, interval& fx )
{
  DerivType xD, fxD;

  DerivOrder = 0;
  xD         = DerivVar(x);
  fxD        = f(xD);
  fx         = fxD.f;
  DerivOrder = 2;
}

//----------------------------------------------------------------------------
// Purpose: Evaluation of function 'f' for argument 'x' in differentiation
//    arithmetic computing the function value and the value of the first
//    derivative.
// Parameters:
//    In : 'f'   : function of 'DerivType'
//         'x'   : argument for evaluation of 'f'
//    Out: 'fx'  : returns the function value 'f(x)'
//         'dfx': returns the first derivative value 'f'(x)'
// Description: This function sets 'DerivOrder' to 1, evaluates 'f(x)' in
//    differentiation arithmetic, and returns the function value and the
//    value of the first derivative.
//----------------------------------------------------------------------------
void dfEval ( ddf_FctPtr f, interval x, interval& fx, interval& dfx )
{
  DerivType xD, fxD;
```

```
  DerivOrder = 1;
  xD         = DerivVar (x);
  fxD        = f(xD);
  fx         = fxD.f;
  dfx        = fxD.df;
  DerivOrder = 2;
}

//----------------------------------------------------------------------------
// Purpose: Evaluation of function 'f' for argument 'x' in differentiation
//    arithmetic computing the function value, the value of the first, and
//    the value of the second derivative.
// Parameters:
//    In : 'f'   : function of 'DerivType'
//         'x'   : argument for evaluation of 'f'
//    Out: 'fx'  : returns the function value 'f(x)'
//         'dfx' : returns the value of the first derivative 'f'(x)'
//         'ddfx': returns the value of the second derivative 'f''(x)'
// Description: This function keeps 'DerivOrder' = 2, evaluates 'f(x)' in
//    differentiation arithmetic, and returns the function value, the value
//    of the first, and the value of the second derivative.
//----------------------------------------------------------------------------
void ddfEval ( ddf_FctPtr f, interval x, interval& fx, interval& dfx,
                                                        interval& ddfx )
{
  DerivType xD, fxD;

  xD   = DerivVar(x);
  fxD  = f(xD);
  fx   = fxD.f;
  dfx  = fxD.df;
  ddfx = fxD.ddf;
}
```

### 5.3.2 Examples

We first illustrate how to write a program using our differentiation arithmetic in Example 5.1:

**Example 5.2** We must define a C++ function with result type *DerivType* for the function $f(x) = x \cdot (4 + x)/(3 - x)$:

```
DerivType f ( DerivType x ) {
  return( x * (4.0 + x) / (3.0 - x) );
}
```

With the declarations

```
DerivType  x, fx;
interval   y, fy, dfy, ddfy;
```

we are able to compute function and derivative values

```
   y = 123.0;
   x = DerivVar(y);
  fx = f(x);
  fy = fValue(fx);       // Function value f(y)
 dfy = dfValue(fx);      // First derivative f'(y)
ddfy = ddfValue(fx);     // Second derivative f''(y)
```

Even easier is

```
y = 123.0;
ddfEval(f,y,fy,dfy,ddfy);
```

**Example 5.3** Newton's method for approximating a zero of a nonlinear function $f : \mathbb{R} \to \mathbb{R}$ requires $f'$. Other methods use derivatives of second or higher order. Halley's method uses the first and the second derivatives of $f$. Starting from an approximation $x^{(0)} \in \mathbb{R}$, Halley's method iterates according to

$$\left.\begin{array}{rcl} a^{(k)} & := & -\dfrac{f(x^{(k)})}{f'(x^{(k)})} \\[2ex] b^{(k)} & := & a^{(k)} \cdot \dfrac{f''(x^{(k)})}{f'(x^{(k)})} \\[2ex] x^{(k+1)} & := & x^{(k)} + \dfrac{a^{(k)}}{1 + \frac{b^{(k)}}{2}} \end{array}\right\} \quad k = 0, 1, 2, \ldots$$

In our sample program for Halley's method, all real operations are replaced by interval operations, because we make use of the interval differentiation arithmetic. The influence of rounding errors is demonstrated directly by the increasing interval diameters. We terminate the iteration if an interval evaluation of $f$ over the current interval iterate $[x]^{(k)}$ contains zero, or if the number of iterations is greater than or equal to the constant value *MaxIter*. This approximate method can *not* guarantee to find and enclose a zero of the function $f$ because we have enclosed only roundoff errors in Halley's method, not truncation errors. We present in Chapter 6 a more sophisticated interval method for finding *all* zeros of a nonlinear function with guarantees and high accuracy.

```
//----------------------------------------------------------------------------
// Example: Differentiation arithmetic (one-dimensional)
// This is an implementation of Halley's method for approximating a zero of
// a twice continuously differentiable function.
//
//    given:        the function f(x) and the starting value x[0]
//    iteration:    x[k+1] := x[k] + a[k]/(1+b[k]/2)            \
//    with:         a[k]   := - f(x[k])/f'(x[k])                 ) k=0,1,2,...
//                  b[k]   := a[k]*f''(x[k])/f'(x[k])           /
//
// All real operations are replaced by interval operations. All function and
// derivative evaluations are calculated by differentiation arithmetic.
// This approximate method can NOT guarantee to find and enclose a zero of
// the function 'f' because we have enclosed only roundoff errors, not
// truncation errors.
//----------------------------------------------------------------------------
#include <ddf_ari.hpp>      // Differentiation arithmetic

const int MaxIter = 100;    // Maximum number of iterations

interval  x, a, b, fx, dfx, ddfx;
int       n;
real      start;
```

```
DerivType f ( DerivType xD )                    // Sample function
  { return( exp(xD) * sin(4.0*xD) ); }

int main ( )
{
  start = 1.25;

  cout << SetPrecision(23,15) << Scientific;   // Output format

  cout << "Halley's method for the function  f(x) = exp(x) * sin(4*x)"
       << endl << endl;
  cout << "Starting value = " << start << endl;
  cout << "Iteration:" << endl;

  x = start;
  ddfEval(f,x,fx,dfx,ddfx);
  n = 0;
  do {
    n++;
    cout << "x               : " << x << endl;
    cout << "f(x)            : " << fx << endl;
    a = - fx / dfx;
    b = ddfx / dfx * a;
    x = x + a / (1.0 + 0.5*b);
    ddfEval(f,x,fx,dfx,ddfx);
  } while( !( in(0.0,fx) || (n >= MaxIter) ) );

  cout << endl
       << "Approximate zero : " << x << endl;
  cout << "Function value   : " << fx << endl << endl;
  cout << "Expected zero    : 1.570796326794896619202..." << endl;

  return 0;
}
```

If we run our sample program for Halley's method, we get the following runtime output.

```
Halley's method for the function  f(x) = exp(x) * sin(4*x)

Starting value =  1.250000000000000E+000
Iteration:
x              : [ 1.250000000000000E+000, 1.250000000000000E+000]
f(x)           : [-3.346974588809691E+000,-3.346974588809689E+000]
x              : [ 1.271022734475337E+000, 1.271022734475338E+000]
f(x)           : [-3.321078714403206E+000,-3.321078714403198E+000]
x              : [ 1.330698806738156E+000, 1.330698806738163E+000]
f(x)           : [-3.100410348120015E+000,-3.100410348119942E+000]
x              : [ 1.463583515060168E+000, 1.463583515060210E+000]
f(x)           : [-1.796959342435823E+000,-1.796959342435113E+000]
x              : [ 1.565154617724422E+000, 1.565154617724544E+000]
f(x)           : [-1.079373694784440E-001,-1.079373694761117E-001]
x              : [ 1.570795749328203E+000, 1.570795749328450E+000]
f(x)           : [-1.111155544687062E-005,-1.111155072141681E-005]

Approximate zero : [ 1.570796326794650E+000, 1.570796326795143E+000]
Function value   : [-4.726632018387835E-012, 4.728548126295277E-012]

Expected zero    : 1.570796326794896619202...
```

Mathematically, Halley's method is converging to the correct root. The floating-point algorithm converges, but grows wider. This shows the importance of using true *interval* algorithms, as opposed to *point* algorithms in interval arithmetic. The algorithm presented in Chapter 6 uses a contractive map to generate a sequence of intervals that both

- is guaranteed to enclose the root, and
- can achieve one ulp accuracy.

### 5.3.3 Restrictions and Hints

The implementation in module *ddf_ari* uses the standard error handling of the interval library employed, if the interval argument of an elementary function does not lie in the domain specified. The same holds for an interval containing zero as second operand of the division operator. These cases can also occur during a function evaluation using differentiation arithmetic because of the known overestimation effects of interval arithmetic (see Section 3.1).

The rules for getting true enclosures in connection with conversion errors discussed in Section 3.7 also apply to interval differentiation arithmetic. If you want to compute enclosures for the values of the derivatives of the function $f$ in Example 5.2 at the point $x = 0.1$, you must insure that the machine interval argument $[x]$ used as argument for the interval function evaluation satisfies $x \in [x]$.

## 5.4 Exercises

**Exercise 5.1** Apply the sample program of page 89 to find positive roots of the function $f(x) = x - \tan x$.

**Exercise 5.2** Implement a *real* differentiation arithmetic by replacing in module *ddf_ari* the data type *interval* by the data type *real*. Then use the new module for Halley's method.

## 5.5 References and Further Reading

The method used in the implementation above is called the *forward method* of automatic differentiation, because all differentiation arithmetic operations can be executed in the same order as the corresponding floating-point operations. Automatic differentiation for gradients and Hessian matrices (see Chapter 12) can optimize the time complexity using the *backward method*, also called fast automatic differentiation (see [18]).

Automatic differentiation methods can also be used to compute interval slopes (see [70]).

Further applications and differentiation arithmetics such as Taylor arithmetics, power series arithmetics, or parallel implementation of differentiation arithmetics can be found in [22], [23], and [75]. A large bibliography on automatic differentiation is given in [12].

# Chapter 6

# Nonlinear Equations in One Variable

One of the most important tasks in scientific computing is the problem of finding zeros (or roots) of nonlinear functions. In classical numerical analysis, root-finding methods for nonlinear functions begin with an approximation and apply an iterative method (such as Newton's or Halley's methods), which hopefully improves the approximation. It is a myth that no numerical algorithm is able to compute *all* zeros of a nonlinear equation with guaranteed error bounds, or even more, that no method is able to give concrete information about the existence and uniqueness of solutions of such a problem.

In this chapter, we consider a method for finding *all* zeros of a nonlinear continuously differentiable function $f : I\!R \rightarrow I\!R$ in a given interval and consider the domain of applicability of that method. It computes tight bounds on the roots, and it delivers information about existence and uniqueness of the computed solutions.

The method we present is an extension of *interval Newton's method* which makes use of the extended interval operations defined in Section 3.3. Interval Newton's method was derived by Moore [67]. Modifications, extensions, and applications are given by Alefeld [3], Hansen [31], and many others.

## 6.1 Theoretical Background

We address the problem of finding all solutions of the equation

$$f(x) = 0 \tag{6.1}$$

for a continuously differentiable function $f : I\!R \rightarrow I\!R$ and $x \in [x]$. Interval Newton's method for solving (6.1) can easily be derived from the mean value form

$$f(m([x])) - f(x^*) = f'(\xi) \cdot (m([x]) - x^*),$$

where $x^*, \xi \in [x]$. If we assume $x^*$ to be a zero of $f$, we get

$$x^* = m([x]) - \frac{f(m([x]))}{f'(\xi)} \in m([x]) - \frac{f(m([x]))}{f'([x])} =: N([x]). \tag{6.2}$$

Hence, every zero of $f$ in $[x]$ also lies in $N([x])$, and therefore in $N([x]) \cap [x]$. Using standard interval arithmetic, interval Newton's method starts with an interval $[x]^{(0)}$ satisfying $0 \notin f'([x]^{(0)})$ and iterates according to

$$[x]^{(k+1)} := [x]^{(k)} \cap N([x]^{(k)}), \quad k = 0, 1, 2, \ldots \tag{6.3}$$

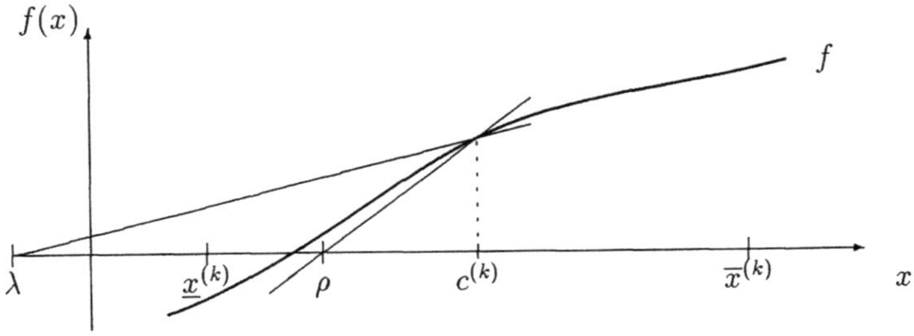


Figure 6.1: Interval Newton step with $0 \notin f'([x]^{(k)})$

The method cannot diverge due to the intersection of $N([x]^{(k)})$ with $[x]^{(k)}$. Like classical Newton's method, interval Newton's method can be geometrically interpreted as drawing two lines from the point $(c^{(k)}, f(c^{(k)}))$ with $c^{(k)} = m([x]^{(k)})$, and intersecting them with the $x$-axis. The lines have the slope $\underline{g}$ (the smallest slope of $f$ in $[x]$) and $\overline{g}$ (the largest slope of $f$ in $[x]$), respectively, where $[\underline{g}, \overline{g}] = f'([x]^{(k)})$. The points of intersection with the $x$-axis, let us call them $\lambda$ and $\rho$, form the new interval $N([x]^{(k)}) = \lambda \underline{\cup} \rho$. Figure 6.1 demonstrates a single interval Newton step with resulting interval

$$[x]^{(k+1)} = [\lambda, \rho] \cap [x]^{(k)} = [\underline{x}^{(k)}, \rho].$$

If the intersection is empty, we know that there is no root of $f$ in $[x]^{(k)}$.

Using extended interval arithmetic as defined in Section 3.3, we are able to treat the case $0 \in f'([x]^{(0)})$ that occurs if there are several zeros in the starting interval $[x]^{(0)}$. In this case, $N([x]^{(k)})$ is given by one or two extended intervals resulting from the interval division. Even though $N([x]^{(k)})$ is not finite, the intersection $[x]^{(k+1)} = N([x]^{(k)}) \cap [x]^{(k)}$ is finite and may be a single interval, the union of two intervals, or the empty set. Then, the next step of interval Newton's iteration must be applied to each of the resulting intervals. In this way it is possible to enclose *all* zeros of $f$ in the starting interval $[x]^{(0)}$.

In Figure 6.2, we illustrate one extended interval Newton step geometrically. Again we draw lines through the point $(c^{(k)}, f(c^{(k)}))$ with $c^{(k)} = m([x]^{(k)})$. The first line with the smallest (negative) slope of $f$ in $[x]^{(k)}$ intersects the $x$-axis in point $\rho$. The line with the largest slope intersects the $x$-axis in point $\lambda$. Therefore, we get

$$N([x]^{(k)}) = [-\infty, \rho] \cup [\lambda, \infty]$$

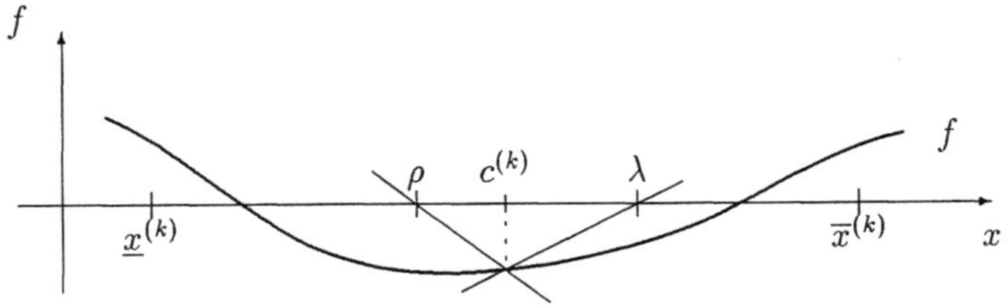


Figure 6.2: Extended interval Newton step with $0 \in f'([x]^{(k)})$

and

$$[x]^{(k+1)} = [\underline{x}^{(k)}, \rho] \cup [\lambda, \overline{x}^{(k)}].$$

The following theorem summarizes the most important properties of interval Newton's method.

**Theorem 6.1** *Let $f : D \subseteq I\!R \to I\!R$ be a continuously differentiable function, and let $[x] \in II\!R$ be an interval with $[x] \subseteq D$. Then*

$$N([x]) := m([x]) - \frac{f(m([x]))}{f'([x])}$$

*has the following properties:*

1. *Every zero $x^* \in [x]$ of $f$ satisfies $x^* \in N([x])$.*
2. *If $N([x]) \cap [x] = \emptyset$, then there exists no zero of $f$ in $[x]$.*
3. *If $N([x]) \overset{\circ}{\subset} [x]$, then there exists a unique zero of $f$ in $[x]$ and hence in $N([x])$.*

The proofs appear in [31], [67], or [70].

**Remark:** The conditions of Theorem 6.1 can be checked on a computer. For example, if $N_\diamond([x])$ denotes the evaluation of $N([x])$ in floating-interval arithmetic and if Condition 3 is satisfied for $N_\diamond$, then we have

$$N([x]) \subseteq N_\diamond([x]) \overset{\circ}{\subset} [x].$$

Thus, the condition is fulfilled for $N([x])$, too. On the other hand, if we cannot fulfill Conditions 2 and 3 of the Theorem, then $[x]$ *may* contain one or more zeros.

## 6.2 Algorithmic Description

Algorithm 6.3 AllZeros consists of two parts. The first part is the extended interval Newton iteration itself, including intermediate checks for the uniqueness of a zero in a computed enclosure. The second part is an additional verification step which tries to verify the uniqueness for enclosures that have not already been marked as enclosing a unique zero.

Algorithm 6.1 is a recursive function for the execution of the extended interval Newton method for the function $f$. The input parameter $[y]$ specifies the actual interval. The input parameter $\varepsilon$ corresponds to the desired relative accuracy or tolerance for the resulting interval enclosures of the zeros. The calling function AllZeros guarantees that $\varepsilon$ is not chosen less than the relative machine accuracy (1 ulp). The input parameter *yUnique* signals whether we have already verified that the incoming interval $[y]$ contains a unique zero.

If $0 \notin f([y])$, then we know that no zero can lie within the interval $[y]$. Hence, a single interval function evaluation of $f$ can guarantee that a complete range of real values cannot contain a zero of $f$. If $f([y])$ contains zero, then the extended interval

Newton step given by (6.3) is applied to $[y]$ resulting in at most two intervals $[V]_i$ (Steps 3 and 4). Interval arithmetic must be used to evaluate $f(c)$ in an implementation in order to bound all rounding errors. Therefore, we use the notation $f_\diamond(c)$ in Step 3 of the algorithm.

If no improvement for $[y]$ is achieved, $[y]$ gets bisected (Step 5). If the Newton step results in only one single interval and if the uniqueness of a zero in $[y]$ is not already proven, then the algorithm checks the Condition 3 of Theorem 6.1 for the resulting interval $[V]_1$, and sets the flag *yUnique* (Step 6).

In Step 7(b) the actual interval $[V]_i$ gets stored in the interval vector $[Zero]$ if the tolerance criterion is satisfied. Otherwise, the function XINewton is recursively called (else-branch of Step 7(b)). The corresponding information on the uniqueness of the zero is stored in the flag vector *Info*. $N$ represents the number of enclosures of zeros stored in $[Zero]$.

Function XINewton terminates when no more recursive calls are necessary, that is if $0 \notin f([y])$ or if $(d_{\text{rel}}([V]_i) \leq \varepsilon)$ for $i = 1, 2$. The bisection step (Step 5) guarantees that this second condition is fulfilled at some stage of the recursion.

**Algorithm 6.1:** XINewton ($f, [y], \varepsilon, yUnique, [Zero], Info, N$)

1. **if** $0 \notin f([y])$ **then return**;
2. $c := m([y])$;
3. $[z] := c - f_\diamond(c)/f'([y])$; {Extended interval Newton step}
4. $[V] := [y] \cap [z]$; {Intersection $[y] \cap [z] = [V]_1 \cup [V]_2$}
5. **if** $[V]_1 = [y]$ **then** $[V]_1 := [\underline{y}, c]$; $[V]_2 := [c, \overline{y}]$; {Bisection}
6. **if** $[V]_1 \neq \emptyset$ **and** $[V]_2 = \emptyset$ **then**
   $yUnique := yUnique$ **or** $([V]_1 \overset{\circ}{\subset} [y])$; {Inner inclusion $\Rightarrow$ uniqueness}
   **else**
   $yUnique := false$;
7. **for** $i := 1$ **to** 2 **do**
   (a) **if** $[V]_i = \emptyset$ **then next**$_i$;
   (b) **if** $(d_{\text{rel}}([V]_i) \leq \varepsilon)$ **then** {Store enclosure of zero and uniqueness info}
       **if** $(0 \in f([V]_i))$ **then**
           $N := N + 1$; $[Zero]_N := [V]_i$; $Info_N := yUnique$;
       **else** {Recursive call of XINewton for $[V]_i$}
       XINewton ($f, [V]_i, \varepsilon, yUnique, [Zero], Info, N$);
8. **return** $[Zero]$, $Info$, $N$;

Algorithm 6.2 describes an additional verification step which checks the uniqueness of the zero enclosed in the interval $[y]$. It can be used for intervals which have not yet been guaranteed to enclose a unique zero. This is done according to Condition 3 of Theorem 6.1 by applying interval Newton steps including an epsilon inflation of the iterates $[y]$, which can help to verify zeros lying on the edge of $[y]$. We use $k_{\max} = 10$ as the maximum number of iterations, $\varepsilon = 0.25$ as the starting value for the epsilon inflation, and a factor of 8 to increase $\varepsilon$ within the iterations. It turned out that these are good values for minimizing the effort if no verification is possible (see also [52]).

**Algorithm 6.2:** VerificationStep $(f, [y], yUnique)$

1. $k_{\max} := 10$; $k := 0$; $[y_{\text{in}}] := [y]$; $\varepsilon := 0.25$; $yUnique := false$; {Initializations}
2. **while** (**not** $yUnique$) **and** $(k < k_{\max})$ **do** {Do $k_{\max}$ loops to achieve inclusion}
   (a) $[y_{\text{old}}] := [y] \bowtie \varepsilon$; {Epsilon inflation of $[y]$}
   (b) **if** $0 \in f'([y_{\text{old}}])$ **then exit**$_{\text{while-loop}}$; {No verification possible}
   (c) $k := k + 1$; $c := m([y_{\text{old}}])$;
   (d) $[y] := c - f_{\diamond}(c)/f'([y_{\text{old}}])$; {Interval Newton step}
   (e) **if** $[y] \cap [y_{\text{old}}] = \emptyset$ **then exit**$_{\text{while-loop}}$; {No root}
   (f) $yUnique := ([y] \overset{\circ}{\subset} [y_{\text{old}}])$; {Inner inclusion $\Rightarrow$ uniqueness}
   (g) $[y] := [y] \cap [y_{\text{old}}]$; {Intersection with old value}
   (h) **if** $[y] = [y_{\text{old}}]$ **then** $\varepsilon := \varepsilon \cdot 8$; {Increase $\varepsilon$}
3. **if not** $yUnique$ **then** $[y] := [y_{\text{in}}]$; {Reset $[y]$ to starting value}
4. **return** $[y]$, $yUnique$;

Algorithm 6.3 now combines Algorithm 6.1 and Algorithm 6.2 to compute enclosures for all zeros of the function $f$ within the input interval $[x]$ and tries to prove the local uniqueness of a zero in each enclosure computed. The desired accuracy (relative diameter) of the interval enclosures is specified by the input parameter $\varepsilon$. 1 ulp accuracy is chosen if the specified value of $\varepsilon$ is too small (for example 0). The enclosures for the zeros of $f$ are returned in the interval vector $[Zero]$. The corresponding information on the local uniqueness of the zeros are returned in the Boolean vector *Info*. The number of enclosures computed is returned in the integer variable $N$.

We use a function called CheckParameters as an abbreviation for the error checks for the parameters of AllZeros which are necessary in an implementation. If no error occurs, AllZeros delivers the $N$ enclosures $[Zero]_i$, $i = 1, 2, \ldots, N$, and for each $i = 1, 2, \ldots, N$,

if $Info_i = true$, then $[Zero]_i$ encloses a locally unique zero of $f$,
if $Info_i = false$, then $[Zero]_i$ may enclose a zero of $f$.

If $N = 0$, then it is guaranteed that there is no zero of $f$ in the starting interval $[x]$.

**Algorithm 6.3:** AllZeros ($f, [x], \varepsilon, [Zero], Info, N, Err$)

1. $Err :=$ CheckParameters;
2. **if** $Err \neq$ "No Error" **then return** $Err$;
3. $N := 0$;
4. Set $\varepsilon$ to "1 ulp accuracy" if $\varepsilon$ is too small;
5. XINewton ($f, [x], \varepsilon, false, [Zero], Info, N$);
6. **for** $i := 1$ **to** $N$ **do**
   **if** $Info_i \neq true$ **then** VerificationStep ($f, [Zero]_i, Info_i$);
7. **return** $[Zero], Info, N, Err$;

### Applicability of the Algorithm

To keep our algorithm and implementation as easy to use as possible, Algorithm 6.1 uses only the very simple stopping criteria

$$d_{\text{rel}}([V]_i) \leq \varepsilon.$$

Hence, the method reduces to a bisection method if the interval Newton steps do not improve the actual interval iterate. For more sophisticated stopping criteria, see [31] for example. On the other hand, if the actual interval iterate $[y]$ satisfies the condition

$$0 \notin f'([y]),$$

then the asymptotic rate of convergence to a zero of $f$ in $[y]$ is quadratic.

The algorithm cannot verify the existence and the uniqueness of a multiple zero $x^*$ of $f$ in $[y]$. Nevertheless, the zero is bounded to the desired accuracy specified by $\varepsilon$. In this case, the corresponding component of the *Info*-vector is set to the value *false*. When the zero lies exactly at a splitting point, that zero is enclosed in two different intervals as a consequence of the bisection of the intervals.

The algorithm is well suited for parallel computers due to its natural parallelism given by the splitting of the intervals by the extended interval Newton step or by the bisection.

## 6.3 Implementation and Examples

### 6.3.1 C++ Program Code

First, we list our implementation of the operations of the extended interval arithmetic needed in the extended interval Newton step (see Section 3.3). Newton's method is the only application we make of extended interval arithmetic. Therefore, we provide only those operations on extended intervals we require for our implementation of Newton's method. An implementation of iteration (6.3) requires only operators for

| | | | | |
|---|---|---|---|---|
| interval | / | interval | $\rightarrow$ | extended interval, |
| real | $-$ | extended interval | $\rightarrow$ | extended interval, and |
| interval | $\cap$ | extended interval | $\rightarrow$ | 0, 1, or 2 intervals. |

Second, we list the implementation of Algorithm 6.3 and its subalgorithms. We use module *ddf_ari* for differentiation arithmetic given in Chapter 5.

#### 6.3.1.1 Module xi_ari

The header file of module *xi_ari* supplies the definition of the class *xinterval* representing a normal interval, a single extended interval, or a pair of extended intervals. The data member *kind* specifies these cases by

| *kind* | interval form ($a, b \in I\!R$) |
|---|---|
| *Finite* | $[a, b], \quad a \leq b,$ |
| *PlusInfty* | $[a, \infty],$ |
| *MinusInfty* | $[-\infty, a],$ |
| *Double* | $[-\infty, a] \cup [b, \infty], \quad a \leq b.$ |

The case $[-\infty, \infty]$ is represented by $[-\infty, a] \cup [a, \infty]$ for simplicity.

```
//----------------------------------------------------------------------------
// File: xi_ari (header)
// Purpose: Definition of an extended interval arithmetic which allows the
//    division by an interval containing zero.
// Global type:
//    KindType          : component type of extended intervals
// Global function:
//    EmptyIntval()     : empty set as irregular interval
// Class xinterval:
//    operators %, -, &: operators of extended interval arithmetic
//    operator =        : assignment operator
//----------------------------------------------------------------------------
#ifndef __XI_ARI_HPP
#define __XI_ARI_HPP

#include <interval.hpp>      // Interval arithmetic
#include <ivector.hpp>       // Interval vector arithmetic

typedef enum { Finite, PlusInfty, MinusInfty, Double } KindType;

extern interval EmptyIntval ( );    // Irregular (empty) interval

class xinterval {                   // Extended intervals according
  private:                          // to the above definition
    KindType  kind;                 //----------------------------
    real      inf, sup;

  public:
    xinterval ( );
    xinterval ( const KindType&, const real&, const real& );
    xinterval ( const xinterval& );
```

```
    xinterval& operator= ( xinterval& );

    friend xinterval operator- ( const real&, const xinterval& );
    friend xinterval operator% ( interval&, interval& );
    friend ivector   operator& ( interval&, const xinterval& );
};
#endif
```

The function *EmptyIntval()* delivers an irregular value of type *interval* (i.e. an interval with infimum greater than supremum) representing the empty set and serving as result of the intersection of two disjoint (extended) intervals.

Since overloaded operators are distinguished by their operands, the operator could not be used for the extended interval division. We used the operator % instead.

The operator & for the intersection of an interval with an extended interval pair of type *xinterval* delivers a result of type *ivector* with two components, where one or both of the interval components may be an empty interval.

```
//----------------------------------------------------------------------------
// File: xi_ari (implementation)
// Purpose: Definition of an extended interval arithmetic which allows the
//    division by an interval containing zero.
// Method: Overloading of operators for arithmetic and lattice operations
//    of data type 'xinterval'.
// Global type:
//    KindType           : component type of extended intervals
// Global function:
//    EmptyIntval()      : empty set as irregular interval
// Class xinterval:
//    operators %, -, &: operators of extended interval arithmetic
//    operator =         : assignment operator
//----------------------------------------------------------------------------
#include <i_util.hpp>      // Interval utility functions
#include <xi_ari.hpp>

//----------------------------------------------------------------------------
// An extended interval 'x', represented by the type 'xinterval', is defined
// according to the following rules (a <= b):
//
// x = [a, b]              :  x.kind = Finite,     x.inf = a, x.sup = b
// x = [a, +oo]            :  x.kind = PlusInfty,  x.inf = a, x.sup undefined
// x = [-oo, a]            :  x.kind = MinusInfty, x.sup = a, x.inf undefined
// x = [-oo, a] v [b, +oo] :  x.kind = Double,     x.inf = b, x.sup = a
// x = [-oo, +oo]          :  x.kind = Double,     x.inf = a, x.sup = a
//
// In this definition, 'v' stands for the set union and 'oo' for infinity.
//----------------------------------------------------------------------------
interval EmptyIntval ( )                        // Irregular (empty) interval
{                                               //---------------------------
  interval x;
  x = _unchecked_interval(999999999.0,-999999999.0);
  return x;
}

//----------------------------------------------------------------------------
// Constructors and assignment operator
//----------------------------------------------------------------------------
xinterval::xinterval ( )
{
  kind = Finite;
  inf  = 0.0;
```

```
  sup  = 0.0;
}

xinterval::xinterval ( const KindType& k, const real& i, const real& s )
{
  kind = k;
  inf  = i;
  sup  = s;
}

xinterval::xinterval ( const xinterval& a )
{
  kind = a.kind;
  inf  = a.inf;
  sup  = a.sup;
}

xinterval& xinterval::operator= ( xinterval& a )
{
  kind = a.kind;
  inf  = a.inf;
  sup  = a.sup;
  return *this;
}

//----------------------------------------------------------------------------
// Subtraction of an extended interval 'B' from a real value 'a'.
//----------------------------------------------------------------------------
xinterval operator- ( const real& a, const xinterval& B )
{
  xinterval D;

  switch (B.kind) {
    case Finite     : D.kind = Finite;                  // D = [D.inf,D.sup]
                      D.inf  = subd(a,B.sup);           //------------------
                      D.sup  = subu(a,B.inf);
                      break;
    case PlusInfty  : D.kind = MinusInfty;                 // D = [inf,+oo]
                      D.sup  = subu(a,B.inf);              //--------------
                      break;
    case MinusInfty : D.kind = PlusInfty;                  // D = [-oo,sup]
                      D.inf  = subd(a,B.sup);              //--------------
                      break;
    case Double     : D.kind = Double;        // D = [-oo,D.sup] v [D.inf,+oo]
                      D.inf  = subd(a,B.sup); //------------------------------
                      D.sup  = subu(a,B.inf);
                      if (D.inf < D.sup) D.inf = D.sup;
                      break;
  } // switch
  return D;
}

//----------------------------------------------------------------------------
// Extended interval division 'A / B' where 0 in 'B' is allowed.
//----------------------------------------------------------------------------
xinterval operator% ( interval& A, interval& B )
{
  interval  c;
  xinterval Q;

  if ( in(0.0,B) ) {
    if ( ( (Inf(A) < 0.0) && (0.0 < Sup(A) ) ) || (A == 0.0) || (B == 0.0) ) {
      Q.kind = Double;                      // Q = [-oo,+oo] = [-oo,0] v [0,+oo]
      Q.sup  = 0.0;                         //----------------------------------
```

```
      Q.inf  = 0.0;
    }
    else if ( (Sup(A) <= 0.0) && (Sup(B) == 0.0) ) {        // Q = [Q.inf,+oo]
      Q.kind = PlusInfty;                                   //---------------
      Q.inf  = divd(Sup(A),Inf(B));
    }
    else if ( (Sup(A) <= 0.0) && (Inf(B) < 0.0) && (Sup(B) > 0.0) ) {
      Q.kind = Double;                         // Q = [-oo,Q.sup] v [Q.inf,+oo]
      Q.sup  = divu(Sup(A),Sup(B));            //------------------------------
      Q.inf  = divd(Sup(A),Inf(B));
    }
    else if ( (Sup(A) <= 0.0) && (Inf(B) == 0.0) ) {        // Q = [-oo,Q.sup]
      Q.kind = MinusInfty;                                  //---------------
      Q.sup  = divu(Sup(A),Sup(B));
    }
    else if ( (Inf(A) >= 0.0) && (Sup(B) == 0.0) ) {        // Q = [-oo,Q.sup]
      Q.kind = MinusInfty;                                  //---------------
      Q.sup  = divu(Inf(A),Inf(B));
    }
    else if ( (Inf(A) >= 0.0) && (Inf(B) < 0.0) && (Sup(B) > 0.0) ) {
      Q.kind = Double;                         // Q = [-oo,Q.sup] v [Q.inf,+oo]
      Q.sup  = divu(Inf(A),Inf(B));            //------------------------------
      Q.inf  = divd(Inf(A),Sup(B));
    }
    else { // if ( (Inf(A) >= 0.0) && (Inf(B) == 0.0) )
      Q.kind = PlusInfty;                                   // Q = [Q.inf,+oo]
      Q.inf  = divd(Inf(A),Sup(B));                         //---------------
    }
  } // in(0.0,B)
  else {  // !in(0.0,B)
    c = A / B;                                              // Q = [C.inf,C.sup]
    Q.kind = Finite;                                        //-----------------
    Q.inf  = Inf(c);
    Q.sup  = Sup(c);
  }

  return Q;
} // operator%

//----------------------------------------------------------------------------
// Intersection of an interval 'X' and an extended interval 'Y'. The result
// is given as a pair (vector) of intervals, where one or both of them can
// be empty intervals.
//----------------------------------------------------------------------------
ivector operator& ( interval& X, const xinterval& Y )
{
  interval H;
  ivector  IS(2);

  IS[1] = EmptyIntval();
  IS[2] = EmptyIntval();

  switch (Y.kind) {
    case Finite     : // [X.inf,X.sup] & [Y.inf,Y.sup]
                      //------------------------------
                      H = _interval(Y.inf,Y.sup);
                      if ( !Disjoint(X,H) ) IS[1] = X & H;
                      break;
    case PlusInfty  : // [X.inf,X.sup] & [Y.inf,+oo]
                      //---------------------------
                      if (Sup(X) >= Y.inf)
                        if (Inf(X) > Y.inf)
                          IS[1] = X;
                        else
```

```
                    IS[1] = _interval(Y.inf,Sup(X));
                  break;
    case MinusInfty : // [X.inf,X.sup] & [-oo,Y.sup]
                  //----------------------------
                  if (Y.sup >= Inf(X))
                    if (Sup(X)<Y.sup)
                      IS[1] = X;
                    else
                      IS[1] = _interval(Inf(X),Y.sup);
                  break;
    case Double     : if ( (Inf(X) <= Y.sup) && (Y.inf <= Sup(X)) ) {
                    IS[1] = _interval(Inf(X),Y.sup);    // X & [-oo,Y.sup]
                    IS[2] = _interval(Y.inf,Sup(X));    // X & [Y.inf,+oo]
                  }
                  else if (Y.inf <= Sup(X)) // [X.inf,X.sup] & [Y.inf,+oo]
                    if (Inf(X) >= Y.inf)    //----------------------------
                      IS[1] = X;
                    else
                      IS[1] = _interval(Y.inf,Sup(X));
                  else if (Inf(X) <= Y.sup) // [X.inf,X.sup] & [-oo,Y.sup]
                    if (Sup(X) <= Y.sup)    //----------------------------
                      IS[1] = X;
                    else
                      IS[1] = _interval(Inf(X),Y.sup);
                  break;
  } // switch

  return IS;
} // operator&
```

#### 6.3.1.2 Module nlfzero

The header file of module *nlfzero* supplies the interface of the function *AllZeros()* for computing all zeros of a continuously differentiable function and the interface of the function *AllZerosErrMsg()* which can be used to get an error message for the error code returned by *AllZeros()*.

```
//----------------------------------------------------------------------------
// File: nlfzero (header)
// Purpose: Computing enclosures for all zeros of a continuously
//    differentiable one-dimensional, real-valued function.
// Global functions:
//    AllZeros()      : computes enclosures for all zeros
//    AllZerosErrMsg(): delivers an error message text
//----------------------------------------------------------------------------
#ifndef __NLFZERO_HPP
#define __NLFZERO_HPP

#include <intvector.hpp>     // Integer vector type
#include <ivector.hpp>       // Interval vector arithmetic
#include <mvi_util.hpp>      // Interval matrix/vector utility functions
#include <ddf_ari.hpp>       // Differentiation arithmetic

const int MaxCount = 10000;  // Maximum count of result components

extern char* AllZerosErrMsg ( int );
extern void  AllZeros ( ddf_FctPtr, interval, real, ivector&, intvector&,
                        int&, int&, int = MaxCount );
#endif
```

The function *AllZeros()* is an implementation of Algorithm 6.3. The static functions *XINewton()* and *VerificationStep()* are implementations of the Algorithms 6.1 and 6.2, respectively. All derivative evaluations are done with the help of the differentiation arithmetic *ddf_ari* (see Chapter 5).

The function *AllZeros()* uses the *DerivType* function *f()* and the starting interval *Start* as input parameters and stores all computed enclosures in the interval vector *ZeroVector* which is also passed to and from the function *XINewton()*. The information about the uniqueness of the roots is stored in the integer vector *InfoVector*. Both result vectors are automatically resized to the length given by the number of zeros located. The function *AllZeros()* has an optional last parameter *MaxNumberOfZeros* by which you can specify the maximum number of zeros to be computed. Its default value is given by the local constant *MaxCount* = 10000 which should be big enough to compute *all* zeros in most cases. *AllZeros()* reports a warning if there are more zeros than specified by parameter *MaxNumberOfZeros*. The value of the desired accuracy *Epsilon* is checked to be greater than or equal to *MinEpsilon* (the relative machine accuracy). If this is not true, then *Epsilon* is reset to 1 ulp accuracy.

```
//----------------------------------------------------------------------------
// File: nlfzero (implementation)
// Purpose: Computing enclosures for all zeros of a continuously
//    differentiable one-dimensional, real-valued function.
// Method: Extended interval Newton's method.
// Global functions:
//    AllZeros()      : computes enclosures for all zeros
//    AllZerosErrMsg(): delivers an error message text
//----------------------------------------------------------------------------
#include <string.h>          // String handling
#include <interval.hpp>      // Interval arithmetic
#include <i_util.hpp>        // Interval utility functions
#include <xi_ari.hpp>        // Extended interval arithmetic
#include <nlfzero.hpp>

static int MaxZeroNo;        // Maximum number of zeros to be computed

//----------------------------------------------------------------------------
// Error codes used in this module.
//----------------------------------------------------------------------------
const int
  NoError      = 0,  // No error occurred
  IllMaxZeroNo = 1,  // Illegal value for the maximum number of zeros
  NotAllZeros  = 2;  // Not all zeros computed

//----------------------------------------------------------------------------
// Error messages depending on the error code.
//----------------------------------------------------------------------------
char* AllZerosErrMsg ( int Err )
{
  static char Msg[80] = "";

  switch (Err) {
    case NoError:
      break;  // Blank message
    case IllMaxZeroNo:
      sprintf(Msg,"Error: Parameter for maximum number of zeros "
                  "must lie in 1,...,%1d!",MaxCount); break;
    case NotAllZeros:
```

```
      sprintf(Msg,"Warning: Not all zeros found due to the user "
                  "limit of %1d zero(s)!",MaxZeroNo); break;
    default:
      strcpy(Msg,"Error: Code not defined!");
  }
  return(Msg);
}

//----------------------------------------------------------------------------
// Purpose: Recursive function for the execution of the extended interval
//    Newton's method for the function 'f'.
// Parameters:
//    In    : 'f'          : must be declared for the 'DerivType' to enable
//                           the internal use of the differentiation
//                           arithmetic 'ddf_ari'
//            'y'          : specifies the starting interval
//            'Epsilon'    : specifies the desired relative accuracy
//                           (interval diameter) of the result intervals
//            'yUnique'    : signals whether it is already verified that the
//                           actual interval 'y' contains a unique zero
//    Out   : 'ZeroVector': stores the enclosures for the zeros of 'f'
//            'InfoVector': stores the corresponding information on the
//                           uniqueness of the zero in each enclosure
//    In/Out: 'ZeroNo'     : represents the total number of enclosures
//                           computed at the end of the recursion
// Description:
//    The function 'XINewton' is recursively called whenever the extended
//    interval division results in a bisection of the actual interval 'y' in
//    two intervals 'y1' and 'y2', and the tolerance condition is not ful-
//    filled yet. Otherwise, the enclosures for the zeros of 'f' are stored
//    in the interval vector 'ZeroVector', the corresponding information on
//    the uniqueness of the zero in each enclosure is stored in the integer
//    vector 'InfoVector'. The maximum number of enclosures to be computed
//    is given by the value of the global variable 'MaxZeroNo'.
//----------------------------------------------------------------------------
static void XINewton ( ddf_FctPtr f,
                       interval   y,
                       real       Epsilon,
                       int        yUnique,
                       ivector&   ZeroVector,
                       intvector& InfoVector,
                       int&       ZeroNo )
{
  if (ZeroNo > MaxZeroNo) return;        // Stop due to user specified limit
                                         //--------------------------------

  xinterval  z;
  interval   fc, fy, dfy;
  ivector    V(2);
  real       c;
  int        i;

  dfEval(f,y,fy,dfy);                    // Compute f(y) and f'(y)
  if ( !in(0.0,fy) ) return;             // Start if 0 in f(y), else do nothing
                                         //------------------------------------
  c  = mid(y);
  fEval(f,_interval(c),fc);         // Compute f(c) and f'(c).
  z  = c - fc % dfy;                // Extended interval Newton step.
  V = y & z;                        // Intersect interval y and extended interval z
                                    // resulting in two intervals V[1] and V[2].
                                    //-------------------------------------------
  if (V[1] == y) {                                        // Stagnation, so 'y'
    V[1] = _interval(Inf(y),c);                           // must be bisected.
    V[2] = _interval(c,Sup(y));                           //------------------
  }
```

```
  if ( (V[1] != EmptyIntval()) && (V[2] == EmptyIntval()) )
    yUnique = yUnique || in(V[1],y);          // Inner inclusion ===> uniqueness
  else                                        //---------------------------------
    yUnique = 0;

  for (i = 1; i <= 2; i++) {
    if (V[i] == EmptyIntval()) continue;     // Next i

    if (RelDiam(V[i]) <= Epsilon) {          // No more Newton steps

        fEval(f,V[i],fy);                     // Compute f(V[i])
        if ( in(0.0,fy) ) {                   // Store enclosure and info
          ZeroNo++;                           //-------------------------

          if (ZeroNo > MaxZeroNo) return;     // Stop due to user
                                              // specified limit
          if (ZeroNo > Ub(ZeroVector))        //-----------------
            { DoubleSize(ZeroVector); DoubleSize(InfoVector); }

          ZeroVector[ZeroNo] = V[i];          // Store enclosure of the zero
          InfoVector[ZeroNo] = yUnique;       // Store uniqueness info
        }                                     //----------------------------
      }
    else // Recursive call of 'XINewton' for interval 'V[i]'
      XINewton(f,V[i],Epsilon,yUnique,ZeroVector,InfoVector,ZeroNo);
  } // for (i ...
} // XINewton

//----------------------------------------------------------------------------
// Purpose: Execution of a verification step including the use of an epsilon
//    inflation.
// Parameters:
//    In    : 'f'      : function of 'DerivType'
//    Out   : 'yUnique': returns 'true' if the verification is successful
//    In/Out: 'y'      : interval enclosure to be verified
// Description: This function checks the uniqueness of the zero enclosed in
//    the variable 'y' by an additional verification step including the use
//    of an epsilon inflation of the iterates.
//----------------------------------------------------------------------------
static void VerificationStep ( ddf_FctPtr f, interval& y, int& unique )
{
  const int kmax = 10;

  interval yIn, yOld, fc, fy, dfy;
  real     c, eps;
  int      k;

  k = 0;  yIn = y;  eps = 0.25; unique = 0;               // Initializations
                                                          //----------------
  while ( !unique && (k < kmax) ) {                 // Do kmax loops to achieve
                                                    // inclusion
    yOld  = Blow(y,eps);                            // Epsilon inflation of 'y'
    dfEval(f,yOld,fy,dfy);                          //-------------------------

    if ( in(0.0,dfy) ) break;                       // No verification possible
                                                    //-------------------------
    k++;  c = mid(yOld);                        // Perform interval Newton step
    fEval(f,_interval(c),fc);                   //-----------------------------
    y = c - fc / dfy;

    if (Disjoint(y,yOld)) break;                    // No verification possible
                                                    //-------------------------
    unique = in(y,yOld);                      // Inner inclusion ===> uniqueness
```

```
    y      = y & yOld;                         // Intersection with old value
    if (y == yOld)                             //--------------------------------
      eps = eps * 8.0;                         // Increase the value of 'eps'
  } // while

  if (!unique) y = yIn;
} // VerificationStep

//----------------------------------------------------------------------------
// Purpose: Computation of enclosures for all zeros of a continuously
//    differentiable one-dimensional, real-valued function.
// Parameters:
//    In : 'f'                : objective function, must be declared for the
//                              'DerivType' to enable the internal use of
//                              the differentiation arithmetic 'ddf_ari'
//         'Start',           : specifies the starting interval
//         'Epsilon'          : specifies the desired relative accuracy
//                              (interval diameter) of the result intervals
//    Out: 'ZeroVector'       : stores and returns the enclosures for the
//                              zeros of 'f'
//         'InfoVector'       : stores the corresponding information on the
//                              uniqueness of the zeros in these enclosures
//         'NumberOfZeros'    : returns the number of enclosures computed
//         'Err'              : returns an error code
//         'MaxNumberOfZeros' : maximum number of zeros to be computed
//                              (default value: 'MaxCount' = compute all)
// Description:
//    The enclosures for the zeros of 'f' are computed by calling function
//    'XINewton'. Then an additional verification step is applied to those
//    enclosures which have not been verified. If an error occurs the value
//    of 'Err' is different from 0. The result vectors 'ZeroVector' and
//    'InfoVector' are automatically resized to standard index range with
//    lower index bounds 1. The maximum number of zeros which can be computed
//    is given by 'MaxNumberOfZeros'. It is bounded by 'MaxCount'. If your
//    system can handle more than 'MaxCount' zeros, you can increase its
//    value.
//----------------------------------------------------------------------------
void AllZeros ( ddf_FctPtr f,
                interval   Start,
                real       Epsilon,
                ivector&   ZeroVector,
                intvector& InfoVector,
                int&       NumberOfZeros,
                int&       Err,
                int        MaxNumberOfZeros )
{
  int       i, MinSize;
  real      MinEpsilon;

  if (1 <= MaxNumberOfZeros && MaxNumberOfZeros <= MaxCount) {
    MaxZeroNo = MaxNumberOfZeros;   // Initialize global static variable

    // Resize the result vectors to 'MinSize' components with standard index
    // range. Use 'MaxNumberOfZeros' if specified and less than 'MaxCount'.
    //------------------------------------------------------------------------
    MinSize = (MaxNumberOfZeros == MaxCount) ? 10 : MaxNumberOfZeros;

    Resize(ZeroVector,MinSize);
    Resize(InfoVector,MinSize);

    // Start extended interval Newton method
    //--------------------------------------
    Err = NoError;  NumberOfZeros = 0;
```

```
      MinEpsilon = succ(1.0) - 1.0;         // Relative machine accuracy (1 ulp)
      if (Epsilon < MinEpsilon)
        Epsilon = MinEpsilon;               // Set 'Epsilon' to 1 ulp accuracy

      XINewton(f,Start,Epsilon,0,ZeroVector,InfoVector,NumberOfZeros);

      if (NumberOfZeros > MaxNumberOfZeros)
        { Err = NotAllZeros;  NumberOfZeros = MaxNumberOfZeros; }
    }
    else {
      Err = IllMaxZeroNo;
      NumberOfZeros = 0;
    }

    // Resize the result vectors to appropriate length if no error
    // has occurred. Otherwise, resize them to minimum dimensions
    // with undefined value.
    //--------------------------------------------------------------
    if (NumberOfZeros == 0) {
      Resize(ZeroVector,1);
      Resize(InfoVector,1);
    }
    else {
      Resize(ZeroVector,NumberOfZeros);      // Values undefined
      Resize(InfoVector,NumberOfZeros);      //-----------------
    }

    for (i = 1; i <= NumberOfZeros; i++)                        // Verification step
      if (InfoVector[i] == 0)                                   // for the enclosures
        VerificationStep(f,ZeroVector[i],InfoVector[i]);        //-------------------
  } // AllZeros
```

### 6.3.2 Example

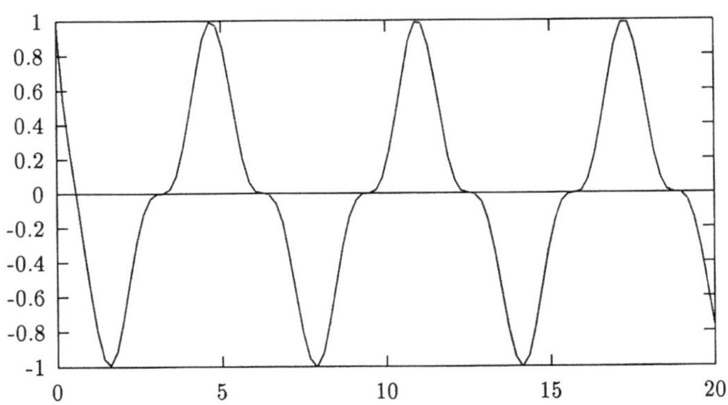


Figure 6.3: Function $f(x) = e^{-3x} - \sin^3 x$

Our sample program uses modules *nlfzero* to compute all zeros of the function

$$f(x) = e^{-3x} - \sin^3 x$$

in a specified starting interval (see Figure 6.3). The first positive root near $x = \frac{\pi}{2}$ is easy to locate numerically. As $x \longrightarrow \infty$, roots of $f$ approach the triple roots of $\sin^3 x$ at $\frac{\pi}{2} + n\pi$. We first have to define the function $f$ for the type *DerivType*. Then we use a function *compute()* with the function $f$ and a string (a description of the function) as parameters. This function prompts you to enter the necessary data for the call of function *AllZeros()*. If we add further function definitions in our program, we can apply this function in the same way using another function as parameter.

```
//----------------------------------------------------------------------------
// Example: Nonlinear equations in one variable
// This program uses module 'nlfzero' to compute the zeros of the function
//
//             exp(-3*x) - power(sin(x), 3)
//
// A starting interval and a tolerance must be entered.
//----------------------------------------------------------------------------
#include <nlfzero.hpp>      // Nonlinear equations
#include <stacksz.hpp>      // To increase stack size for some
                            // special C++ compilers

DerivType f ( DerivType x )                  // Sample function
  { return exp(-3.0*x) - power(sin(x),3); }

//----------------------------------------------------------------------------
// Function for prompting and reading information to call the function
// 'AllZeros()'. This function must be called with the function 'f'
// and a string 'Name' containing a textual description of that function.
//----------------------------------------------------------------------------
void compute ( ddf_FctPtr f, const char *Name )
{
  interval  SearchInterval;
  real      Tolerance;
  ivector   Zero;
  intvector Unique;
  int       NumberOfZeros, i, Error;

  cout << SetPrecision(23,15) << Scientific;   // Output format

  cout << "Computing all zeros of the function  " << Name << endl;
  cout << "Search interval       : ";
  cin  >> SearchInterval;
  cout << "Tolerance (relative) : ";
  cin  >> Tolerance;
  cout << endl;

  AllZeros(f,SearchInterval,Tolerance,Zero,Unique,NumberOfZeros,Error);

  for ( i = 1; i <= NumberOfZeros; i++) {
    cout << Zero[i] << endl;
    if (Unique[i])
      cout << "encloses a locally unique zero!" << endl;
    else
      cout << "may contain a zero (not verified unique)!" << endl;
  }

  cout << endl << NumberOfZeros << " interval enclosure(s)" << endl;

  if (Error) cout << endl << AllZerosErrMsg(Error) << endl;
}
```

```
//--------------------
// Body of the program
//--------------------
int main ( )
 { compute(f, "EXP(-3x)-POWER(SIN(x),3)"); return 0;}
```

If we execute this program, we get the following runtime output:

```
Computing all zeros of the function  EXP(-3x)-POWER(SIN(x),3)
Search interval      : [0,20]
Tolerance (relative) : 1e-10

[ 5.885327439797041E-001, 5.885327439840196E-001]
encloses a locally unique zero!
[ 3.096363932410612E+000, 3.096363932410676E+000]
encloses a locally unique zero!
[ 6.285049273382585E+000, 6.285049273382588E+000]
encloses a locally unique zero!
[ 9.424697254681708E+000, 9.424697254772081E+000]
encloses a locally unique zero!
[ 1.256637410168852E+001, 1.256637410169038E+001]
encloses a locally unique zero!
[ 1.570796311724486E+001, 1.570796311724936E+001]
encloses a locally unique zero!
[ 1.884955592762931E+001, 1.884955592835263E+001]
encloses a locally unique zero!

7 interval enclosure(s)
```

The seven interval enclosures are disjoint, so we can conclude that there are seven roots of our sample function within the specified starting interval.

### 6.3.3 Restrictions and Hints

The function $f$ whose root we seek must be expressible as a finite sequence of arithmetic operations and elementary functions supported by the differentiation arithmetic module *ddf_ari* described in Chapter 5.

*AllZeros()* may report a warning that not all zeros were found due to a user limit for the number of zeros to be computed. In this case, you may call *AllZeros()* again to find additional zeros. You may then use a smaller starting interval, namely the original starting interval *without* the area already fully treated which is given by the interval hull of the computed solutions. This works, because the recursive implementation of our method treats the starting interval from the left to the right.

The method is not very fast if a very small value of $\varepsilon$ (*Epsilon*) is used and the interval Newton step does not improve the actual iterates because of rounding and overestimation effects of the machine interval arithmetic. In this case the method is equivalent with a bisection method.

In *XINewton()*, the evaluation of the function with differentiation arithmetic can cause a runtime error if the interval argument of an elementary function does not lie in the domain specified for this interval function or if a division by an interval containing zero occurs, even though the domain is mathematically valid. For example, $\sqrt{1 + x \cdot x}$ for $x \in [-2, 2]$ attempts to take the square root of a negative number due to the overestimation effects of interval arithmetic (see Section 3.1). To get rid of these errors, you may rewrite $f$ in a different (but mathematically equivalent) form ($\sqrt{1+x^2}$ in this example), or you may try to split the starting interval in several parts and call *AllZeros()* for these parts.

The rules for getting true enclosures in connection with conversion errors (see Section 3.7) also apply here. If you want to compute enclosures for the zeros of the function $f(x) = (x - \frac{1}{3}) \cdot (x - 0.1)$, for example, be careful that the values $\frac{1}{3}$ and 0.1 are enclosed in machine intervals when computing interval evaluations of $f$. A corresponding implementation of $f$ could be

```
DerivType f ( DerivType x )
{
  interval one = interval(1.0,1.0);
  return (x - one/3.0) * (x - one/10.0);
}
```

## 6.4 Exercises

**Exercise 6.1** Use the function *AllZeros()* to compute the zero of the function $f(x) = (x-1)^3$ with starting intervals $[-3,4]$ and $[-3,5]$, respectively, to demonstrate the behavior of our implementation when computing multiple zeros or zeros lying on the splitting point of the bisection step.

**Exercise 6.2** In [15, Section 2.1], Dennis and Schnabel write:

> "*It is unlikely, that there will ever be a wonderful general purpose computer routine that would tell us for* $f_1(x) = x^4 - 12x^3 + 47x^2 - 60x$, $f_2(x) = f_1(x) + 24$, *and* $f_3(x) = f_1(x) + 24.1$ *the roots of* $f_1$ *are* $x = 0$, *3, 4, and 5, the real roots of* $f_2$ *are* $x = 1$ *and* $x = 0.888305\ldots$, *and* $f_3$ *has no real roots. In general, the question of existence and uniqueness are beyond the capabilities one can expect of algorithms that solve nonlinear problems.*"

Use our *AllZeros()* routine to compute the zeros of $f_1$, $f_2$, and $f_3$ in the starting interval $[-3,8]$ to demonstrate that the wonderful routine exists!

**Exercise 6.3** Use the function *AllZeros()* to verify an approximation of a zero of the function $f(x) = \sin\frac{1}{x}$. Compute the approximation using an arbitrary method of your choice (e.g. Newton's method). Then apply *AllZeros()* to a neighborhood interval of this approximation.

## 6.5 References and Further Reading

The method we discussed in this chapter is an *a priori* method because the iteration starts with a (possibly large) interval enclosing all the solutions which have to be found. Here, the iterates of the method are subintervals of the previous iterates.

There are also methods for finding (and bounding) one single solution of a nonlinear equation called *a posteriori* methods. These methods start with an approximation of a zero and apply a test function for a neighborhood interval of the approximation to verify a zero within that interval. Our method presented in this chapter can also be applied to verify such an approximation, if we start the process with a small interval containing the approximation (Exercise 6.3).

Many authors have dealt with the problem of computing enclosures of zeros of nonlinear functions. For further references in the field of *a priori* methods, see [3], [8], [31], [68], [69], [70], [77], or [78], for example. For *a posteriori* methods, see [11], [42], [43], [56], [59], [61], [66], [70], [83], [84], or [85], for example.

# Chapter 7

# Global Optimization

We want to find the global minimum in an interval $[x]$ of a function $f$ that may have many local minima. We want to compute the minimum value of $f$ and the point(s) at which the minimum value is attained. This is a very difficult problem for classical methods because narrow, deep valleys may escape detection. In contrast, the interval method presented here evaluates $f$ on a continuum of points, including those points that are not finitely representable, so valleys, no matter how narrow, are recognized with certainty. Further, interval techniques often can reject large regions in which the optimum can be guaranteed not to lie, so they can be faster overall than classical methods for many problems.

In classical numerical analysis, global optimization methods proceed by iteration starting from some approximate trial points. Classical optimization methods sample the objective function at only a finite number of points and cannot guarantee that the function does not have some unexpectedly small values in between these trial points.

In this chapter, we consider an algorithm based on the method of Hansen [28] for computing all solutions of the global unconstrained optimization problem. The algorithm computes enclosures for all global minimizers in a given interval and for the global minimum value of a twice continuously differentiable function.

## 7.1 Theoretical Background

We address the problem of finding all solutions $x^*$ of

$$\min_{x \in [x]} f(x) \tag{7.1}$$

for a twice continuously differentiable function $f : I\!R \to I\!R$ and $[x] \in II\!R$. We wish to find the set of all global minimizers $x^*$ and the global minimum value $f^* = f(x^*)$. The existence of a global minimizer is assured since $f$ is a continuous function on a compact set, but need not be unique. In fact, the minimizers may include continua of points.

Our algorithm for problem (7.1) is based on the method of Hansen (see [28] or [31]) and the modifications of Ratz (see [79]). Starting from the initial interval $[x]$, our algorithm subdivides $[x]$ and stores the subintervals $[y] \subset [x]$ in a list $L$. Subintervals which are guaranteed not to contain a global minimizer of $f$ are discarded from that list, while the other subintervals get subdivided again until the desired accuracy (relative diameter) of the intervals in the list is achieved. The key to the speed of this method is its use of several tests to discard subintervals in which no minimizer may occur:

- midpoint test,
- monotonicity test,
- concavity test, and
- interval Newton step.

In the following sections, we consider each of these methods in detail. We use the notation $\underline{f_y}$ as abbreviation for the lower interval bound of the interval function evaluation $[f_y] := f_{[]}([y])$ for $[y] \in I\mathbb{R}$.

### 7.1.1 Midpoint Test

If we are able to determine an upper bound $\widetilde{f}$ for the global minimum value $f^*$, then we can delete all subintervals $[y]$ for which

$$\underline{f_y} > \widetilde{f} \geq f^*. \tag{7.2}$$

The midpoint test first determines or improves such an upper bound for $f^*$. Initially, let $\widetilde{f} = +\infty$. We choose an interval $[y]$ out of the list $L$ which satisfies $\underline{f_y} \leq \underline{f_z}$ for all intervals $[z]$ in the list $L$. That is, $[y]$ has the smallest lower bound for the range of $f$. Hence, it is a likely candidate to contain a minimizer. Let $c = m([y])$ (or any other point in $[y]$), and compute $\widetilde{f} = \min\{f(c), \widetilde{f}\}$. Such an upper bound can also be computed on a computer when rounding errors occur: we compute $f_{\diamond}(c)$ and use the upper bound of the resulting interval as the possibly new value $\widetilde{f}$.

Now, with a possibly improved (decreased) value of $\widetilde{f}$, we can discard all intervals $[z]$ from the list $L$ for which $\widetilde{f} < \underline{f_z}$. The midpoint test is relatively inexpensive, and it often allows us to discard from consideration large portions of the original interval $[x]$. Figure 7.1 illustrates this procedure, which deletes the intervals $[y]_2$,

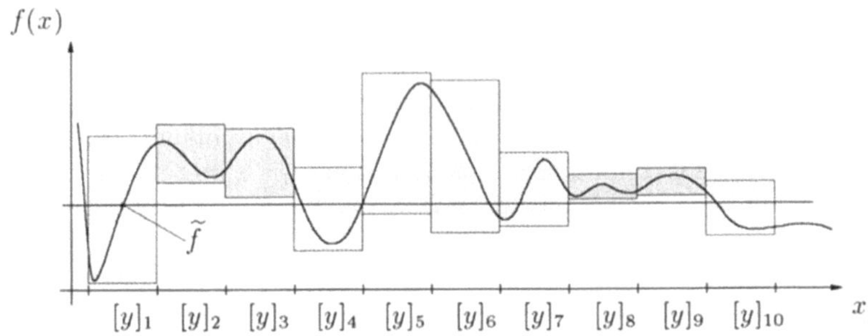


Figure 7.1: Midpoint test

$[y]_3$, $[y]_8$, and $[y]_9$ in this special case. The test remains valid if an arbitrary $c' \in [y]$ is used instead of $c = m([y])$. Our algorithm could be extended by using a local approximate search to find a $c' \in [y]$ that is likely to give a smaller upper bound for $f$ than $c$ gives. In Figure 7.1, we evaluate $f$ at $c = m([y]_1)$ to get $\widetilde{f}$. A value of $c$ near the left end of $[y]_1$ would have yielded a more effective $\widetilde{f}$. We pursue this modification in the Exercises.

The value $\widetilde{f}$ is also used when entering newly subdivided intervals $[w]$ in our list $L$. If we know that $[w]$ satisfies $\underline{f_w} > \widetilde{f}$, then $[w]$ cannot contain a global minimizer. Thus, we must only enter intervals $[w]$ that satisfy $\underline{f_w} \leq \widetilde{f}$ in the list $L$.

### 7.1.2 Monotonicity Test

The monotonicity test determines whether the function $f$ is *strictly monotone* in an entire subinterval $[y] \subset [x]$. If $f$ is strictly monotone in $[y]$, then $[y]$ cannot contain a global minimizer in its interior. Further, a global minimizer can only lie on a boundary point of $[y]$ if this point is also a boundary point of $[x]$. Therefore, if $f$ satisfies

$$0 \notin f'([y]), \tag{7.3}$$

then the subinterval $[y]$ can be deleted (with the exception of boundary points of $[x]$).

Figure 7.2 demonstrates the monotonicity test for four subintervals of $[x]$. In this special case, $[y]_1$ can be reduced to the boundary point $\underline{x}$, $[y]_2$ remains unchanged, and $[y]_3$ can be deleted. Since $f$ is monotonically increasing in $[y]_4$, this entire interval can also be deleted because we are looking for a minimum. In this example, the monotonicity test has reduced the list $L$ from four elements $[y]_1$, $[y]_2$, $[y]_3$, and $[y]_4$ to two elements $[\underline{x}]$ and $[y]_2$. It appears that applying the midpoint test with $[\underline{x}]$ might also discard $[y]_2$, leaving the unique solution $x^* = \underline{x}$ at a cost of only four interval evaluations of $f'$ and two interval evaluations of $f$.

The monotonicity test costs only one interval evaluation of $f'([y])$. As illustrated in Figure 7.2, it often allows us to discard large portions of the original interval $[x]$ from further consideration.

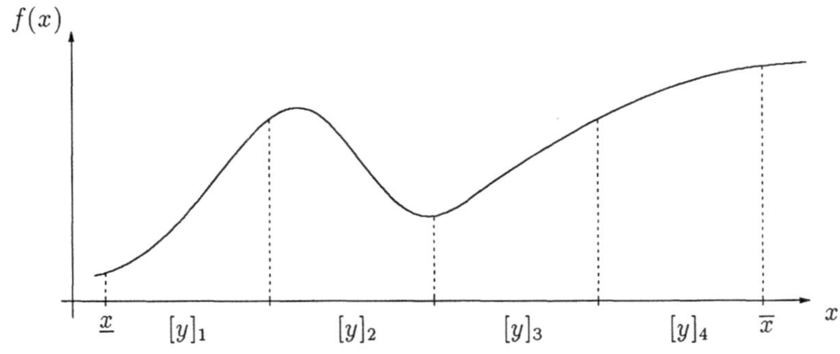


Figure 7.2: Monotonicity test

### 7.1.3 Concavity Test

The concavity test (non-convexity test) examines the concavity of $f$. If $f$ is *not convex* in a subinterval $[y] \subset [x]$, then $[y]$ cannot contain a global minimizer in its interior. Further, a global minimizer can only lie on a boundary point if this point is also a boundary point of $[x]$.

A function $f$ is convex in $[y]$ if the second derivative is positive or zero. Therefore, if $f$ satisfies

$$\overline{f''([y])} < 0, \tag{7.4}$$

then $f$ is concave over $[y]$, and the subinterval $[y]$ can be deleted (with the exception of boundary points of $[x]$). One of the reasons to include second-order derivatives in our automatic differentiation module in Chapter 5 was to provide the computational tool to apply the concavity test to non-trivial functions $f$.

Figure 7.3 demonstrates the concavity test for four subintervals of $[x]$. In this special case, $[y]_1$ can be reduced to the boundary point $\underline{x}$, $[y]_2$ remains unchanged, $[y]_3$ can be deleted, and $[y]_4$ can be reduced to the boundary point $\overline{x}$. In this example, the concavity test has reduced the list $L$ from four elements $[y]_1$, $[y]_2$, $[y]_3$, and $[y]_4$ to a much smaller subset of $[x]$ ($[\underline{x}]$, $[y]_2$, $[\overline{x}]$) at a cost of only four interval evaluations of $f''$.

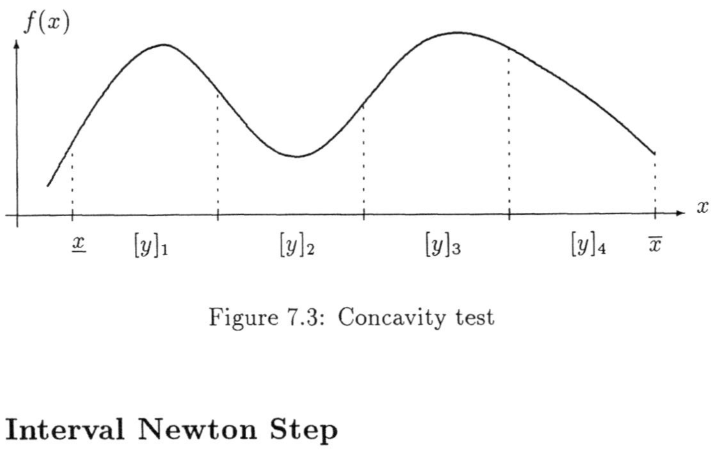


Figure 7.3: Concavity test

### 7.1.4 Interval Newton Step

In our global optimization method, we apply one step of the extended interval Newton's method (see Chapter 6) to the problem

$$f'(y) = 0, \quad y \in [y]. \tag{7.5}$$

Let

$$N'([y]) := m([y]) - \frac{f'(m([y]))}{f''([y])}, \tag{7.6}$$

and compute

$$[y_N] := [y] \cap N'([y]), \tag{7.7}$$

where $[y_N]$ is a single interval, the union of two intervals, or the empty set. Thus, the interval Newton step may eliminate parts of $[y]$ which cannot contain a stationary point of $f$. We do not apply several iterations of interval Newton because it would be inefficient to compute all stationary points which not necessarily are minimizers.

An interval Newton step is computationally more expensive than the midpoint, monotonicity, or concavity tests. However, it is often very effective in reducing the widths of subintervals to the point where overestimation by interval arithmetic is small enough that one of the less expensive tests can succeed.

### 7.1.5 Verification

We have to check two conditions to verify the existence and uniqueness of a local minimizer within an interval of the final list our method produces. The first condition is

$$N'([y]) \overset{\circ}{\subset} [y], \tag{7.8}$$

which guarantees the existence and uniqueness of a stationary point of $f$, i.e. a zero of $f'$ in $[y]$ (cf. Theorem 6.1). The second condition is

$$\underline{f''([y])} > 0, \tag{7.9}$$

which guarantees that $f$ is convex over $[y]$ (recall that we are assuming in this chapter that $f$ is twice continuously differentiable). We know of no way to verify the uniqueness of a global minimizer in general. The global minimizer can even be one or more continua of points. Hence, we settle for attempting to verify that intervals we compute as candidates for containing a global minimizer contain unique local minimizers. Failure to verify the uniqueness of a local minimizer in a subinterval is not grounds for discarding that subinterval from the list of candidates.

In fact, our method produces a final list containing enclosures for *locally unique candidates for global minimizers.* If we have in the final list exactly one subinterval $[y]$ in which we can validate a local minimizer, then we have validated a unique global minimizer in the starting interval $[x]$. If we have two or more subintervals validated to contain unique local minimizers, then the best we can say is that each contains a candidate for a global minimizer, which need not be unique. In this case, the global minimizer could not be a continuum of points.

## 7.2 Algorithmic Description

The algorithm AllGOp1 (Algorithm 7.4) for determining <u>all</u> <u>g</u>lobal <u>op</u>timizers of a <u>1</u>-dimensional twice continuously differentiable function $f$ in an interval $[x]$ consists of two parts. The first part is the subdivision method including the tests and the extended interval Newton steps described in Section 7.1. The second part is a verification step, which tries to verify the uniqueness of a local minimizer within the remaining intervals of the work list.

The algorithm for the execution of the extended interval Newton step (Algorithm 7.1), yields a pair of intervals $[v]$ and the number $p$, specifying the number of non-empty interval components of $[v]$.

**Algorithm 7.1:** NewtonStep$(f, [y], [v], p)$

1. $c := m([y]);\quad p := 0;$
2. $[z] := c - f'_{\diamond}(c)/f''([y]);$ {Extended interval Newton step}
3. $[v] := [y] \cap [z];$ {Intersection $[y] \cap [z] = [v]_1 \cup [v]_2$ }
4. **if** $[v]_1 \neq \emptyset$ **then** $p := p + 1;$
5. **if** $[v]_2 \neq \emptyset$ **then** $p := p + 1;$
6. **return** $[v], p;$

Algorithm 7.2 manages the list $L$ of pending subintervals that may contain global minimizers $x^*$. Subintervals are removed from the list and placed in an accepted list when they satisfy relative or absolute error acceptance criteria. Subintervals are also removed from the list by the midpoint, monotonicity, concavity tests, or by the interval Newton step. Subintervals are added to the pending list when an element from the list is bisected or when the extended interval Newton step yields two candidate intervals.

Our algorithm stores the subdivided intervals $[y]$ and the lower bound of the interval function evaluations $[f_y] := f_{[]}([y])$ as pairs $([y], \underline{f_y})$. Pairs are stored in the list sorted in increasing order of lower bounds $\underline{f_y}$. Therefore, a newly computed pair is stored in the list $L$ according to the ordering rule (cf. [79]):

$$\left.\begin{array}{ll} \bullet \text{ either} & \underline{f_w} \leq \underline{f_y} < \underline{f_z} \text{ holds,} \\ \bullet \text{ or} & \underline{f_y} < \underline{f_z} \text{ holds, and } ([y], \underline{f_y}) \text{ is the first element of the list,} \\ \bullet \text{ or} & \underline{f_w} \leq \underline{f_y} \text{ holds, and } ([y], \underline{f_y}) \text{ is the last element of the list,} \\ \bullet \text{ or} & ([y], \underline{f_y}) \text{ is the only element of the list,} \end{array}\right\} \quad (7.10)$$

where $([w], \underline{f_w})$ is the predecessor and $([z], \underline{f_z})$ is the successor of $([y], \underline{f_y})$ in $L$.

That is, the second components of the list elements may not decrease, and a new pair is entered behind all other pairs with the same second component. Thus, the first element of the list has the smallest second component, and we can directly use the corresponding interval to compute $f(c)$ for the improvement of $\tilde{f}$ in performing the midpoint test. Because of the ordering of the list, we can also save some work when deleting elements in the midpoint test. When we have found the first element to be deleted, we can delete the rest of the list. At each step of our algorithm, $\tilde{f}$ is the best known upper bound for the global minimum $f^*$, and $\underline{f_y}$ from the first element of the list is the best known lower bound for $f^*$. Hence in some applications, we might halt the algorithm when $\tilde{f} - \underline{f_y} < \varepsilon_f$ (absolute tolerance) or when $\tilde{f} - \underline{f_y} < |f_{[]}([y])|\varepsilon_f$ (relative tolerance).

Given the list $L$ and a list element $E$, we use the following notations in our algorithms:

| Notation | Meaning |
|---|---|
| $L := \{\}$ | Initialization by an empty list |
| $L := E$ | Initialization by a single element |
| $L := L + E$ | Enter element $E$ in $L$ in sorted order according to (7.10) |
| $L := L - E$ | Discard element $E$ from $L$ |
| $E :=$ Head $(L)$ | Set $E$ to the first element of $L$ |
| MultiDelete $(L, \tilde{f})$ | Discard all elements from $L$ satisfying Condition (7.2) |
| Length $(L)$ | Delivers the number of elements in $L$ |

In GlobalOptimize, we first compute an upper bound for the global minimum value and we do some initializations. Then in Step 3, we separately treat the two boundary points of the starting interval $[x]$. We enter them into the work list if they are candidates for minimizers. Step 4 is the main iteration. Here, we first do a bisection of the actual interval $[y]$. Then in Step 4(b), we apply the monotonicity test, a function value check using the centered form, the concavity test, and the extended interval Newton step to the bisected intervals $[u]_1$ and $[u]_2$. The interval Newton step may result in at most two intervals. We have to handle them both in Step 4(b)vii, where we again apply a monotonicity test and a function value check with centered forms. If the actual interval $[v]_j$ is still a candidate for a minimizer, we store it in $L$.

In Step 4(d), we remove the first element from the list $L$, i.e. the element of $L$ with the smallest lower bound of the interval function evaluation, and we perform the midpoint test. Then, we check the tolerance criterion for the new actual interval. If the desired accuracy is achieved, we store this interval in the result list $L_{\text{res}}$. Otherwise, we go to the bisection step.

When the iteration stops because the list $L$ is empty, we compute a final enclosure $[f^*]$ for the global minimum value in Step 5, and we return $L_{\text{res}}$ and $[f^*]$. Algorithm GlobalOptimize terminates because the elements of $L$ move to $L_{\text{res}}$ if $(d_{\text{rel}}([y]) \leq \varepsilon)$ or if $(d_{\text{rel}}([f^*]) \leq \varepsilon)$. The bisection step (Step 4(a)) guarantees that the first condition is fulfilled at some stage of the iteration.

**Algorithm 7.2:** GlobalOptimize $(f, [x], \varepsilon, L_{\text{res}}, [f^*])$

1. $c := m([x])$; $[f_c] := f_\Diamond(c)$; $\tilde{f} := \overline{f_c}$; {Compute upper bound for $f^*$}
2. $[y] := [x]$; $L := \{\}$; $L_{\text{res}} := \{\}$;
3. {Treat boundary points separately}

   $[a] := f_\Diamond(\underline{y})$; $[b] := f_\Diamond(\overline{y})$; $\tilde{f} := \min\{\tilde{f}, \overline{a}, \overline{b}\}$

   **if** $\tilde{f} \geq \underline{a}$ **then** $L := L + ([\underline{y}, \underline{y}], \underline{a})$;

   **if** $\tilde{f} \geq \underline{b}$ **then** $L := L + ([\overline{y}, \overline{y}], \underline{b})$;
4. **repeat** {Start iteration}

   (a) $[u]_1 := [\underline{y}, c]$; $[u]_2 := [c, \overline{y}]$; {Bisect $[y]$}

(b) **for** $i := 1$ **to** $2$ **do**

i. **if** $0 \notin f'([u]_i)$ **then next**$_i$; {Monotonicity test}

ii. $[f_u] := (f(c) + f'([u]_i) \cdot ([u]_i - c)) \cap f([u]_i)$; {Centered form}

iii. **if** $\widetilde{f} < \underline{f_u}$ **then next**$_i$;

iv. $[h] := f''([u])$;

v. **if** $\overline{h} < 0$ **then next**$_i$; {Concavity test}

vi. NewtonStep$(f, [u]_i, [v], p)$; {Extended interval Newton step}

vii. **for** $j := 1$ **to** $p$ **do**

A. **if** $0 \notin f'([v]_j)$ **then next**$_j$; {Monotonicity test}

B. $c_v := m([v]_j)$;

C. $[f_v] := (f(c_v) + f'([v]_j) \cdot ([v]_j - c_v)) \cap f([v]_j)$; {Centered form}

D. **if** $\widetilde{f} \geq \underline{f_v}$ **then** $L := L + ([v]_j, \underline{f_v})$; {Store $[v]_j$}

(c) *Bisect* := *false*;

(d) **while** $(L \neq \{\,\})$ **and** (**not** *Bisect*) **do**

i. $([y], \underline{f_y})$ := Head$(L)$; $L := L - ([y], \underline{f_y})$; $c := m([y])$;

ii. $\widetilde{f} := \min\{\widetilde{f}, f(c)\}$; MultiDelete$(L, \widetilde{f})$; {Midpoint test}

iii. $[f^*] := [\underline{f_y}, \widetilde{f}]$;

iv. **if** $(d_{\text{rel}}([f^*]) \leq \varepsilon)$ **or** $(d_{\text{rel}}([y]) \leq \varepsilon)$ **then**

$L_{\text{res}} := L_{\text{res}} + ([y], \underline{f_y})$; {Accept this subinterval}

**else**

*Bisect* := *true*;

**until** (**not** *Bisect*);

5. $([y], \underline{f_y})$ := Head$(L_{\text{res}})$; $[f^*] := [\underline{f_y}, \widetilde{f}]$;

6. **return** $L_{\text{res}}, [f^*]$;

Algorithm 7.3 describes a verification step that checks the uniqueness of a local minimizer enclosed in the interval $[y]$. The function tries to do a "zero check" for the derivative $f'$ according to condition 3 of Theorem 6.1 by applying interval Newton steps including an epsilon inflation of the iterates $[y]$. We also check the condition $f''(y) > 0$ for all $y \in [y]$ to verify that $f$ is convex in $[y]$. We use $k_{\max} = 10$ as the maximum number of iterations, $\varepsilon = 0.25$ as the starting value for the epsilon inflation, and a factor of 8 to increase $\varepsilon$ within the iterations. It turned out that these are good values for minimizing the effort if no verification is possible (see also [52]).

**Algorithm 7.3:** VerificationStep $(f, [y], yUnique)$

1. $yUnique := (\underline{y} = \overline{y})$; {Point interval $\Rightarrow$ uniqueness of local minimizer}
2. **if** $yUnique$ **then return** $[y], yUnique$;

3. $k_{\max} := 10$; $k := 0$; $[y_{\text{in}}] := [y]$; $\varepsilon := 0.25$; {Initializations}
4. **while** (**not** $yUnique$) **and** ($k < k_{\max}$) **do** {Do $k_{\max}$ loops to achieve inclusion}
   (a) $[y_{\text{old}}] := [y] \bowtie \varepsilon$; {Epsilon inflation of $[y]$}
   (b) $[h] := f''([y_{\text{old}}])$;
   (c) **if** $\underline{h} \leq 0$ **then exit**$_{\text{while-loop}}$; {No verification possible}
   (d) $k := k + 1$; $c := m([y_{\text{old}}])$;
   (e) $[y] := c - f'_{\diamond}(c)/f''([y_{\text{old}}])$; {Interval Newton step}
   (f) **if** $[y] \cap [y_{\text{old}}] = \emptyset$ **then exit**$_{\text{while-loop}}$; {No verification possible}
   (g) $yUnique := ([y] \overset{\circ}{\subset} [y_{\text{old}}])$; {Inner inclusion $\Rightarrow$ local uniqueness}
   (h) $[y] := [y] \cap [y_{\text{old}}]$; {Intersection with old value}
   (i) **if** $[y] = [y_{\text{old}}]$ **then** $\varepsilon := \varepsilon \cdot 8$; {Increase $\varepsilon$}
5. **if not** $yUnique$ **then** $[y] := [y_{\text{in}}]$;
6. **return** $[y]$, $yUnique$;

Algorithm 7.4 now combines these procedures to compute enclosures for all global minimizers $x^*$ of the function $f$ and for the global minimum value $f^*$ within the input interval $[x]$ and tries to prove the uniqueness of the local minimizers within the computed enclosures. The desired accuracy (relative diameter) of the interval enclosures is specified by the input parameter $\varepsilon$. 1 ulp accuracy is chosen if the specified value of $\varepsilon$ is too small (for example 0). The enclosures for the global minimizers of $f$ are returned in the interval vector $[Opt]$, the corresponding information on the local uniqueness of the optimizers is returned in the Boolean vector *Info*. The number of enclosures computed is returned in the integer variable $N$.

We use a function called CheckParameters as an abbreviation for the error checks for the parameters of AllGOp1 which are necessary in an implementation. If no error occurs, AllGOp1 delivers the $N$ enclosures $[Opt]_i$, $i = 1, 2, \ldots, N$, and for each $i = 1, 2, \ldots, N$,

if $Info_i = true$, then $[Opt]_i$ encloses a unique local minimizer of $f$,
if $Info_i = false$, then $[Opt]_i$ may enclose a local or global minimizer of $f$.

**Algorithm 7.4:** AllGOp1 ($f, [x], \varepsilon, [Opt], Info, N, [f^*], Err$)

1. $Err$ := CheckParameters;
2. **if** $Err \neq$ "No Error" **then return** $Err$;
3. Set $\varepsilon$ to "1 ulp accuracy" if $\varepsilon$ is too small;
4. GlobalOptimize ($f, [x], \varepsilon, L_{\text{res}}, [f^*]$);
5. $N$ := Length ($L_{\text{res}}$);
6. $([Opt], Info) := L_{\text{res}}$; {Transfer list to result vectors}

7. **for** $i := 1$ **to** $N$ **do**

    VerificationStep ($f, [Opt]_i, Info_i$);

8. **return** $[Opt]$, *Info*, $N$, $[f^*]$, *Err*;

## Applicability of the Algorithm

We have assumed that $f$ is twice continuously differentiable. However, we can apply our algorithm to functions that are only once continuously differentiable or to functions that are not differentiable if we leave out some parts of the algorithm. If we do not use the interval Newton step (Step 4(b)vi in our algorithm), i.e. if we leave out Steps 4(b)iv, 4(b)v, and 4(b)vi, and replace them by the sequence

$$[v]_1 := [u]_i; \quad p := 1;$$

then our method works for functions that are only once differentiable. Algorithm 7.3 cannot be applied in this case. We can even do problems in nonsmooth global optimization. If we replace Step 4(b) by the sequence

**for** $i := 1$ **to** $2$ **do**
  $[f_u] := f([u]_i)$;
  **if** $\widetilde{f} \geq \underline{f_u}$ **then** $L := L + ([u]_i, \underline{f_u})$;

then we can apply our method, consisting of subdividing and the midpoint test, to functions that are *not* differentiable.

The closer the upper bound $\widetilde{f}$ is to the global minimum value $f^*$, the more intervals we can delete in the midpoint test (Step 4(d)ii of Algorithm 7.2). Thus, the method can be improved by incorporating a floating-point approximate local search procedure in an attempt to decrease the value $\widetilde{f}$. See [31] or [78] for the description of such local search procedures. We explore this extension in the Exercises.

For a multiple zero $x^*$ of $f'$, the algorithm cannot verify the existence and the uniqueness of $x^*$ in the enclosing result interval. Nevertheless, the zero of $f'$, which is possibly a global minimizer, is bounded to the desired accuracy specified by $\varepsilon$. In this case, the corresponding component of the *Info*-vector has the value *false*.

As a consequence of the splitting (bisecting) of the intervals, it may happen that a minimizer lying exactly at the splitting point is enclosed in two different intervals. A sophisticated supplement to our method avoiding this can be found in [79].

The algorithm is well suited for parallel computers due to its natural parallelism given by the splitting of the intervals by the extended interval Newton step or by the bisection.

# 7.3 Implementation and Examples

## 7.3.1 C++ Program Code

We begin by describing our implementation of the operations needed for handling lists and list elements (pairs). Then, we describe the implementation of Algorithm 7.4 and its subalgorithms.

### 7.3.1.1 Module lst1_ari

The header file of module *lst1_ari* supplies the definition of the class *Pair1* representing a pair of an interval and a real value. The type *Pair1Ptr* is provided to manage lists of such pairs. The constant pointer *EmptyList* denotes an empty list.

```
//----------------------------------------------------------------------------
// File: lst1_ari (header)
// Purpose: Definition of a list arithmetic used in connection with an
//    interval bisection method in global optimization for storing pairs of
//    an interval and a real value.
// Class Pair1:
//    Pair1()        : constructors
//    ~Pair1()       : destructor
//    operator =     : assignment operator
//    _Pair1()       : for explicit generating a pair
//    GetInt()       : get the interval component
//    GetFyi()       : get the function value
// Global type:
//    Pair1Ptr       : list of pairs
// Global constant:
//    EmptyList      : the empty list (NULL pointer)
// Global functions and operators:
//    operator +     : adding a new element to a list
//    Head()         : access functions for the head of a list
//    Length()       : access function to length of list
//    FreeAll()      : free complete list
//    MultiDelete()  : deletes several elements in a list
//    DelHead()      : deletes the first element of a list
//    ListToVector(): transfer a list into a dynamic interval vector
//----------------------------------------------------------------------------
#ifndef __LST1_ARI_HPP
#define __LST1_ARI_HPP

#include <intvecto.hpp>      // Integer vector type
#include <ivector.hpp>       // Interval vector arithmetic

#define EmptyList NULL       // The empty list
                             //---------------

struct                       // Structure used for a list of pairs
  Pair1Elmt;                 //-----------------------------------

typedef                      // Pointer to a list of pairs
  Pair1Elmt* Pair1Ptr;       //---------------------------

class Pair1 {                // Pair of an interval 'intv' and
  private:                   // the real value 'inf(f(intv))'.
    interval intv;           //-------------------------------
    real     fyi;
```

```
  public:
    Pair1 ( ) { };
    Pair1 ( Pair1& );

    Pair1& operator= ( Pair1& );

    friend Pair1    _Pair1 ( interval&, const real& );
    friend interval GetInt ( Pair1& );
    friend real     GetFyi ( Pair1& );
};

extern void      MultiDelete  ( Pair1Ptr&, const real& );
extern void      ListToVector ( Pair1Ptr, ivector&, intvector& );
extern Pair1Ptr operator+     ( Pair1Ptr, Pair1 );

extern void      FreeAll      ( Pair1Ptr& );
extern int       Length       ( Pair1Ptr );
extern Pair1     Head         ( Pair1Ptr );
extern void      DelHead      ( Pair1Ptr& );
#endif
```

A list of pairs is realized as a linearly linked list. The static variable *FreeList* is used to hold the list of freed elements. The static functions *NewPP()*, *Free()*, and the global function *FreeAll()* work on that list to prevent the creation of garbage in memory. The friend functions *_Pair1()*, *GetInt()*, and *GetFyi()* are transfer and access functions for pairs.

The operator + enters a new list element $p$ in the list *List* according to Condition (7.10). The function *MultiDelete()* deletes all elements $p$ in *List* for which $p.fyi > fmax$. This function assumes that the list elements are ordered according to Condition (7.10). Function *Head()* delivers the first pair of *List*, whereas *DelHead()* deletes the first pair of *List*. Function *Length()* delivers the number of elements in *List*. The void function *ListToVector()* transfers a list of pairs to the result vectors *Opt* and *Info* according to Step 6 of Algorithm 7.4.

```
//----------------------------------------------------------------------------
// File: lst1_ari (implementation)
// Purpose: Definition of a list arithmetic used in connection with an
//    interval bisection method in global optimization for storing pairs of
//    an interval and a real value.
// Method: Overloading of functions and operators of class type 'Pair1' (list
//    element) and data type 'Pair1Ptr' (list). See module 'lst_ari' for the
//    definition of a class 'Pair' for pairs with a vector component 'fyi'.
//    The suffix 1 for the type 'Pair1' is needed to avoid a duplicate name
//    error when linking all toolbox modules to a library.
// Class Pair1:
//    Pair1()        : constructors
//    ~Pair1()       : destructor
//    operator =     : assignment operator
//    _Pair1()       : for explicit generating a pair
//    GetInt()       : get the interval component
//    GetFyi()       : get the function value
// Global type:
//    Pair1Ptr       : list of pairs
// Global functions and operators:
//    operator +     : adding a new element to a list
//    Head()         : access functions for the head of a list
//    Length()       : access function to length of list
//    FreeAll()      : free complete list
//    MultiDelete() : deletes several elements in a list
```

```
//    DelHead()      : deletes the first element of a list
//    ListToVector(): transfer a list into a dynamic interval vector
//----------------------------------------------------------------------------
#include <interval.hpp>     // Interval arithmetic
#include <intvecto.hpp>     // Integer vector type
#include <lst1_ari.hpp>

struct Pair1Elmt {         // Type for a list of pairs
  Pair1    P;              //-----------------------
  Pair1Ptr N;
};

static Pair1Ptr            // Local variable storing list of free
  FreeList = EmptyList;    // elements (automatic garbage recycling)
                           //--------------------------------------

//----------------------------------------------------------------------------
// Constructors and assignment operator
//----------------------------------------------------------------------------
Pair1::Pair1 ( Pair1& a )
  { intv = a.intv; fyi  = a.fyi; }

Pair1& Pair1::operator= ( Pair1& a )
{
  intv = a.intv;
  fyi  = a.fyi;
  return *this;
}

//----------------------------------------------------------------------------
// Transfer and access functions for pairs
//----------------------------------------------------------------------------
Pair1 _Pair1 ( interval& i, const real& r )
{
  Pair1 p;
  p.intv = i;
  p.fyi  = r;
  return p;
}

interval GetInt ( Pair1& p )          // Get int-component
  { return p.intv; }                  //------------------

real GetFyi ( Pair1& p )              // Get fyi-component
  { return p.fyi; }                   //------------------

//----------------------------------------------------------------------------
// Functions for generating and freeing of list elements (pairs)
//----------------------------------------------------------------------------
static void NewPP ( Pair1Ptr& pp )     // 'NewPP' generates a new list element
{                                      // or gets one from 'FreeList'.
  if (FreeList == EmptyList)           //-----------------------------------
    { pp = new Pair1Elmt; pp->N = NULL; }
  else
    { pp = FreeList; FreeList = FreeList->N; pp->N = NULL; }
}

static void Free ( Pair1Ptr& pp )            // 'Free' enters one element of a
{                                            // list in the 'FreeList'.
  if (pp != NULL)                            //-----------------------------
    { pp->N = FreeList; FreeList = pp; pp = NULL; }
}

void FreeAll ( Pair1Ptr& List )              // 'FreeAll' enters all elements
```

```
{                                          // of 'List' in the 'FreeList'
  Pair1Ptr H;                              //------------------------------

  if (List != EmptyList) {
    H = List;
    while (H->N != NULL) H = H->N;
    H->N = FreeList; FreeList = List; List = NULL;
  }
}

//----------------------------------------------------------------------------
// Operator + enters the pair 'P' on the list 'List' in such a way that after
// entering, one of the four following condition holds:
//   1) O.fyi <= P.fyi < Q.fyi,
//   2)          P.fyi < Q.fyi  and  'P' is the first element of 'List',
//   3) O.fyi <= P.fyi          and  'P' is the last  element of 'List',
//   4)                              'P' is the only  element of 'List',
// where 'O' is the preceding and 'Q' is the succeeding element of 'P' in
// the resulting list.
//----------------------------------------------------------------------------
Pair1Ptr operator+ ( Pair1Ptr List, Pair1 p )
{
  Pair1Ptr H, HN;
  int     ready, alreadyIn;

  if (List == EmptyList) {                           // List is empty, so new
    NewPP(H); H->P = p; H->N = NULL;                 // list contains only P.
    return H;                                        //----------------------
  }
  else if (GetFyi(List->P) > GetFyi(p)) {            // p becomes new first
    NewPP(H); H->P = p; H->N = List;                 // element of the list.
    return H;                                        //----------------------
  }
  else {
    H = List; HN = H->N; ready = 0;
    alreadyIn = (GetInt(H->P) == GetInt(p));

    while ( !(ready || alreadyIn) ) {                 // Search for the right
      if (HN == NULL)                                 // position to enter P.
        ready = 1;                                    //---------------------
      else if (GetFyi(HN->P) > GetFyi(p))
        ready = 1;
      else {
        H = HN; HN = H->N;
        alreadyIn = (GetInt(H->P) == GetInt(p));
      }
    } // while

    if (!alreadyIn)                                      // Enter p between H
      { NewPP(H->N); H = H->N; H->P = p; H->N = HN; }    // and HN.
                                                         //------------------
    return List;
  }
} // operator+

//----------------------------------------------------------------------------
// 'MultiDelete' deletes all elements 'P' in 'List' for which the condition
// 'P.fyi > fmax' holds.  This function assumes that the 'fyi' components of
// the list elements are sorted in increasing order (see operator +).
//----------------------------------------------------------------------------
void MultiDelete ( Pair1Ptr& List, const real& fmax )
{
  Pair1Ptr DelPrev, Del;
  int     ready;
```

```
  if (List != EmptyList) {
    if (GetFyi(List->P) > fmax)                  // All list elements fulfill
      { Del  = List; List = EmptyList; }         // 'P.fyi > fmax'.
    else {                                       //--------------------------
      DelPrev = List; Del = DelPrev->N; ready = (Del == NULL);

      while (!ready) {
        if (Del == NULL)
          ready = 1;
        else if (GetFyi(Del->P) > fmax)
          { ready = 1; DelPrev->N = NULL; }
        else
          { DelPrev = Del; Del = Del->N; }
      }
    }
    FreeAll(Del);
  }
} // Multidelete

Pair1 Head ( Pair1Ptr List )                    // Delivers first pair of the
  { return List->P; }                           // list, i.e. the pair P with
                                                // the smallest value P.fyi.
                                                //---------------------------

void DelHead ( Pair1Ptr& List )                  // Deletes the first pair of
{                                                // the List
  Pair1Ptr Del;                                  //--------------------------

  Del = List; List = List->N; Free(Del);
}


int Length ( Pair1Ptr List )                   // 'Length' delivers the number
{                                              // of elements in list 'List'
  int     i = 0;                               //----------------------------

  while (List != EmptyList)
    { i++; List = List->N; }
  return i;
}

//----------------------------------------------------------------------------
// Copies the interval components of every pair of the 'List' to the interval
// vector 'Vec' and allocates the intvector 'bv'. For an empty list the
// vectors 'Vec' and 'bv' are sized to unit length with undefined components.
//----------------------------------------------------------------------------
void ListToVector ( Pair1Ptr List, ivector& Vec, intvector& bv )
{
  int     Len, i;
  Pair1Ptr H;

  Len = Length(List);
                                                // Resize the vectors:
  if (Len == 0)                                 //    - to minimum length with
    { Resize(Vec,1); Resize(bv,1); }            //      undefined components
  else                                          // or
    { Resize(Vec,Len); Resize(bv,Len); }        //    - to correct length
                                                //---------------------------
  H = List;
  for (i = 1; i <= Len; i++)
    { Vec[i] = GetInt(H->P); H = H->N; }
}
```

### 7.3.1.2 Module gop1

The header file of module *gop1* supplies the interface of the function *AllGOp1()* for computing all global minimizers and the global minimum value of a twice continuously differentiable function and the interface of the function *AllGOp1ErrMsg()* which can be used to get an error message for the error code returned by *AllGOp1()*.

```
//----------------------------------------------------------------------------
// File: gop1 (header)
// Purpose: Computing enclosures for all global minimizers and for the global
//    minimum value of a twice continuously differentiable one-dimensional,
//    scalar valued function.
// Global functions:
//    AllGOp1()      : computes enclosures for all global optimizers
//    AllGOp1ErrMsg(): delivers an error message text
//----------------------------------------------------------------------------
#ifndef __GOP1_HPP
#define __GOP1_HPP

#include <interval.hpp>       // Interval arithmetic
#include <intvector.hpp>      // Integer vector type
#include <ivector.hpp>        // Interval vector arithmetic
#include <ddf_ari.hpp>        // Differentiation arithmetic

const int MaxCount = 10000;   // Maximum count of result components

extern char* AllGOp1ErrMsg ( int );
extern void  AllGOp1 ( ddf_FctPtr, interval, real, ivector&, intvector&,
                       int&, interval&, int&, int = MaxCount );
#endif
```

The function *AllGOp1()* is an implementation of Algorithm 7.4. The static functions *NewtonStep()*, *GlobalOptimize()*, and *VerificationStep()* are implementations of the Algorithms 7.1 to 7.3. All derivative evaluations are done with the help of the differentiation arithmetic *ddf_ari* described in Chapter 5.

The function *AllGOp1()* uses the *DerivType* function *f()* and the starting interval *Start* as input parameters and stores all computed enclosures in the interval vector *OptiVector*. The information about the uniqueness of the optimizers is stored in the integer vector *InfoVector*. When the working list of pairs is transferred to the result vectors by calling function *ListToVector()*, both vectors *OptiVector* and *InfoVector* are automatically resized to the length given by the number of minimizers in the list. The function *AllGOp1()* has an optional last parameter *MaxNumberOfOptis* by which you can specify the maximum number of optimizers to be computed. Its default value is given by the local constant *MaxCount* = 10000 which should be big enough to compute *all* optimizers in most cases. *AllGOp1()* reports a warning if there are more optimizers than specified by parameter *MaxNumberOfOptis*. The value of the desired accuracy *Epsilon* is checked to be greater than or equal to *MinEpsilon* (the relative machine accuracy). If this is not true, then *Epsilon* is reset to 1 ulp accuracy.

```
//----------------------------------------------------------------------------
// File: gop1 (implementation)
// Purpose: Computing enclosures for all global minimizers and for the global
//    minimum value of a twice continuously differentiable one-dimensional,
//    scalar valued function.
```

```
// Method: Bisection method combined with midpoint, monotonicity, concavity
//    test and extended interval Newton step.
// Global functions:
//    AllGOp1()       : computes enclosures for all global optimizers
//    AllGOp1ErrMsg(): delivers an error message text
//----------------------------------------------------------------------------
#include <string.h>          // String handling
#include <i_util.hpp>        // Interval utility functions
#include <xi_ari.hpp>        // Extended interval arithmetic
#include <lst1_ari.hpp>      // List arithmetic
#include <gop1.hpp>

static int MaxOptiNo;        // Maximum number of optimizers to be computed

//----------------------------------------------------------------------------
// Error codes used in this module.
//----------------------------------------------------------------------------
const int
  NoError      = 0,  // No error occurred
  IllMaxOptiNo = 1,  // Illegal value for the maximum number of optimizers
  NotAllOptis  = 2;  // Not all optimizers computed

//----------------------------------------------------------------------------
// Error messages depending on the error code.
//----------------------------------------------------------------------------
char* AllGOp1ErrMsg ( int Err )
{
  static char Msg[160] = "";

  switch (Err) {
    case NoError:
      break;  // Blank message
    case IllMaxOptiNo:
      sprintf(Msg,"Error: Parameter for maximum number of optimizers "
                  "must lie in 1,...,%1d!",MaxCount); break;
    case NotAllOptis:
      sprintf(Msg,"Warning: Not all optimizers found due to the user "
                          "limit of %1d optimizer(s).\n"
                  "        The enclosure of the global minimum value "
                          "could not be optimal!",MaxOptiNo); break;
    default:
      strcpy(Msg,"Error: Code not defined!");
  }
  return(Msg);
}

//----------------------------------------------------------------------------
// Purpose: Execution of one extended interval Newton step for the derivative
//    of function 'f'.
// Parameters:
//    In : 'f'    : must be declared for the 'DerivType' to enable the
//                  internal use of the differentiation arithmetic 'ddf_ari'
//         'Y'    : specifies the starting interval
//         'ddfY' : f''(Y), already computed outside of 'NewtonStep'
//    Out: 'V'    : Pair of intervals V[1] and V[2]
//         'p'    : Number of valid intervals in V (0, 1, or 2)
// Description:
//    One extended interval Newton step for 'Y' is executed resulting in 'p'
//    interval(s) 'V[1]' and 'V[2]' which can be empty.
//----------------------------------------------------------------------------
static void NewtonStep ( ddf_FctPtr f,
                         interval   Y,
                         interval   ddfY,
                         ivector&   V,
```

```
                           int&        p )
{
  real      c;
  interval fC, dfC;

  c = mid(Y); dfEval(f,_interval(c),fC,dfC);      // Midpoint evaluation of 'f'
  V = Y & (c - dfC % ddfY)     ;                  // Execution of Newton step
  if (V[1] != EmptyIntval()) p = 1; else p = 0;   // Fix number of non-empty
  if (V[2] != EmptyIntval()) p++;                 // intervals
}                                                 //---------------------------

//----------------------------------------------------------------------------
// Purpose: Execution of the global optimization method including a bisection
//    method, midpoint test, monotonicity test, concavity test, and extended
//    interval Newton steps.
// Parameters:
//    In : 'f'          : must be declared for the 'DerivType' to enable the
//                        internal use of the differentiation arithmetic
//                        'ddf_ari'
//         'Start:      : specifies the starting interval
//         'Epsilon'    : specifies the desired relative accuracy
//    Out: 'ResultList': stores the candidates for enclosure of a global
//                        minimizer
//         'ListLength': length of the result list, that is the number of
//                        candidates in the result list
//         'Minimum'    : stores the enclosure of the global minimum value
// Description:
//    The function manages the list 'L' of pending subintervals that may
//    contain global minimizers. Subintervals are removed from the list and
//    placed in the accepted list 'ResultList' when they satisfy relative
//    error acceptance criteria. Subintervals are also removed from the list
//    by the midpoint, monotonicity, concavity tests, or by the interval
//    Newton steps. Subintervals are added to the pending list when an element
//    from the list is bisected or when the extended interval Newton step
//    yields two candidate intervals. 'ResultList' returns the list of
//    enclosures of the global minimizers. The maximum number of enclosures
//    to be computed is given by the value of the global variable 'MaxOptiNo'.
//    'Minimum' returns the enclosure of the global minimum value.
//----------------------------------------------------------------------------
static void GlobalOptimize ( ddf_FctPtr f,
                             interval   Start,
                             real       Epsilon,
                             Pair1Ptr&  ResultList,
                             int&       ListLength,
                             interval&  Minimum )
{
  Pair1     PairY;                         // For the pair (Y, inf(f(Y))
  interval  Y, fY, dfY, ddfY, BdP;         // Initial box, f(Y), f'(Y), f''(Y)
  ivector   U(2), V(2);                    // Subboxes of Y and V
  interval  fU, fV, fC, fCV, fBdP;         // Function evaluations of f
  interval  dfU, dfV, ddfU;                // Derivative evaluations of f
  real      fmax;                          // Upper bound for minimum value
  real      c, cV;                         // Midpoint of interval Y and V
  Pair1Ptr  WorkList;                      // List of pairs
  int       i, j, p;                       // Control variables
  int       Bisect;                        // Flag for iteration and algorithm

  c = mid(Start); fEval(f,_interval(c),fC); // Compute upper bound for minimum
  fmax = Sup(fC);                           //--------------------------------

  if ( !UlpAcc(Start,1) ) {                                  // Start method
    Y = Start;                                               //-------------
    WorkList = EmptyList;
    ResultList = EmptyList;
```

```
ListLength = 0;

BdP = Inf(Y);                              // Treat boundary points separately
for (i = 1; i <= 2; i++) {                 //---------------------------------
  fEval(f,BdP,fBdP);
  if (Sup(fBdP) < fmax) fmax = Sup(fBdP);
  if (fmax >= Inf(fBdP))
    WorkList = WorkList + _Pair1(BdP,Inf(fBdP));
  BdP = Sup(Y);
}

do {                                                       // Start iteration
  U[1] = _interval(Inf(Y),c);                              // Bisect 'Y'
  U[2] = _interval(c,Sup(Y));                              //----------------
  for (i = 1; i <= 2; i++) {
    dfEval(f,U[i],fU,dfU);                              // Compute dfu = f'(U)
                                                        //--------------------
    if ( !in(0.0,dfU) ) continue;      // Monot. test: if not 0 in dfU stop
                                       //----------------------------------
    fU = (fC + dfU * (U[i] - c)) & fU;                 // Centered form 'f(U)'
                                                       //---------------------
    if (fmax < Inf(fU)) continue;

    ddfEval(f,U[i],fU,dfU,ddfU);                        // Compute dfu = f'(U)
                                                        //--------------------
    if (Sup(ddfU) < 0.0) continue;      // Concavity test: if 0 > ddfU stop
                                        //---------------------------------
    NewtonStep(f,U[i],ddfU,V,p);           // Extended interval Newton step
                                           //------------------------------
    for (j = 1; j <= p; j++) {
      dfEval(f,V[j],fV,dfV);
      if ( !in(0.0,dfV) ) continue;    // Monot. test: if not 0 in dfV stop
                                       //----------------------------------
      cV = mid(V[j]);                                   // Try centered form
      fEval(f,_interval(cV),fCV);                       // to get better en-
      fV = (fCV + dfV * (V[j] - cV)) & fV;              // closure of 'f(U)'
                                                        //------------------
      if (fmax >= Inf(fV))                                        // Store V
        WorkList = WorkList + _Pair1(V[j],Inf(fV));               //--------
    } // for j ...
  } // for i ...

  Bisect = 0;                                              // Get next 'Y' of
  while ( (WorkList != EmptyList) && !Bisect ) {           // the work list
    PairY = Head(WorkList); DelHead(WorkList);             //----------------

    Y = GetInt(PairY); c = mid(Y); fEval(f,_interval(c),fC);
    if (Sup(fC) < fmax) fmax = Sup(fC);
    MultiDelete(WorkList,fmax);

    Minimum = _interval(GetFyi(PairY),fmax);
    if ( (RelDiam(Minimum) <= Epsilon) || (RelDiam(Y) <= Epsilon) ) {

      ListLength++;                   // Checking termination criterion. Stop
                                      // due to user specified maximum for
      if (ListLength > MaxOptiNo) {// the number of optimizers.
        FreeAll(WorkList);            //-------------------------------------
        Minimum = _interval(GetFyi(Head(ResultList)),fmax);
        return;
      }

      ResultList = ResultList + PairY;
    }
    else
```

```
        Bisect = 1;
      } // while

    } while (Bisect);
  } // if ( !UlpAcc(Start,1) )
  else {                                                // Store starting interval
    fEval(f,Start,fY);                                  //------------------------
    ResultList = EmptyList + _Pair1(Start,Inf(fY));
    ListLength = 1;
  }
                                                        // Compute good enclosure
  Minimum = _interval(GetFyi(Head(ResultList)),fmax);// of the global minimum
} // GlobalOptimize                                     //-----------------------

//----------------------------------------------------------------------------
// Purpose: Execution of a verification step including the use of an epsilon
//    inflation.
// Parameters:
//    In    : 'f'       : function of 'DerivType'
//    Out   : 'yUnique': returns 'true' if the verification is successful
//    In/Out: 'y'       : interval enclosure to be verified
// Description: This function checks the uniqueness of the local minimizer
//    enclosed in the interval variable 'y' by a verification step including
//    the use of an epsilon inflation of the iterates.
//----------------------------------------------------------------------------
static void VerificationStep ( ddf_FctPtr f, interval& y, int& yUnique )
{
  const
    int kmax = 10;  // Maximum number of iterations

  interval  yIn, yOld, fm, dfm, fy, dfy, ddfy;
  real      m, eps;
  int       k;

  yUnique = (Inf(y) == Sup(y));                         // y is a point interval
  if (yUnique) return;                                  //-----------------------

  yIn = y; k = 0; eps = 0.25;                                  // Initializations
                                                               //----------------
  while ( !yUnique && (k < kmax) ) {      // Do kmax loops to achieve inclusion
                                          //-----------------------------------
    yOld = Blow(y,eps);                               // Epsilon inflation of 'y'
    ddfEval(f,yOld,fy,dfy,ddfy);                      //-------------------------

    if (Inf(ddfy) <= 0.0) break;       // No verification of a minimum possible
                                       //--------------------------------------
                                                 // Perform interval Newton step
    k++; m = mid(yOld);                          //-----------------------------
    dfEval(f,_interval(m),fm,dfm);
    y = (m - dfm / ddfy);

    if (Disjoint(y,yOld)) break;                      // No verification possible
                                                      //-------------------------
    yUnique = in(y,yOld);             // Inner inclusion ===> unique zero of f'
    y = y & yOld;                     // Intersection with old value
    if (y == yOld)                    //---------------------------------------
      eps = eps * 8.0;                           // Increase the value of 'eps'
                                                 //----------------------------
  } // while

  if (!yUnique) y = yIn;
} // VerificationStep

//----------------------------------------------------------------------------
```

```
// Purpose: Computation of enclosures for all global minimizers and for the
//    global minimum value of a twice continuously differentiable one-dimen-
//    sional, scalar-valued function.
// Parameters:
//    In : 'f'                  : objective function, must be declared for the
//                                'DerivType' to enable the internal use of
//                                the differentiation arithmetic 'ddf_ari'
//         'Start',             : specifies the starting interval
//         'Epsilon'            : specifies the desired relative accuracy
//                                (interval diameter) of the result intervals
//         'MaxNumberOfOptis': maximum number of optimizers to be computed
//                                (default value: 'MaxCount' = compute all)
//    Out: 'OptiVector'         : stores and returns the enclosures for the
//                                global optimizers of 'f'
//         'InfoVector'         : stores the corresponding information on the
//                                uniqueness of the local optimizers in these
//                                enclosures
//         'NumberOfOptis'      : returns the number of enclosures computed
//         'Minimum'      '     : returns the enclosure for the minimum value
//         'Err'                : returns an error code
// Description:
//    The enclosures for the global minimizers of 'f' are computed by calling
//    function 'GlobalOptimize'. Then a verification step is applied.
//    The enclosures for the global minimizers of 'f' are stored in the
//    interval vector 'OptiVector', the corresponding information on the
//    uniqueness of the local minimizers in these enclosures is stored in the
//    integer vector 'InfoVector'. The number of enclosures computed is
//    returned in the integer variable 'NumberOfOptis'. The enclosure for the
//    global minimum value is returned in the variable 'Minimum'. If an error
//    occurs, the value of 'Err' is different from 0. The result vectors
//    'OptiVector' and 'InfoVector' are automatically resized to standard
//    index range with lower index bounds 1. The maximum number of optimizers
//    which can be computed is given by 'MaxNumberOfOptis'. It is bounded by
//    'MaxCount'. If your system can handle more than 'MaxCount' optimizers,
//    you can increase its value.
//----------------------------------------------------------------------------
void AllGOp1 ( ddf_FctPtr  f,
               interval    Start,
               real        Epsilon,
               ivector&    OptiVector,
               intvector&  InfoVector,
               int&        NumberOfOptis,
               interval&   Minimum,
               int&        Err,
               int         MaxNumberOfOptis )
{
  int        i;
  real       MinEpsilon;
  Pair1Ptr   ResultList;

  if (1 <= MaxNumberOfOptis && MaxNumberOfOptis <= MaxCount) {
    // Start global optimization method
    //---------------------------------
    Err = NoError;
    MaxOptiNo = MaxNumberOfOptis;           // Initialize global static variable

    MinEpsilon = succ(1.0) - 1.0;           // Relative machine accuracy (1 ulp)
    if (Epsilon < MinEpsilon)
      Epsilon = MinEpsilon;                 // Set 'Epsilon' to 1 ulp accuracy

    GlobalOptimize(f,Start,Epsilon,ResultList,NumberOfOptis,Minimum);

    if (NumberOfOptis > MaxNumberOfOptis)
      { Err = NotAllOptis;  NumberOfOptis = MaxNumberOfOptis; }
```

```
    ListToVector(ResultList,OptiVector,InfoVector);

    for (i = 1; i <= NumberOfOptis; i++)                        // Verification step
      VerificationStep(f,OptiVector[i],InfoVector[i]);          // for the enclosure
                                                                // intervals
    FreeAll(ResultList);                                        //------------------
  }
  else {
    Err = IllMaxOptiNo;
    NumberOfOptis = 0;
  }
}
```

### 7.3.2 Examples

As a sample program, we wish to use *AllGOp1()* to compute all global minimizers and the global minimum of the function

$$f(x) = (x + \sin x) \cdot \mathrm{e}^{-x^2},$$

which is the test function $f_4$ in [93], and of the function of Shubert

$$g(x) = -\sum_{k=1}^{5} k \sin((k+1)x + k),$$

which is the test function $f_3$ in [93]. Figures 7.4 and 7.5 show the plots of these two functions. Within the interval $[-10, 10]$, $f$ has the global minimizer $-6.7957...$, and the global minimizers of $g$ occur at $-6.7745...$, $-0.49139...$, and $5.7917...$. According

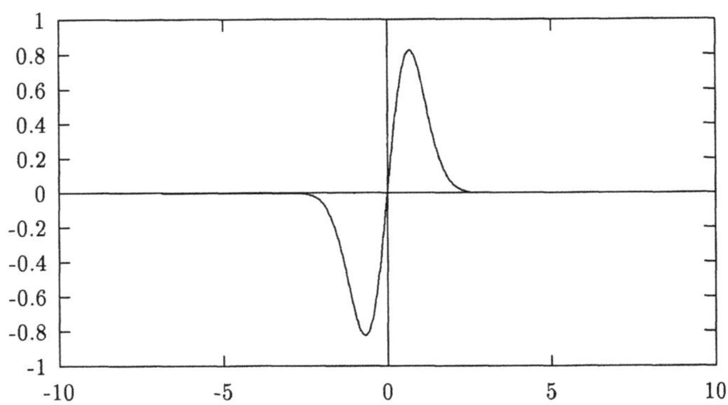


Figure 7.4: Function $f$ (test function $f_4$ in [93])

to [93], both functions are difficult to minimize and represent practical problems. The large number of local optimizers of $g$ make it very difficult for an approximation method to find the global optimizers.

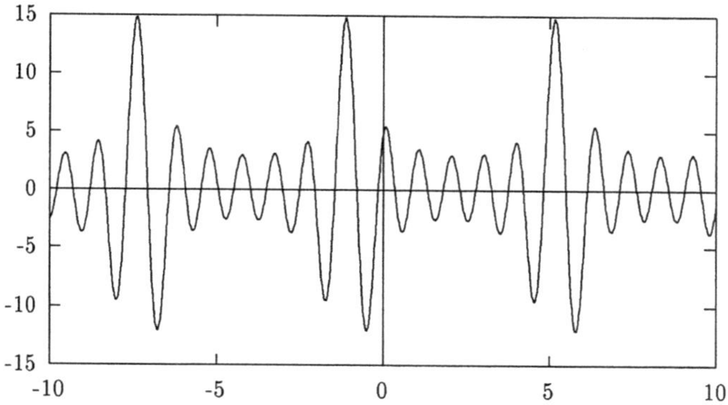


Figure 7.5: Shubert's function $g$ (test function $f_3$ in [93])

We first have to define the functions $f$ and $g$ for the type *DerivType*. Then we use a function *compute()* with the function $f$ and a string (a description of the function) as parameters. This function prompts you to enter the necessary data for the call of function *AllGOp1()*. If we add further function definitions in our program, we can apply this function in the same way using another function as parameter.

```
//----------------------------------------------------------------------------
// Example: Global optimization (one-dimensional)
// This program uses module 'gop1' to compute the global optimizers of the
// functions
//
//    f(x) = (x + sin(x)) * exp(-sqr(x)).
//
// and
//                 5
//    g(x) = - sum ( k * sin((k+1)*x + k) )
//                k=1
//
// A starting interval and a tolerance must be entered.
//----------------------------------------------------------------------------
#include <gop1.hpp>      // Global optimization

DerivType f ( DerivType x )
  { return (x + sin(x)) * exp(-sqr(x)); }

DerivType g ( DerivType x )
{
  DerivType  s;
  int        k;

  s = DerivConst(0.0);
  for (k = 1; k <= 5; k++)
    s = s + double(k) * sin ( double(k+1) * x + double(k) );
  return -s;
}

//----------------------------------------------------------------------------
// Function for printing and reading information to call the function
// 'AllGOp1'. It must be called with the function parameter 'f' and a
// string 'Name' containing a textual description of that function.
```

```
//------------------------------------------------------------------------
void compute ( ddf_FctPtr f, char *Name )
{
  interval   SearchInterval, Minimum;
  real       Tolerance;
  ivector    Opti;
  intvector  Unique;
  int        NumberOfOptis, i, Error;

  cout << SetPrecision(23,15) << Scientific;   // Output format

  cout << "Computing all global minimizers of the function  " << Name << endl;
  cout << "Search interval       : "; cin >> SearchInterval;
  cout << "Tolerance (relative) : "; cin >> Tolerance;
  cout << endl;

  AllGOp1(f,SearchInterval,Tolerance,
          Opti,Unique,NumberOfOptis,Minimum,Error);

  for (i = 1; i <= NumberOfOptis; i++) {
    cout << Opti[i] << endl;
    if (Unique[i])
      cout << "encloses a locally unique candidate for a global minimizer!";
    else
      cout << "may contain a local or global minimizer!";
    cout << endl;
  }

  if (NumberOfOptis != 0)
    cout << endl << Minimum << endl
         << "encloses the global minimum value!" << endl;
  cout << endl << NumberOfOptis << " interval enclosure(s)" << endl;

  if (Error)
    cout << endl << AllGOp1ErrMsg(Error) << endl;
  else if ( (NumberOfOptis == 1) && Unique[1] )
    cout << endl << "We have validated that there is "
                    "a unique global optimizer!" << endl;
}

int main ( )
{
  compute(f,"(x + SIN(x))*EXP(-x↑2)");
  cout << endl << endl;
  compute(g,"-SUM(k*SIN((k+1)*x+k),k,1,5)");
  return 0;
}
```

If we execute this program, we get the following runtime output:

```
Computing all global minimizers of the function  (x + SIN(x))*EXP(-x^2)
Search interval       : [-10,10]
Tolerance (relative) : 1e-12

[-6.795786600198817E-001,-6.795786600198813E-001]
encloses a locally unique candidate for a global minimizer!

[-8.242393984760770E-001,-8.242393984760765E-001]
encloses the global minimum value!

1 interval enclosure(s)

We have validated that there is a unique global optimizer!
```

```
Computing all global minimizers of the function  -SUM(k*SIN((k+1)*x+k),k,1,5)
Search interval      : [-10,10]
Tolerance (relative) : 1e-12

[ 5.791794470920271E+000, 5.791794470920272E+000]
encloses a locally unique candidate for a global minimizer!
[-6.774576143438902E+000,-6.774576143438900E+000]
encloses a locally unique candidate for a global minimizer!
[-4.913908362593147E-001,-4.913908362593144E-001]
encloses a locally unique candidate for a global minimizer!

[-1.203124944216720E+001,-1.203124944216713E+001]
encloses the global minimum value!

3 interval enclosure(s)
```

Thus, we know that there is only one global minimizer of $f$ and three locally unique minimizers which are good candidates for global minimizers of $g$ within the specified starting interval $[-10, 10]$.

### 7.3.3 Restrictions and Hints

The objective function $f$ must be expressible as a finite sequence of arithmetic operations and elementary functions supported by the differentiation arithmetic module *ddf_ari*.

If *AllGOp1()* reports a warning that not all optimizers were found due to a user limit for the number of optimizers to be computed, then you may call *AllGOp1()* again to find additional optimizers.

The method is not very fast if a very small value of $\varepsilon$ (*Epsilon*) is used, if the interval Newton step does not improve the actual iterates, *and* if the different tests do not discard intervals any more because of rounding and the overestimation effects of the machine interval arithmetic. Under these circumstances, the method is equivalent with a bisection method.

In *GlobalOptimize()*, the evaluation of the function in differentiation arithmetic can cause a runtime error if the interval argument of an elementary function does not lie in the domain specified for this interval function or if a division by an interval containing zero occurs. This also may be due to the known overestimation effects of interval arithmetic (see Section 3.1). To get rid of these errors, the user may try to split the starting interval in several parts and call *AllGOp1()* for these parts.

Note that the rules for getting true enclosures in connection with conversion errors (see Section 3.7) also apply here. For example, the user who wishes to compute enclosures for the minimizers of the function $f(x) = (x - \frac{1}{3})^2 \cdot (x - 0.1)^2$, must be careful that the values $\frac{1}{3}$ and 0.1 are enclosed in machine intervals when computing interval evaluations of $f$. An appropriate coding of $f$ is

```
DerivType f ( DerivType x )
{
  interval one = interval(1.0,1.0);
  return sqr(x - one/3.0) * sqr(x - one/10.0);
}
```

## 7.4 Exercises

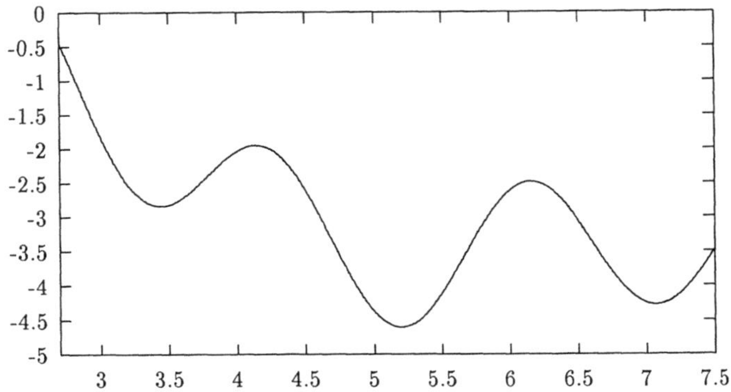


Figure 7.6: Function $h$ (test function $f_1$ in [93]) with several local minima

**Exercise 7.1** Use our function *AllGOp1()* to compute the global minimizer and the global minimum value of the function

$$h(x) = \sin x + \sin \frac{10x}{3} + \ln x - \frac{84x}{100}.$$

This function is the test function $f_1$ from [93] shown in Figure 7.6 with starting interval $[2.7, 7.5]$. It is a rather simple problem with several local minima.

Find an interval $[a, b]$ such that you can prove that the unique global (on $[a, b]$) minimizer located by *AllGOp1()* is also the global minimizer in $I\!R$.

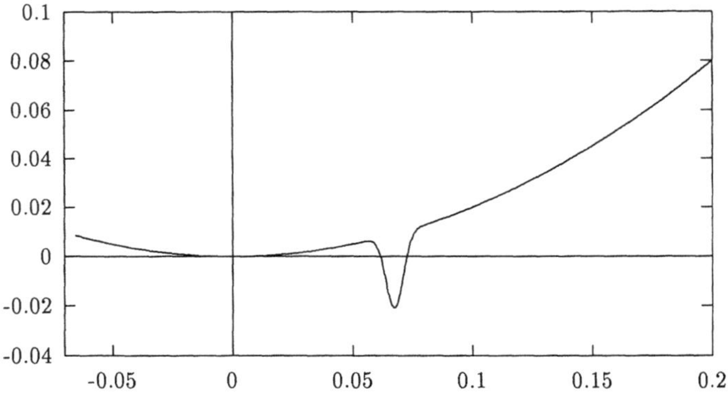


Figure 7.7: Test function $r$ with a sharp valley

**Exercise 7.2** Use our function *AllGOp1()* to compute the global minimizer and the global minimum value of the function

$$r(x) = 2x^2 - \frac{3e^{-(a(x-b))^2}}{c}$$

with starting interval $[-10, 10]$. The global minimum of $r$ is at a sharp valley for $a = 200$, $b = 0.0675$, and $c = 100$ as shown in Figure 7.7. Find a set of values $a$, $b$, and $c$ for which an approximate optimization routine available to you fails. Hint: Make the valley very sharp, rather far from the local minimizer near $x = 0$, and deep enough to cut below the $x$-axis. Can you trick *AllGOp1()* similarly?

**Exercise 7.3** Modify module *go1* to use a local approximate search strategy to locate a point $c$ that may be more effective than the midpoint in eliminating subintervals from the list $L$ (cf. Section 7.1.1). Be careful that your strategy does not leave the subinterval $[y]$. Is the resulting algorithm faster or slower than the one presented here?

## 7.5 References and Further Reading

The method we discussed in this chapter is an *a priori* method because the iteration starts with a (possibly large) interval enclosing all the solutions which have to be found. Here, the iterates of the method are subintervals of the previous iterates. There are also methods for finding (and bounding) one single *local* optimizer called *a posteriori* methods. These methods start with an approximation of a minimizer of $f$ and apply a test procedure for a neighborhood interval of the approximation to verify that a zero of $f'$ lies within that interval.

Our global optimization method can benefit from an approximation of a global minimizer in the way that we use that approximation to compute the value $\tilde{f}$ incorporated in the midpoint test. But we have to to do the same work for validating that the global minimizer has been found as for computing its value by our method. That is, a global optimization method with automatic result verification cannot be an *a posteriori* method.

For the approximation of a minimizer, there are also a huge number of classical optimization methods for local and global optimization available without any verification of the result. For an overview on such approximation methods see [93] or [96], for example.

The method presented in this chapter can easily be extended to become more efficient and faster. For more sophisticated extensions, see [28], [29], [30], [31], [39], [78], [79], and [80].

# Chapter 8

# Evaluation of Arithmetic Expressions

The evaluation of arithmetic expressions using floating-point arithmetic may lead to unpredictable results due to an accumulation of roundoff errors. As an example, the evaluation of $x + 1 - x$ for $x > 10^{20}$ using the standard floating-point format on almost every digital computer yields the wrong result 0. Since the evaluation of arithmetic expressions is a basic task in digital computing, we should have a method to evaluate a given expression to an arbitrary accuracy. We will develop such a method for real expressions composed of the operations $+$, $-$, $\cdot$, $/$, and $\uparrow$, where $\uparrow$ denotes exponentiation by an integer. The method is based on the principle of iterative refinement as discussed in Section 3.8.

## 8.1 Theoretical Background

There is a very simple method for evaluating a real arithmetic expression to a desired accuracy. Supposed we have a multiple-precision interval arithmetic. We could start evaluating the expression with single precision and check the relative diameter of the resulting interval. If the width is greater than the desired accuracy, we would repeat the evaluation with doubled precision, then in a triple-precision format, and so on. When we finally succeed, we would return the enclosing interval rounded to the target format specified by the user. Here, we suppose that the target format is the standard floating-point format. The method holds two principal disadvantages. First, we would have to implement a multiple-precision interval arithmetic. Secondly, an already computed approximation would be discarded instead of being used for the next step when restarting the evaluation with an increased precision. In the sequel, we will discuss a more sophisticated approach avoiding these disadvantages. See Fischer, Haggenmüller, and Schumacher [19] for a detailed description of the method.

### 8.1.1 A Nonlinear Approach

Let $f(x) = f(x_1, \ldots, x_n)$ be an arithmetic expression composed of $n$ operands $x_1, \ldots, x_n$ and $m$ operations. We define the set of legal operations by $\{+, -, \cdot, /, \uparrow\}$, where $\uparrow$ denotes exponentiation by an integer. We show how the evaluation of $f(x)$ is equivalent to finding the solution of a special nonlinear system of equations, which is similar to the *code lists* used in automatic differentiation techniques (cf. Rall [72]).

The procedure of transforming $f(x)$ into such a system is straight forward. Every operation of $f$ defines a separate equation. The order of these equations is defined by the conventional rules of algebra for left-to-right evaluation.

**Example 8.1** Let $f(x) = \frac{(x_1+x_3)^4(x_2-x_4)^{-3}}{x_5 x_5}$. Thus, we have $n = 5$ and $m = 7$. If we denote the intermediate results of the left-to-right evaluation by $x_{n+1}$ to $x_{n+m}$, then we get the following system of nonlinear equations

$$\begin{array}{rcl} x_6 & = & x_1 + x_3 \\ x_7 & = & x_6^4 \\ x_8 & = & x_2 - x_4 \\ x_9 & = & x_8^{-3} \\ x_{10} & = & x_7 \cdot x_9 \\ x_{11} & = & x_5 \cdot x_5 \\ x_{12} & = & x_{10}/x_{11} \end{array} \quad \Longleftrightarrow \quad \begin{array}{rcl} x_6 - (x_1 + x_3) & = & 0 \\ x_7 - x_6^4 & = & 0 \\ x_8 - (x_2 - x_4) & = & 0 \\ x_9 x_8^3 - 1 & = & 0 \\ x_{10} - x_7 \cdot x_9 & = & 0 \\ x_{11} - x_5 \cdot x_5 & = & 0 \\ x_{11} \cdot x_{12} - x_{10} & = & 0 \end{array}$$

with $x_{12} = x_{n+m} = f(x)$. Note that $x_{n+1}, \ldots, x_{n+m}$ are the unknowns of the system.

Once an approximate solution $(\widetilde{x}_{n+1}, \ldots, \widetilde{x}_{n+m})^{\mathrm{T}}$ of the nonlinear system is computed, the absolute error of any intermediate result is given by $\Delta x_\nu = x_\nu - \widetilde{x}_\nu$, $\nu = n+1, \ldots, n+m$. Central to our approach is the computation of enclosures of all $\Delta x_\nu$. Thus, we get

$$x_\nu - \widetilde{x}_\nu \in [\Delta x]_\nu, \quad \nu = n+1, \ldots, n+m.$$

In particular, this yields $f(x) = x_{n+m} \in \widetilde{x}_{n+m} + [\Delta x]_{n+m}$.

To find enclosures of the absolute error of the intermediate results, we use the special form of the nonlinear equations. Each of the equations has the form

$$g(x_i, x_j, x_k) = 0, \quad i \leq j < k, \tag{8.1}$$

where $x_k$ is the new intermediate result. In fact, $g$ has five special forms that will be individually considered shortly. If $g$ has continuous partial derivatives (ours do), then expanding $g$ about $\widetilde{x}_k$ yields

$$g(x_i, x_j, x_k) = g(x_i, x_j, \widetilde{x}_k) + \frac{\partial g}{\partial x_k}(x_i, x_j, \xi_k)\Delta x_k,$$

with $\xi_k$ lying in the hull of $x_k$ and $\widetilde{x}_k$. Similarly, we expand $g(x_i, x_j, \widetilde{x}_k)$ about $\widetilde{x}_j$

$$g(x_i, x_j, \widetilde{x}_k) = g(x_i, \widetilde{x}_j, \widetilde{x}_k) + \frac{\partial g}{\partial x_j}(x_i, \xi_j, \widetilde{x}_k)\Delta x_j,$$

and then $g(x_i, \widetilde{x}_j, \widetilde{x}_k)$ about $\widetilde{x}_i$ to get

$$0 = g(\widetilde{x}_i, \widetilde{x}_j, \widetilde{x}_k) + \frac{\partial g}{\partial x_k}(x_i, x_j, \xi_k)\Delta x_k + \frac{\partial g}{\partial x_j}(x_i, \xi_j, \widetilde{x}_k)\Delta x_j + \frac{\partial g}{\partial x_i}(\xi_i, \widetilde{x}_j, \widetilde{x}_k)\Delta x_i.$$

Here, we have $\xi_\nu \in x_\nu \,\underline{\cup}\, \widetilde{x}_\nu$ for $\nu = i, j, k$. Hence, we obtain the following representation of the absolute error of the $k$-th intermediate result:

$$\Delta x_k = \frac{-g(\widetilde{x}_i, \widetilde{x}_j, \widetilde{x}_k) - \frac{\partial g}{\partial x_j}(x_i, \xi_j, \widetilde{x}_k)\Delta x_j - \frac{\partial g}{\partial x_i}(\xi_i, \widetilde{x}_j, \widetilde{x}_k)\Delta x_i}{\frac{\partial g}{\partial x_k}(x_i, x_j, \xi_k)}. \tag{8.2}$$

We will now use (8.2) to derive explicit rules for computing enclosures of $\Delta x_k$. We will see that each of the partial derivatives in Equation (8.2) is easily computable for each of the five special forms of $g$ we consider below. For this, we assume that enclosures $[\Delta x]_i$ and $[\Delta x]_j$ of $\Delta x_i$ and $\Delta x_j$ are already known. Actually, this is no restriction since $[\Delta x]_\nu = 0$ for the exactly known input parameters i.e. for $\nu = 1, \ldots, n$. See Exercise 8.6 for the treatment of interval input parameters.

**Addition and Subtraction:** Here, we have $g(x_i, x_j, x_k) = x_k - (x_i \pm x_j)$. Applying Equation (8.2), we get the following enclosure of $\Delta x_k$

$$\Delta x_k = (\widetilde{x}_i \pm \widetilde{x}_j - \widetilde{x}_k) \pm \Delta x_j + \Delta x_i \in \underbrace{\Diamond(\widetilde{x}_i \pm \widetilde{x}_j - \widetilde{x}_k) \pm [\Delta x]_j + [\Delta x]_i}_{[\Delta x]_k}. \tag{8.3}$$

**Multiplication:** For $g(x_i, x_j, x_k) = x_k - x_i x_j$, Equation (8.2) gives an expression where the unknown $x_i$ is replaced by its enclosure $\widetilde{x}_i + [\Delta x]_i$. Thus, we get

$$\begin{aligned} \Delta x_k &= (\widetilde{x}_i\widetilde{x}_j - \widetilde{x}_k) + x_i \Delta x_j + \widetilde{x}_j \Delta x_i \\ &\in \underbrace{\Diamond(\widetilde{x}_i\widetilde{x}_j - \widetilde{x}_k) + (\widetilde{x}_i + [\Delta x]_i)[\Delta x]_j + \widetilde{x}_j[\Delta x]_i}_{[\Delta x]_k}. \end{aligned} \tag{8.4}$$

**Division:** Here, we have $g(x_i, x_j, x_k) = x_j x_k - x_i$. Applying Equation (8.2) and replacing $x_j$ by $\widetilde{x}_j + [\Delta x]_j$ yields

$$\Delta x_k = \frac{(\widetilde{x}_i - \widetilde{x}_j\widetilde{x}_k) - \widetilde{x}_k \Delta x_j + \Delta x_i}{x_j} \in \underbrace{\frac{\Diamond(\widetilde{x}_i - \widetilde{x}_j\widetilde{x}_k) - \widetilde{x}_k \Delta x_j + \Delta x_i}{\widetilde{x}_j + [\Delta x]_j}}_{[\Delta x]_k}. \tag{8.5}$$

**Exponentiation:** We consider exponentiation by an integer in three cases. For a zero exponent, we have $x_k = 1$. Hence, we get $[\Delta x]_k = 0$. For positive exponents $n > 0$, we have $g(x_j, x_k) = x_k - x_j^n$. Applying Equation (8.2) and enclosing $\xi_j$ by $x_j \,\underline{\cup}\, \widetilde{x}_j \in (\widetilde{x}_j + [\Delta x]_j) \,\underline{\cup}\, \widetilde{x}_j = \widetilde{x}_j + (0 \,\underline{\cup}\, [\Delta x]_j)$, we get the following enclosure of $\Delta x_k$

$$\Delta x_k = (\widetilde{x}_j^n - \widetilde{x}_k) + n\xi_j^{n-1}\Delta x_j \in \underbrace{\Diamond(\widetilde{x}_j^n - \widetilde{x}_k) + n(\widetilde{x}_j + (0 \,\underline{\cup}\, [\Delta x]_j))^{n-1}[\Delta x]_j}_{[\Delta x]_k}. \tag{8.6}$$

For negative exponents $n < 0$, we use $g(x_j, x_k) = x_k x_j^{-n} - 1$. In a similar way as for positive exponents, we get

$$\Delta x_k = \frac{\Diamond(1 - \widetilde{x}_k \widetilde{x}_j^{-n}) + n\widetilde{x}_k \xi_j^{-(n+1)}[\Delta x]_j}{(\widetilde{x}_j + [\Delta x]_j)^{-n}}$$

$$\in \underbrace{\frac{\Diamond(1 - \widetilde{x}_k \widetilde{x}_j^{-n}) + n\widetilde{x}_k(\widetilde{x}_j + (0 \,\underline{\cup}\, [\Delta x]_j))^{-(n+1)}[\Delta x]_j}{(\widetilde{x}_j + [\Delta x]_j)^{-n}}}_{[\Delta x]_k}. \tag{8.7}$$

**Elementary functions:** We give the enclosure (8.8) for an elementary function $s(t) : I\!R \to I\!R$ for completeness, but our implementation does not support the accurate evaluation of expressions involving elementary functions. For this special case, we have $g(x_j, x_k) = x_k - s(x_j)$. Thus, we get

$$\Delta x_k = s(\widetilde{x}_j) - \widetilde{x}_k + s'(\xi_j)[\Delta x]_j \in \underbrace{\Diamond(s(\widetilde{x}_j) - \widetilde{x}_k) + s'(\widetilde{x}_j + (0 \,\underline{\cup}\, [\Delta x]_j))[\Delta x]_j}_{[\Delta x]_k}. \tag{8.8}$$

We point out that it is essential for the convergence of the algorithms described below to enclose the terms for $-g(\widetilde{x}_i, \widetilde{x}_j, \widetilde{x}_k)$ in intervals as tight as possible. We have indicated this by the $\Diamond(\ldots)$ notation. That is, these terms should be evaluated with only one final interval rounding.

## 8.2 Algorithmic Description

The explanations of the previous section motivate the following simple iterative refinement scheme. Let $f$ be an arithmetic expression depending on $n$ real-valued parameters $x_1, \ldots, x_n$. Initially, set $\widetilde{x}_\nu = x_\nu$ for $\nu = 1, \ldots, n$, and compute an approximation of $f(x)$ saving the $m$ intermediate results $\widetilde{x}_{n+1}, \ldots, \widetilde{x}_{n+m}$. The order of the intermediate results is given by the conventional rules for left-to-right evaluation. Set $[\Delta x]_\nu = 0$ for the exactly known parameters $x_1, \ldots, x_n$. Start the iteration by computing enclosures of the absolute error of each intermediate result using Equations (8.3)–(8.8). The procedure is finished if the relative diameter of $[y] = \widetilde{x}_{n+m} + [\Delta x]_{n+m}$ is less than the desired accuracy $\varepsilon > 0$. Otherwise, improve the approximation of each intermediate result by adding the midpoint of the corresponding error interval. Repeat the procedure to get a, hopefully, better $[y]$. The iteration fails if the desired accuracy $\varepsilon$ cannot be satisfied after $k_{\max}$ steps, where $k_{\max}$ is only restricted by the memory available on an actual machine.

**Algorithm 8.1:** NaiveEval $(f, x, \varepsilon, [y], k, Err)$

1. {Initialization: Max. number of iteration steps, error code, input parameters}
   $k_{\max} := 10$; $Err :=$ "No error";
   $\widetilde{x}_\nu^{(0)} := x_\nu$; $[\Delta x]_\nu := 0$; $(\nu = 1, \ldots, n)$

2. {Computation of a first approximation}
   Compute $\tilde{x}_\nu^{(0)}$; $(\nu = n+1, \ldots, n+m)$
3. {Iterative refinement}
   $k := 0$;
   **repeat**
   Compute $[\Delta x]_\nu$; $(\nu = n+1, \ldots, n+m)$
   $[y] := \tilde{x}_{n+m}^{(k)} + [\Delta x]_{n+m}$;
   $Success := (d_{\text{rel}}([y]) \leq \varepsilon)$;
   **if** (**not** Success) **then** {Add correction terms}
   $\tilde{x}_\nu^{(k+1)} := \tilde{x}_\nu^{(k)} + m([\Delta x]_\nu)$; $(\nu = n+1, \ldots, n+m)$
   $k := k + 1$;
   **until** *Success* **or** $(k \geq k_{\max})$;
4. {Return: Enclosure, corrections needed, and error message}
   **if** (**not** *Success*) **then**
   *Err* := "Cannot satisfy the desired accuracy, $k_{\max}$ is to small!";
   **return** $[y]$, $k$, *Err*;

In general, Algorithm 8.1 may not yield an enclosure that is sufficiently tight if a finite precision arithmetic is used. To satisfy a desired accuracy for the result, it is often necessary to compute some or all intermediate results with an increased precision. For this reason, we will introduce a special kind of multi-precision arithmetic. Instead of adding the correction term $m([\Delta x]_\nu)$ to get a new approximation of an intermediate result (Step 3), we will store its value separately. An intermediate result is stored in the staggered correction form $\tilde{x}_\nu = \sum_{\mu=0}^{p} \tilde{x}_\nu^{(\mu)}$, where $p$ is the actual degree of precision. See Stetter [88] for more details on the staggered correction representation. For our purposes, every component of this representation is assumed to be an element of a number screen $R(b, l, e_{\min}, e_{\max})$ as introduced in Section 3.5. Some of these components may have overlapping mantissae, so a staggered representation is not unique as Example 8.2 demonstrates.

**Example 8.2** Let $t = 0.123456$ and $R = R(10, 3, -5, +5)$ be a decimal screen with three significant digits. There are different staggered representations of $t$ using components of $R$. For instance, we have

$$\begin{aligned} t &= 0.103 \cdot 10^0 + 0.203 \cdot 10^{-1} + 0.156 \cdot 10^{-3} \\ &= 0.133 \cdot 10^0 - 0.961 \cdot 10^{-2} + 0.660 \cdot 10^{-4}. \end{aligned}$$

We will now modify Algorithm 8.1 to employ a staggered representation of the intermediate results. For reasons of simplicity, we also use a staggered representation for the input parameters. Since these values are exactly known, we set $\tilde{x}_\nu^{(0)} = x_\nu$ and $\tilde{x}_\nu^{(\mu)} = 0$ for $\mu > 0$ and $\nu = 1, \ldots, n$. We could extend this algorithm to support user entry of the parameters $x_1, \ldots, x_n$ in a staggered correction format, but we do not do so here (see Exercise 8.5). In Algorithm 8.2, the actual degree of precision is denoted by $p$. We use the same $p$ for each intermediate result. When the algorithm has finished, then $p$ is a kind of condition number for the problem.

Large values of $p$ signal that the evaluation problem was poorly conditioned. The representation of the intermediate results in the staggered format has its effects on the computation of the enclosures for the absolute error. Thus, we will refer the reader to Algorithms 8.3–8.6 for a detailed description of how to compute these intervals.

**Algorithm 8.2:** Eval $(f, x, \varepsilon, [y], p, Err)$

1. {Initialization: Max. number of corrections, error code, input parameters}
   $p_{\max} := 10$; $Err :=$ "No error";
   $\tilde{x}_\nu^{(0)} := x_\nu$; $\tilde{x}_\nu^{(\mu)} := 0$; $[\Delta x]_\nu := 0$; $(\mu = 1, \ldots, p_{\max},\ \nu = 1, \ldots, n)$
2. {Computation of a first approximation}
   Compute $\tilde{x}_\nu^{(0)}$; $(\nu = n+1, \ldots, n+m)$
3. {Iterative refinement}
   $p := 0$;
   **repeat**
   Compute $[\Delta x]_\nu$; $(\nu = n+1, \ldots, n+m)$ using Algorithms 8.3–8.6
   $[y] := \left( \sum_{\mu=0}^{p} \tilde{x}_{n+m}^{(\mu)} + [\Delta x]_{n+m} \right)$;
   $Success := (d_{\text{rel}}([y]) \le \varepsilon)$;
   **if** (**not** Success) **and** $(p < p_{\max})$ **then** {Store next staggered component}
   $\tilde{x}_\nu^{(p+1)} := m([\Delta x]_\nu)$; $(\nu = n+1, \ldots, n+m)$
   $p := p + 1$;
   **until** *Success* **or** $(p \ge p_{\max})$;
4. {Return: Enclosure, corrections needed, and error message}
   **if** (**not** *Success*) **then**
   $Err :=$ "Cannot satisfy the desired accuracy, $p_{\max}$ is to small!";
   **return** $[y]$, $p$, $Err$;

As mentioned earlier, we do not support elementary functions. This is because would need elementary functions that accept arguments in a staggered correction format. An implementation of such functions is beyond the scope of this book (see Krämer [53]). This is also the reason for we do not give an algorithmic description of the exponentiation by an integer for arguments in a staggered correction format. To implement Equations (8.6) and (8.7) accurately, we would have to implement powers of operands in a staggered correction format. To avoid this, we will consider an integer exponentiation as a sequence of multiplications or divisions until the end of this chapter (see Algorithm A.1 in the appendix for details). Hence, the number of intermediate results increases by formally substituting these operations.

Let us now describe the algorithms for computing $[\Delta x]_\nu$ for the basic operations $+$, $-$, $\cdot$, and $/$. In order to simplify the notation slightly, we identify the parameters $x_i$, $x_j$, and $x_k$ as $x$, $y$, and $z$, respectively.

**Algorithm 8.3:** Add $(\widetilde{x}, \widetilde{y}, \widetilde{z}, [\Delta x], [\Delta y], [\Delta z])$

1. {Evaluation of (8.3) for operands in stagg. corr. format of precision $p$}

$$[\Delta z] := \Diamond \Big( \sum_{\mu=0}^{p} \widetilde{x}^{(\mu)} + \sum_{\mu=0}^{p} \widetilde{y}^{(\mu)} - \sum_{\mu=0}^{p} \widetilde{z}^{(\mu)} + [\Delta x] + [\Delta y] \Big);$$

2. **return** $[\Delta z]$;

**Algorithm 8.4:** Sub $(\widetilde{x}, \widetilde{y}, \widetilde{z}, [\Delta x], [\Delta y], [\Delta z])$

1. {Evaluation of (8.3) for operands in stagg. corr. format of precision $p$}

$$[\Delta z] := \Diamond \Big( \sum_{\mu=0}^{p} \widetilde{x}^{(\mu)} - \sum_{\mu=0}^{p} \widetilde{y}^{(\mu)} - \sum_{\mu=0}^{p} \widetilde{z}^{(\mu)} + [\Delta x] - [\Delta y] \Big);$$

2. **return** $[\Delta z]$;

**Algorithm 8.5:** Mul $(\widetilde{x}, \widetilde{y}, \widetilde{z}, [\Delta x], [\Delta y], [\Delta z])$

1. {Evaluation of (8.4) for operands in stagg. corr. format of precision $p$}

$$[\Delta z] := \Diamond \Big( \sum_{\mu=0}^{p} \sum_{\nu=0}^{p} \widetilde{x}^{(\mu)} \cdot \widetilde{y}^{(\nu)} - \sum_{\mu=0}^{p} \widetilde{z}^{(\mu)} + \sum_{\mu=0}^{p} \widetilde{x}^{(\mu)} \cdot [\Delta y] + \sum_{\mu=0}^{p} \widetilde{y}^{(\mu)} \cdot [\Delta x] + [\Delta x] \cdot [\Delta y] \Big);$$

2. **return** $[\Delta z]$;

**Algorithm 8.6:** Div $(\widetilde{x}, \widetilde{y}, \widetilde{z}, [\Delta x], [\Delta y], [\Delta z])$

1. {Evaluation of (8.5) for operands in stagg. corr. format of precision $p$}

$$[u] := \Diamond \Big( \sum_{\mu=0}^{p} \widetilde{x}^{(\mu)} - \sum_{\mu=0}^{p} \sum_{\nu=0}^{p} \widetilde{z}^{(\nu)} \cdot \widetilde{y}^{(\mu)} - \sum_{\mu=0}^{p} \widetilde{z}^{(\mu)} \cdot [\Delta y] + [\Delta x] \Big);$$

$$[v] := \Diamond \Big( \sum_{\mu=0}^{p} \widetilde{y}^{(\mu)} + [\Delta y] \Big);$$

2. **return** $[\Delta z] := [u]/[v]$;

## 8.3 Implementation and Examples

### 8.3.1 C++ Program Code

In this section, we give the implementation of the module *expreval* for the accurate evaluation of real arithmetic expressions. The module is based on a data type

*Staggered* that holds components for both the approximation and the error interval. We give the definition of a special arithmetic including the operators $+$, $-$, $\cdot$, $/$, and $\uparrow$. A little programming trick implicitly transforms an expression into a system of nonlinear equations so that evaluating the expression using this special arithmetic automatically yields a user-specified accuracy for the result. We stress that those parts of the algorithms of the previous section marked by $\Diamond(\ldots)$ are implemented using accurate evaluation of dot product expressions as described in Section 2.3.

#### 8.3.1.1 Module expreval

The header file of module *expreval* provides the definition of the classes *Staggered* and *StaggArray*. The operators $+$, $-$, $\cdot$, and $/$ are declared for both operands of type *Staggered* and for mixed operands of type *Staggered* and *real*. The interface file also provides the declaration of an assignment operator for converting *real* data to data of type *Staggered* and the declaration of the function *Power()*. *Eval()* is the function for evaluating an arithmetic expression accurately. The function *EvalErrMsg()* can be used to get an error message for the error code returned by *Eval()*.

```
//----------------------------------------------------------------------------
// File: expreval (header)
// Purpose: Computation of an enclosure of the value of a real arithmetic
//    expression composed of the operations +, -, *, /, and ↑, where ↑
//    denotes exponentiation by an integer.
// Type:
//    Stagg_FctPtr          : pointer for a function of type 'Staggered'
// Class Staggered (staggered data representation):
//    Staggered()           : constructors
//    operators =           : assignment of a real argument
//    operators +, -, *, /: both operands of type 'Staggered' or one of
//                            type 'Staggered' and one of type 'real'
//    Power()               : argument of type 'Staggered', exponent of type
//                            'integer' (Exponentiation by an integer)
//    Eval()                : main function for evaluation of an expression
//    EvalErrMsg()          : to get an error message text
// Class StaggArray (arrays of type 'Staggered'):
//    StaggArray()          : constructors
//    ~StaggArray()         : destructor
//    operator []           : component access
//----------------------------------------------------------------------------
#ifndef __EXPREVAL_HPP
#define __EXPREVAL_HPP

#include <interval.hpp>      // Interval arithmetic
#include <i_util.hpp>        // Interval utility functions
#include <rvector.hpp>       // Real vector arithmetic

extern char* EvalErrMsg ( int );

class Staggered;
class StaggArray;

typedef Staggered  (*Stagg_FctPtr)(StaggArray&);

class Staggered {
  private:
    rvector   Val;
    interval  Err;
```

```
    friend void InitEntry       ( real );
    friend void UpdateError     ( interval );
    friend void UpdateStaggComp ( int );

  public:
    Staggered ( );
    Staggered ( Staggered& );

    Staggered& operator= ( const real& );
    Staggered& operator= ( Staggered& );

    friend Staggered  operator+ ( Staggered& , Staggered& );
    friend Staggered  operator- ( Staggered& , Staggered& );
    friend Staggered  operator* ( Staggered& , Staggered& );
    friend Staggered  operator/ ( Staggered& , Staggered& );

    friend Staggered  operator+ ( Staggered& , const real& );
    friend Staggered  operator- ( Staggered& , const real& );
    friend Staggered  operator* ( Staggered& , const real& );
    friend Staggered  operator/ ( Staggered& , const real& );

    friend Staggered  operator+ ( const real& , Staggered& );
    friend Staggered  operator- ( const real& , Staggered& );
    friend Staggered  operator* ( const real& , Staggered& );
    friend Staggered  operator/ ( const real& , Staggered& );

    friend Staggered Power ( Staggered& , const int );

    friend void Eval ( Stagg_FctPtr, rvector, real, real&, interval&, int&,
                                                                      int& );

}; // class Staggered

class StaggArray {
  private:
    Staggered *SA;
    int       Dim;

  public:
    StaggArray  ( );
    StaggArray  ( int );
    ~StaggArray ( );
    StaggArray  ( StaggArray& );

    Staggered& operator[] ( int );
}; // class StaggArray
#endif
```

The function *Power()* is an implementation of the operator ↑ using the binary shift method as described in Algorithm A.1 of the appendix. The function *Eval()* is an implementation of Algorithm 8.2.

As mentioned above, we use a programming trick to transform an arithmetic expression implicitly into an equivalent system of nonlinear equations. We first convert all parameters in an expression $f$ to operands of type *Staggered*. When $f$ is evaluated for the first time using the special staggered arithmetic, every operator appends its result into a linear linked list. We may access these values if we restart the evaluation of $f$ a second time, this time for computing enclosures of the error intervals according to the Algorithms 8.3–8.6. This list now may be used to update the staggered components of the intermediate results as described in Algorithm 8.2. This trick works because the code list for $f$ is always evaluated in exactly the same order each time $f$ is evaluated. All functions needed for allocation and management of the internal list are declared private, so they are hidden for programs which include the module.

Another simple trick is used to avoid an abnormal program termination caused by a division by zero. If a zero denominator occurs while executing the operator /, the flag *DivByZero* is set. Since all operators check this flag before starting execution, this terminates the evaluation without forcing a runtime exception. An update of those intermediate results computed before terminating the evaluation will result in an improved approximation for the denominator. Thus, it is our hope to overcome this critical point of evaluation with the next iteration step. Any other exceptions are handled by the standard exception handling of the C++ runtime system or the interval library employed.

```
//----------------------------------------------------------------------------
// File: expreval (implementation)
// Purpose: Computation of an enclosure of the value of a real arithmetic
//     expression composed of the operations +, -, *, /, and ↑, where ↑
//     denotes exponentiation by an integer.
// Method: Iterative refinement using a defect correction mechanism. The
//     successively computed corrections for the result are stored in a
//     staggered correction format.
// Class Staggered (staggered data representation):
//     Staggered()           : constructors
//     operators =           : assignment of a real argument
//     operators +, -, *, /: both operands of type 'Staggered' or one of
//                             type 'Staggered' and one of type 'real'
//     Power()               : argument of type 'Staggered', exponent of type
//                             'integer' (Exponentiation by an integer)
//     Eval()                : main function for evaluation of an expression
//     EvalErrMsg()          : to get an error message text
// Class StaggArray (arrays of type 'Staggered'):
//     StaggArray()          : constructors
//     ~StaggArray()         : destructor
//     operator []           : component access
//----------------------------------------------------------------------------
#include <stdlib.h>           // For function 'exit()'
#include <string.h>           // String handling
#include <idot.hpp>           // Interval dotprecision type
#include <i_util.hpp>         // Interval utility functions
#include <expreval.hpp>

const int
  MaxStaggPrec = 10;        // Maximum number of staggered corrections which can
                            // be stored in a variable of type 'Staggered'.
                            // Recompile this module if its value is changed.
                            //------------------------------------------------

typedef struct IR           // Type for an intermediate result
{                           //--------------------------------
  Staggered  Entry;         // Staggered entry
  IR         *Next;         // Next entry
} IntRes, *IntResPtr;

//----------------------------------------------------------------------------
// Static variables globally used within other functions this module
//----------------------------------------------------------------------------
```

```
static int
  DivByZero,       // Error flag
  InitFlag,        // Signals the initialization process
  ActStaggPrec;    // Actual length of the staggered format

static IntResPtr
  HeadPtr,         // Head of the list of intermediate results
  ActPtr,          // Pointer to the actual intermediate result
  FreePtr;         // Pointer to freed intermediate results

//----------------------------------------------------------------------------
// Error codes used in this module.
//----------------------------------------------------------------------------
const int
  NoError   = 0,   // No error occurred
  ItFailed  = 1,   // Maximal number of defect corrections exceeded
  ZeroDiv   = 2;   // Division by zero that could not be
                   // removed by iterative refinement

//----------------------------------------------------------------------------
// Error messages depending on the error code.
//----------------------------------------------------------------------------
char* EvalErrMsg ( int Err )
{
  static char Msg[80] = "";

  if (Err != NoError) {
    char Hlp[60];

    switch (Err) {
      case ItFailed:
        sprintf(Hlp,"Maximal number of defect corrections (=%1d) exceeded",
                MaxStaggPrec); break;
      case ZeroDiv:
        strcpy(Hlp,"Division by zero occurred"); break;
      default:
        strcpy(Hlp,"Code not defined");
    }
    sprintf(Msg,"Error: %s!",Hlp);
  }
  return(Msg);
} // EvalErrMsg


//----------------------------------------------------------------------------
// Constructors, destructors, component access for the classes 'Staggered'
// and 'StaggArray
//----------------------------------------------------------------------------
Staggered::Staggered ( )                                        // Constructor
{                                                               //------------
  Resize(Val,0,MaxStaggPrec);
  for (int i = 0; i <= MaxStaggPrec; i++) Val[i] = 0.0;
  Err = 0.0;
}

Staggered::Staggered ( Staggered& x )                      // Copy constructor
{                                                          //-----------------
  Resize(Val,0,MaxStaggPrec);
  for (int i = 0; i <= MaxStaggPrec; i++) Val[i] = x.Val[i];
  Err = x.Err;
}

StaggArray::StaggArray ( )                                      // Constructor
{                                                               //------------
```

```
  Dim = 0;  SA = NULL;
}

StaggArray::StaggArray ( int n )                              // Constructor
{                                                             //------------
  if (n < 1) {
    cerr << "Lower bound is < 1 in 'StaggArray'-constructor!" << endl;
    exit(-1);
  }
  if ( !(SA = new Staggered[n]) ) {
    cerr << "Not enough memory for variable of type 'StaggArray'!" << endl;
    exit(-1);
  }

  Dim = n;
}

StaggArray::StaggArray ( StaggArray& x )                 // Copy constructor
{                                                        //-----------------
  if ( (Dim < 1) || (x.SA == NULL) ) {
    cerr << "Illegal input parameter in 'StaggArray(StaggArray&)'!" << endl;
    exit(-1);
  }
  if ( !(SA = new Staggered[x.Dim]) ) {
    cerr << "Not enough memory for variable of type 'StaggArray'!" << endl;
    exit(-1);
  }

  Dim = x.Dim;
  for (int i = 0; i < Dim; i++) SA[i] = x.SA[i];
}

StaggArray::~StaggArray ( )                                    // Destructor
{                                                              //-----------
  Dim = 0;
  delete [] SA;
  SA = NULL;
}

Staggered& StaggArray::operator[] ( int i )              // Component access
{                                                        //-----------------
  if ( (i < 1) || (i > Dim) ) {
    cerr << "Illegal index access on variable of type 'StaggArray'!" << endl;
    exit(-1);
  }
  return( SA[--i] );
}

//----------------------------------------------------------------------------
// Since the intermediate results of operations for staggered data must be
// iteratively updated, it is necessary to store these data. For this
// purpose, a linear linked list is used. Its head is accessed via 'HeadPtr',
// whereas its actual entry is accessed via 'ActPtr'. 'FreePtr' is a pointer
// to a list of already allocated but actually unused entries. To prevent
// creation of garbage in memory, the entries of this list are used first
// when allocating a new intermediate result. All functions for the handling
// of the list of intermediate results are locally defined.
//----------------------------------------------------------------------------
static void InitList ( )           // Initialize list of intermediate results
{                                  //----------------------------------------
  DivByZero = 0;
  if (HeadPtr != NULL)       // Use list of freed entries
    FreePtr = HeadPtr;
  else                       // A list was not yet allocated
```

```
    FreePtr = NULL;
  HeadPtr = NULL;
  ActPtr = NULL;
}

static void ResetList ( )                   // Reset error flag and set the actual
{                                           // pointer to the head of the list.
  DivByZero = 0;                            //-----------------------------------
  ActPtr = HeadPtr;
}

static void AllocEntry ( IntResPtr& p )             // Allocate memory for a new
{                                                   // entry. Use a previously
  if (FreePtr != NULL) {                            // created entry from the list
    p = FreePtr;                                    // of freed entries if any.
    FreePtr = FreePtr->Next;                        //---------------------------
  }
  else {
    p = new IntRes;
    p->Next = NULL;
  }
}

void InitEntry ( real Approx )                // Get a new list entry and initial-
{                                             // ize its first staggered component
  IntResPtr p;                                // with the value 'Approx'.
                                              //---------------------------------
  AllocEntry(p);
  p->Entry.Val[0] = Approx;
  p->Next = NULL;
  if (HeadPtr == NULL)
    { HeadPtr = p;  ActPtr = p; }
  else
    { ActPtr->Next = p;  ActPtr = p; }
}

void UpdateError ( interval Error )                  // Update the error component
{                                                    // of the actual entry.
  ActPtr->Entry.Err = Error;                         //--------------------------
  ActPtr = ActPtr->Next;
}

void UpdateStaggComp ( int i )                        // Update the i-th staggered
{                                                     // component of all list
  ActPtr = HeadPtr;                                   // entries by the midpoint
  while (ActPtr != NULL) {                            // of the error interval.
    ActPtr->Entry.Val[i] = mid(ActPtr->Entry.Err);//-------------------------
    ActPtr->Entry.Err = 0.0;
    ActPtr = ActPtr->Next;
  }
}

//----------------------------------------------------------------------------
// Assingment operators used to convert real input data to the staggered type
// and to assign a variables of type 'Staggered'. A real operand is assumed
// to be exact! It must not have been rounded. In particular, 'r' cannot be
//    - a real value such as 0.1 which is not exactly representable in the
//      internal binary format,
//    - rounded by conversion to the internal binary data format during
//      input,
//    - any arithmetic expression which cannot be evaluated exactly.
//----------------------------------------------------------------------------
Staggered& Staggered::operator= ( const real& r )
{
```

```
  int i;

  Val[0] = r;
  for (i = 1; i <= MaxStaggPrec; i++) Val[i] = 0.0;
  Err = 0.0;
  return *this;
}

Staggered& Staggered::operator= ( Staggered& x )
{
  int i;

  for (i = 0; i <= MaxStaggPrec; i++) Val[i] = x.Val[i];
  Err = x.Err;
  return *this;
}

//---------------------------------------------------------------------------
// Arithmetic operators +, -, *, and / for operands of type 'Staggered'. A
// division by an interval containing zero will be avoided and is marked by
// setting the 'DivByZero' flag. If 'DivByZero' is set, all succeeding opera-
// tions are not executed. The evaluation is stopped at this point but may be
// restarted after updating a new staggered component by the midpoints of the
// error enclosures computed so far.
//---------------------------------------------------------------------------
// In the comments below the notations #*(...) and ##(...) are used to indi-
// cate the evaluation of the expression specified in round brackets using
// the exact scalar product. I.e. each component of the result is computed
// with only one final rounding. The symbol #* holds for rounding to nearest
// whereas ## holds for rounding to the smallest enclosing interval. An exact
// scalar product may be implemented using dotprecision accumulators.
//---------------------------------------------------------------------------
Staggered operator+ ( Staggered& x , Staggered& y )
{
  int           i;
  Staggered     z;
  idotprecision IAccu;

  if (!DivByZero) {
    if (InitFlag) {           // Initialize first staggered component
      z.Val[0] = x.Val[0] + y.Val[0];
      InitEntry(z.Val[0]);
    }
    else {
      z = ActPtr->Entry;    // Get actual values of z

      // Error:  dz := ##( x + y - z + dx + dy )
      //------------------------------------------
      IAccu  = x.Err;
      IAccu += y.Err;
      for (i = 0; i <= ActStaggPrec; i++) {
        IAccu += x.Val[i];
        IAccu += y.Val[i];
        IAccu -= z.Val[i];
      }
      rnd(IAccu,z.Err);
      UpdateError(z.Err);
    }
  } // if (!DivByZero)
  return(z);
} // operator+

Staggered operator- ( Staggered& x , Staggered& y )
{
```

```
  int           i;
  Staggered     z;
  idotprecision IAccu;

  if (!DivByZero) {
    if (InitFlag) {           // Initialize first staggered component
      z.Val[0] = x.Val[0] - y.Val[0];
      InitEntry(z.Val[0]);
    }
    else {
      z = ActPtr->Entry;   // Get actual values of z

      // Error:  dz := ##( x - y - z + dx - dy )
      //---------------------------------------
      IAccu  = x.Err;
      IAccu -= y.Err;
      for (i = 0; i <= ActStaggPrec; i++) {
        IAccu += x.Val[i];
        IAccu -= y.Val[i];
        IAccu -= z.Val[i];
      }
      rnd(IAccu,z.Err);
      UpdateError(z.Err);
    }
  } // if (!DivByZero)
  return(z);
} // operator-

Staggered operator* ( Staggered& x , Staggered& y )
{
  int           i, j;
  Staggered     z;
  idotprecision IAccu;

  if (!DivByZero) {
    if (InitFlag) {           // Initialize first staggered component
      z.Val[0] = x.Val[0] * y.Val[0];
      InitEntry(z.Val[0]);
    }
    else {
      z = ActPtr->Entry;   // Get actual values of z

      // Error:  dz := ##( x*y - z + y*dx + x*dy + dx*dy )
      //-------------------------------------------------
      IAccu = 0.0;
      accumulate(IAccu,x.Err,y.Err);
      for (i = 0; i <= ActStaggPrec; i++) {
        IAccu -= z.Val[i];
        for (j = 0; j <= ActStaggPrec; j++)
          accumulate(IAccu,x.Val[i],y.Val[j]);
      }
      for (i = 0; i <= ActStaggPrec; i++) {
        accumulate(IAccu,y.Val[i],x.Err);
        accumulate(IAccu,x.Val[i],y.Err);
      }
      rnd(IAccu,z.Err);
      UpdateError(z.Err);
    }
  } // if (!DivByZero)
  return(z);
} // operator*

Staggered operator/ ( Staggered& x , Staggered& y )
{
```

```
  int           i, j;
  Staggered     z;
  interval      num, denom;
  idotprecision IAccu;

  if (!DivByZero) {
    if (InitFlag)            // Initialize first staggered component
      if (y.Val[0] != 0.0) {
        z.Val[0] = x.Val[0] / y.Val[0];
        InitEntry(z.Val[0]);
      }
      else {
        DivByZero = 1;
        InitEntry(0.0);
      }
    else {
      z = ActPtr->Entry;    // Get actual values of z

      // Error:  dz := ##( x - z*y + dx - z*dy )  /  ##( y + dy )
      //------------------------------------------------------
      IAccu = x.Err;
      for (i = 0; i <= ActStaggPrec; i++) {
        IAccu += x.Val[i];
        for (j = 0; j <= ActStaggPrec; j++)
          accumulate(IAccu,-z.Val[i],y.Val[j]);
      }
      for (i = 0; i <= ActStaggPrec; i++)
        accumulate(IAccu,-z.Val[i],y.Err);
      rnd(IAccu,num);
      IAccu = y.Err;
      for (i = 0; i <= ActStaggPrec; i++)
        IAccu += y.Val[i];
      rnd(IAccu,denom);

      if ( in(0.0,denom) )
        { z.Err = 0.0; DivByZero = 1; }
      else
        z.Err = num / denom;
      UpdateError(z.Err);
    }
  } // if (!DivByZero)
  return(z);
} // operator/

//----------------------------------------------------------------------------
// Arithmetic operators for different operands the one of type 'real' and the
// other one of type 'Staggered'. All these operators are implemented by
// first coercing both operands to 'Staggered' type and then calling the
// corresponding operators for the type 'Staggered'.
//----------------------------------------------------------------------------
Staggered operator+ ( const real& x, Staggered& y )
{
  Staggered z;
  z = x; return(z+y);
}

Staggered operator+ ( Staggered& x , const real& y )
{
  Staggered z;
  z = y; return(x+z);
}

Staggered operator- ( const real& x, Staggered& y )
{
```

```
  Staggered z;
  z = x; return(z-y);
}

Staggered operator- ( Staggered& x , const real& y )
{
  Staggered z;
  z = y; return(x-z);
}

Staggered operator* ( const real& x, Staggered& y )
{
  Staggered z;
  z = x; return(z*y);
}

Staggered operator* ( Staggered& x , const real& y )
{
  Staggered z;
  z = y; return(x*z);
}

Staggered operator/ ( const real& x, Staggered& y )
{
  Staggered z;
  z = x; return(z/y);
}

Staggered operator/ ( Staggered& x , const real& y )
{
  Staggered z;
  z = y; return(x/z);
}

//----------------------------------------------------------------------------
// Power function for integer exponents using the binary shift method. Thus
// the number of successive multiplications is reduced from n to log(2,n).
// Note: Since x↑n is considered to be a monomial we define x↑0 := 1.
//----------------------------------------------------------------------------
Staggered Power ( Staggered& x, const int n )
{
  int        m;
  Staggered  p, z;

  if (!DivByZero) {
    p = 1.0;
    if (n != 0) {
      m = (n > 0) ? n : -n;

      p = 1.0; z = x;               // Binary shift method
      while (m > 0) {               //--------------------
        if (m % 2 == 1) p = p * z;
        m = m / 2;                  // integer division!
        if (m > 0) z = z * z;
      }

      if (n < 0) p = 1.0 / p;
    }
  } // if (!DivByZero)
  return p;                  // Note: 'p' is undefined if 'DivByZero' is set
} // Power

//----------------------------------------------------------------------------
// Purpose: The function 'Eval()' may be used for the computation of an
```

```
//    enclosure of the value of a real arithmetic expression composed of the
//    operations +, -, *, /, and ↑, where ↑ denotes exponentiation by an
//    integer.
// Parameters:
//    In : 'f'         : a function of type 'Staggered' whose arguments are
//                       passed in an array of type 'Staggered'
//         'Arg'       : the real-valued arguments of 'f' stored as
//                       components of a real vector
//         'Eps'       : desired accuracy
//    Out: 'Approx'    : result computed with standard floating-point
//                       arithmetic
//         'Encl'      : verified enclosure of the result
//         'StaggPrec' : number of corrections needed
//         'Err'       : error code
// Description:
//    The expression 'f' is evaluated for the real arguments which are stored
//    sequentially in the vector 'Arg'. Initially, the real arguments are
//    converted to arguments of type 'Staggered'. When 'f' is evaluated for
//    the first time using the special staggered arithmetic, the list of
//    intermediate results is initialized. Each time 'f' is evaluated again,
//    the error of every intermediate result is enclosed. The midpoints of
//    these enclosures are used to update the intermediate results. The
//    iteration is finished if the error of the last intermediate result
//    (= value of 'f') is less than the desired accuracy. Otherwise, the
//    iteration is halted after 'MaxStaggPrec' steps.
//----------------------------------------------------------------------------
void Eval ( Stagg_FctPtr f,
            rvector      Arg,
            real         Eps,
            real&        Approx,
            interval&    Encl,
            int&         StaggPrec,
            int&         Err )
{
  int           n = Ub(Arg)-Lb(Arg)+1;
  int           i, Success;
  StaggArray    x(n);
  Staggered     StaggRes;
  idotprecision IAccu;

  for (i = 1; i <= n; i++) x[i] = Arg[i];    // Initialize arguments

  InitList();                     // Initialize list for intermediate results.
  InitFlag = 1;                   // So far, no staggered corrections computed.
  StaggRes = f(x);                // Compute first staggered component.

  // In general, the first component of 'StaggRes' will now hold the usual
  // floating-point approximation, except when a division by zero occurred.
  // In the latter case, a signaling NaN (= Not a Number, for more details
  // see the C-XSC reference) is returned.
  //------------------------------------------------------------------------
  if (DivByZero)
    Approx = SignalingNaN;
  else
    Approx = StaggRes.Val[0];
                                                  // Initial approximations are
  InitFlag = 0; ActStaggPrec = 0;                 // already computed.
  do {
    ResetList();                 // Compute new enclosures of the absolute error
    StaggRes = f(x);             // of any intermediate result.

    // Compute an enclosure of f(x) by ##( StaggRes + StaggRes.Err)
    //-------------------------------------------------------------
    IAccu = StaggRes.Err;
```

```
      for (i = 0; i <= ActStaggPrec; i++) IAccu += StaggRes.Val[i];
      rnd(IAccu,Encl);

      Success = ( !DivByZero && (RelDiam(Encl) <= Eps) );

      // Increment actual staggered precision and store the next component of
      // the intermediate results by updating the list of intermediate results.
      //---------------------------------------------------------------------
      if ( !Success && (ActStaggPrec <= MaxStaggPrec) ) {
        ActStaggPrec++;
        UpdateStaggComp(ActStaggPrec);
      }

    } while ( !(Success || (ActStaggPrec == MaxStaggPrec)) );
    StaggPrec = ActStaggPrec;

    if (Success)                // Set error code
      Err = NoError;
    else if (DivByZero)
      Err = ZeroDiv;
    else
      Err = ItFailed;
  } // Eval
```

### 8.3.2 Examples

The examples of this section demonstrate how to use the function *Eval()* for an accurate black box evaluation of arithmetic expressions. The only thing we have to do is to define a function of type *Staggered* including the expression to be evaluated. By passing the parameters of the expression in an array of type *StaggArray*, we are able to define functions with a variable number of arguments. Note that the arguments are accessed starting with index 1.

**Example 8.3** In our first example, we want to evaluate the real rational function

$$f(x,y) = \frac{1}{y^6 - 3xy^5 + 5x^3y^3 - 3x^5y - x^6}.$$

Let $a_i$ and $a_{i+1}$ be two successive Fibonacci numbers defined by

$$a_0 = 0, \quad a_1 = 1, \quad a_i = a_{i-1} + a_{i-2}, \; i = 2, 3, \ldots$$

Then we always get $f(a_i, a_{i+1}) = (-1)^i$. In the program listed below, $f$ is defined as a function of type *Staggered*. To shorten the definition of its body, we first copy its parameters to auxiliary variables x and y. The main function accepts the parameters and the desired accuracy, evaluates $f$, and reports the results. If *Eval()* succeeds, an enclosure of the function value and, for reasons of comparison, the value of a naive floating-point evaluation are listed.

```
//----------------------------------------------------------------------------
// Example for the evaluation of an arithmetic expression using function
// 'Eval()'. Evaluate the following expression:
//
```

```
//    f(x,y) := 1 / ( y↑6 - 3*x*y↑5 + 5*x↑3*y↑3 - 3*x↑5*y - x↑6 ).
//
// With x and y being successive Fibonacci numbers defined by
//
//    a_0 := 0,  a_1 := 1,  a_i := a_i-1 + a_i-2, i >= 2,
//
// we always get f(a_i,a_i+1) = (-1)↑i. For instance a_66 = 27777890035288
// and a_67 := 44945570212853 are such two successive Fibonacci numbers.
//----------------------------------------------------------------------------
#include <expreval.hpp>     // Expression evaluation
#include <stacksz.hpp>      // To increase stack size for some
                            // special C++ compilers

Staggered f ( StaggArray& v )
{
  Staggered x, y, z;

  x = v[1];
  y = v[2];

  z = 1.0 / ( Power(y,6) - 3.0*x*Power(y,5) + 5.0*Power(x,3)*Power(y,3) -
              3.0*Power(x,5)*y - Power(x,6) );
  return(z);
}

int main ( )
{
  real     Eps, Approx;
  int      StaggPrec, Err;
  rvector  Arg(2);
  interval Encl;

  cout << SetPrecision(23,15) << Scientific;   // Output format

  cout << "Evaluation of 1 / ( y↑6 - 3*x*y↑5 + 5*x↑3*y↑3 - 3*x↑5*y - x↑6 )"
       << endl << endl;

  cout << "Enter the arguments:" << endl;
  cout << "   x = " ; cin >> Arg[1];
  cout << "   y = ";  cin >> Arg[2];
  cout << endl;

  cout << "Desired accuracy:" << endl;
  cout << "   Eps = " ; cin >> Eps;
  cout << endl;

  Eval(f,Arg,Eps,Approx,Encl,StaggPrec,Err);

  if (!Err) {
    cout << "Floating-point evaluation:   " << Approx << endl;
    cout << "Interval enclosure:         " << Encl << endl;
    cout << "Defect corrections needed:    " << StaggPrec << endl;
  }
  else
    cout << EvalErrMsg(Err) << endl;
  return 0;
}
```

A sample output is given below. The IEEE floating-point format used for computation (see Section 3.7) has a precision of about 16 decimals. Thus, it does not make sense to demand a value less than $10^{-16}$ for the relative accuracy $\varepsilon$. We also point out that the result computed by the naive floating-point evaluation has neither the same order of magnitude nor any correct digits in the mantissa!

```
Evaluation of 1 / ( y^6 - 3*x*y^5 + 5*x^3*y^3 - 3*x^5*y - x^6 )

Enter the arguments:
   x = 27777890035288
   y = 44945570212853

Desired accuracy:
   Eps = 1e-20

Floating-point evaluation:     3.165189248586599E-066
Interval enclosure:          [ 1.000000000000000E+000, 1.000000000000000E+000]
Defect corrections needed:     5
```

**Example 8.4** Our second example deals with the evaluation of a second order difference quotient. Let $f(x)$ be a real function, twice differentiable at $x$. Then its second order difference quotient, defined by

$$D_f(x,h) = \frac{f(x-h) - 2f(x) + f(x+h)}{h^2}, \quad h \neq 0, \tag{8.9}$$

is a good approximation of $f''(x)$ for small values of $h$. It is well known that the evaluation of $D_f$ using floating-point arithmetic is unstable. The instability is caused by cancellation while evaluating the numerator of $D_f$. Therefore, the evaluation of $D_f$ for small values of $h$ is a good example for demonstrating the benefits of our algorithm. The sample program listed below may be used to evaluate $D_f$ for a function $f$ defined by

$$f(x) = 540\frac{x^4 - 23x^3 + 159x^2 - 2x + 45}{x^3 + 18x^2 + 501x + 20}.$$

The exact value at $x = 1$ is $f''(1) = 36$. The coding of the sample program is similar to Example 8.3, except that the definition of $D_f$ is realized by nested function calls.

```
//----------------------------------------------------------------------------
// Example for the evaluation of an arithmetic expression using void function
// 'Eval()'. The second order difference quotient for a real function f(x) is
// defined by Df(x,h) := (f(x-h) - 2f(x) + f(x+h)) / h↑2. This quotient is
// used to approximate the second derivative of
//
//    f(x) := 540 (x↑4 - 23x↑3 + 159x↑2 - 2x + 45) / (x↑3 + 18x↑2 + 501x + 20)
//
// at x = 1. Since f is twice differentiable, Df(1,h) should tend to
// f''(1) = 36 if h tends to zero.
//----------------------------------------------------------------------------
#include <expreval.hpp>     // Expression evaluation

Staggered f ( Staggered x )
{
  return( 540.0 * (Power(x,4) - 23.0*Power(x,3) + 159.0*Power(x,2) -
                   2.0*x + 45.0) /
                  (Power(x,3) + 18.0*Power(x,2) + 501.0*x + 20.0) );
}
```

```
Staggered Df ( StaggArray& v )
{
  Staggered x, h;

  x = v[1];
  h = v[2];
  return( (f(x-h) - 2.0*f(x) + f(x+h)) / (h*h) );
}

int main ( )
{
  real     Eps, Approx;
  int      StaggPrec, Err;
  rvector  Arg(2);
  interval Encl;

  cout << SetPrecision(23,15) << Scientific;   // Output format

  cout << "Evaluation of the second order difference quotient " << endl
       << "   Df(x,h) = (f(x-h) - 2f(x) + f(x+h)) / h↑2" << endl
       << "for the function" << endl
       << "   f(x) = 540(x↑4 - 23x↑3 + 159x↑2 - 2x + 45) / "
       << "(x↑3 + 18x↑2 + 501x + 20)" << endl
       << "Note: f''(1) = 36." << endl << endl;

  cout << "Enter the arguments:" << endl;
  cout << "   x = " ; cin >> Arg[1];
  cout << "   h = ";  cin >> Arg[2];
  cout << endl;

  cout << "Desired accuracy:" << endl;
  cout << "   Eps = " ; cin >> Eps;
  cout << endl;

  Eval(Df,Arg,Eps,Approx,Encl,StaggPrec,Err);

  if (!Err) {
    cout << "Floating-point evaluation:   " << Approx << endl;
    cout << "Interval enclosure:          " << Encl << endl;
    cout << "Defect corrections needed:   " << StaggPrec << endl;
  }
  else
    cout << EvalErrMsg(Err) << endl;
  return 0;
}
```

An output of the sample program is given below. Again, the result computed by naive floating-point evaluation has neither the same order of magnitude nor any correct digits in the mantissa. We stress that the result interval is verified to be an enclosure of the value of the second order difference quotient, but it is *not* proved to be an enclosure of the value of the second derivative at $x = 1$. To compute a verified enclosure of $f''$, we should add an enclosure for the truncation error in Equation (8.9).

```
Evaluation of the second order difference quotient
   Df(x,h) = (f(x-h) - 2f(x) + f(x+h)) / h^2
for the function
   f(x) = 540(x^4 - 23x^3 + 159x^2 - 2x + 45) / (x^3 + 18x^2 + 501x + 20)
Note: f''(1) = 36.

Enter the arguments:
```

```
    x = 1
    h = 1e-8

Desired accuracy:
    Eps = 1e-15

Floating-point evaluation:      2.842170943040400E+002
Interval enclosure:           [ 3.600000000000000E+001, 3.600000000000002E+001]
Defect corrections needed:      1
```

### 8.3.3 Restrictions, Hints, and Improvements

Since our method is a method for evaluating *real* expressions, we must be careful that the parameters used for evaluation are exactly representable on the computer's number screen. Actually, this is no restriction if the method is used to evaluate an expression whose input parameters come from previous computations. If an exact representation of real constants such as 0.1 is desired, one might use an appropriate representation as a fraction of integers. For more details about conversion of input data, see the remarks in Section 3.7.

There is another problem we should be wary of when defining an expression of type *Staggered*. Remember that the type of an operator is defined by the types of its operands. As an example, consider the expression $(r_1 + r_2) \cdot x$, where $r_1$ and $r_2$ are operands of type *real* and $x$ is an operand of type *Staggered*. Since both operands $r_1$ and $r_2$ are of type *real*, the sum is computed using the standard operator for a real addition instead of using the operator for addition of type *Staggered*. To be sure that for any operation the corresponding operator of type *Staggered* is used, we recommend converting each real operand of an expression by introducing auxiliary variables of type *Staggered*. In the examples of the previous section, any operators within the definition part of the functions are operators of type *Staggered*.

For the implementation of Algorithm 8.2, we made extensive use of the accurate evaluation of dot product expressions (see Section 2.3). The algorithm can also be applied if a multi-precision interval arithmetic is available and is used instead.

## 8.4 Exercises

**Exercise 8.1** Check by paper and pencil how Algorithm 8.2 works for $x + y - x$ with $x = 10^{100}$ and $y = 1$. How many corrections $p$ are needed?

**Exercise 8.2** Let $f(x, y) = (x^2)^2 - (2y^2)^2 - (2y)^2$ and $g(x, y) = (x^2)^2 - (2y)^2(y^2+1)$. Both expressions are equivalent. Use the module *expreval* to get enclosures of $f$ and $g$ at $x = 665857$ and $y = 470832$. Compare the results with the values computed by ordinary floating-point arithmetic. For these special arguments, the exact result is 1.

**Exercise 8.3** Evaluate the expression

$$f(x, y) = \frac{1}{107751}(1682xy^4 + 3x^3 + 29xy^2 - 2x^5 + 832)$$

at $x = 192119201$ and $y = 35675640$. Use the module *expreval* to get an enclosure of $f(x, y)$. Compare the result with the value computed by ordinary floating-point arithmetic. For these special arguments the exact result is 1783.

**Exercise 8.4** Use the module *expreval* to evaluate the sample polynomials of Chapter 4 on page 65.

**Exercise 8.5** Extend function *Eval()* to accept arguments to $f$ in a staggered correction form and to return the value of $f$ as a *Staggered* type.

**Exercise 8.6** Extend the module *expreval* to support interval-valued expressions. For an interval parameter $[x]_\nu$, $\nu \in \{1, \dots, n\}$, use $\widetilde{x}_\nu := m([x]_\nu)$ and $[\Delta x]_\nu := [x]_\nu - \widetilde{x}_\nu$ for initialization. The method is effective for small intervals, but does not necessarily yield the desired accuracy. See Fischer, Haggenmüller, and Schumacher [19] for more details.

## 8.5 References and Further Reading

A first algorithm for the verified evaluation of arithmetic expressions was given by Böhm [10]. Compared to the method described above, this algorithm is based on a transformation to an equivalent system of linear equations, so it could not handle elementary functions. A summary of the nonlinear approach is found in the papers of Fischer, Haggenmüller, and Schumacher [19], [20]. A more sophisticated approach which works by computing an upper bound for the condition of the problem, and then delivers the precision necessary to achieve the desired accuracy was given by Hammer [25]. It includes an implementation in PASCAL–XSC ([50], [51]). including standard functions and a method for avoiding the problem of data conversion. A different approach which is based on the principles of reverse methods was given by Fischer [18]. In contrast to our algorithm, this approach yields different degrees of precision for computing different intermediate results. Finally, we remark that there are commercial implementations of the nonlinear approach discussed above which come with the subroutine libraries ACRITH of IBM and ARITHMOS of SIEMENS (see [36], [86]).

# Chapter 9

# Zeros of Complex Polynomials

We consider the complex polynomial $p : \mathbb{C} \to \mathbb{C}$ defined by

$$p(z) = \sum_{i=0}^{n} p_i z^i, \qquad p_i \in \mathbb{C}, \quad i = 0, \ldots, n, \qquad p_n \neq 0. \tag{9.1}$$

The Fundamental Theorem of algebra asserts that this polynomial has $n$ zeros counted by multiplicity. Finding these roots is a non trivial problem in numerical mathematics. Most algorithms deliver only approximations of the exact zeros without any or with only weak statements concerning the accuracy.

In this section, we describe an algorithm that computes verified enclosures of the roots of a complex polynomial by enclosing the zeros in narrow bounds. The coefficients of the deflated polynomial also are enclosed.

## 9.1 Theoretical Background

### 9.1.1 Description of the Problem

The algorithm described in Section 9.2 is based on the fact that the roots of the complex polynomial (9.1) of degree $n$ match the eigenvalues of the companion matrix

$$A = \begin{pmatrix} 0 & \ldots & 0 & -p_0/p_n \\ 1 & & & -p_1/p_n \\ & \ddots & & \vdots \\ & & 1 & -p_{n-1}/p_n \end{pmatrix} \in \mathbb{C}^{n \times n} \tag{9.2}$$

since $(-1)^n \cdot p_n \cdot \det(A - zI) = p(z)$, where $I$ is the identity matrix of dimension $n$. Hence, the problem of finding a zero of the complex polynomial $p$ is equivalent to finding an eigenvalue $z^*$ of the matrix $A \in \mathbb{C}^{n \times n}$. We solve the eigenvalue problem

$$Aq^* = z^* q^* \quad \text{or} \quad (A - z^* I)q^* = 0, \qquad z^* \in \mathbb{C}, \qquad q^* \in \mathbb{C}^n, \tag{9.3}$$

where $q^*$ is an eigenvector corresponding to the eigenvalue $z^*$ consisting of the coefficients $q_0^*, q_1^*, \ldots, q_{n-1}^*$ of the deflated polynomial $q^*(z) = \sum_{i=0}^{n-1} q_i^* z^i = \frac{p(z)}{z - z^*}$.

We get a verified enclosure of an eigenvalue of the companion matrix and therefore of a zero of the complex polynomial by using Schauder's fixed-point theorem (see [33]), which is a generalization of Brouwer's fixed-point theorem described in Section 3.8. Additionally, the coefficients of the deflated polynomial are enclosed.

The coefficients of the deflated polynomial may be determined recursively by Horner's evaluation of the polynomial $p$ at the point $z^*$:

$$\begin{aligned} q_{n-1}^* &:= p_n \\ q_{i-1}^* &:= q_i^* z^* + p_i, \quad i = n-1, \ldots, 1. \end{aligned}$$

From the definition of $q^*$, we can scale the vector $q^* = (q_0^*, q_1^*, \ldots, q_{n-1}^*)^T$ by fixing $q_{n-1}^* := p_n$, thus avoiding the division by $p_n$ in the matrix $A$.

We have a system of nonlinear equations in the $n$ unknowns $q_0^*, q_1^*, \ldots, q_{n-2}^*$, and $z^*$. Let the vector $q$ be the first $n-1$ components of the desired eigenvector $q := (q_0^*, q_1^*, \ldots, q_{n-2}^*)^T$. The eigenvalue $z^*$ often is stored as the $n^{th}$ component of a vector $\binom{q}{z}$.

## 9.1.2 Iterative Approach

We introduce the following *a posteriori* method to determine the zeros of a complex polynomial. Let $f : \mathbb{C}^n \to \mathbb{C}^n$ be a nonlinear, differentiable function. A well known strategy to solve a nonlinear system $f(x) = 0$ is the simplified Newton iteration using the fixed-point form of the problem. Let a starting approximation $x^{(0)}$ be given, let $R = f'(x^{(0)})^{-1}$, and iterate according to

$$x^{(k+1)} = x^{(k)} - R \cdot f(x^{(k)}) =: g(x^{(k)}), \quad k = 0, 1, 2, \ldots \tag{9.4}$$

If $x^{(0)}$ is close to the fixed-point $x^*$, the sequence of $x^{(k)}$ for $k \to \infty$ approaches the fixed-point $x^* = g(x^*)$ with $f(x^*) = 0$.

To apply the simplified Newton iteration to the eigenvalue problem (9.3), we let $x = (q, z)^T = (q_0, q_1, \ldots, q_{n-2}, z)^T$, $q_{n-1} := p_n = const$, and define the function $f : \mathbb{C}^n \to \mathbb{C}^n$ as

$$f(x) = f\left(\begin{pmatrix} q \\ z \end{pmatrix}\right) = (A - zI)\begin{pmatrix} q \\ p_n \end{pmatrix}.$$

Then matrix $R \approx J_f^{-1}$ is an approximation to the inverse of the Jacobian matrix

$$\begin{aligned} J_f &= f'(x) = \left[(A - zI)\begin{pmatrix} q \\ p_n \end{pmatrix}\right]' \\ &= (A - zI)\begin{pmatrix} 1 & & & \\ & \ddots & & \\ & & 1 & \\ & & & 0 \end{pmatrix} - \begin{pmatrix} 0 & \cdots & 0 & q_0 \\ \vdots & & \vdots & \vdots \\ \vdots & & \vdots & q_{n-2} \\ 0 & \cdots & 0 & p_n \end{pmatrix} \qquad (9.5) \\ &= \begin{pmatrix} -z & & & -q_0 \\ 1 & \ddots & & \vdots \\ & \ddots & -z & -q_{n-2} \\ & & 1 & -p_n \end{pmatrix}. \end{aligned}$$

Because of the special shape of $J_f$, its inverse $R$ can be determined directly by eliminating the diagonal elements according to Gauss:

$$\begin{pmatrix} 1 & z & & \\ & 1 & & \\ & & \ddots & \\ & & & 1 \end{pmatrix} \cdots \begin{pmatrix} 1 & & & \\ & \ddots & & \\ & & 1 & z \\ & & & 1 \end{pmatrix} \cdot J_f = \begin{pmatrix} 0 & \dots & 0 & -w_0 \\ 1 & & & -w_1 \\ & \ddots & & \vdots \\ & & 1 & -w_{n-1} \end{pmatrix} \tag{9.6}$$

$$\text{with} \quad \begin{aligned} w_{n-1} &:= p_n \\ w_i &:= q_i + z \cdot w_{i+1}, \quad i = n-2, \dots, 0. \end{aligned}$$

With the known inverse of the right-hand side, we immediately get

$$R = J_f^{-1} = \begin{pmatrix} -w_1/w_0 & 1 & & \\ \vdots & & \ddots & \\ -w_{n-1}/w_0 & & & 1 \\ -1/w_0 & 0 & \dots & 0 \end{pmatrix} \cdot \begin{pmatrix} 1 & z & & \\ & 1 & & \\ & & \ddots & \\ & & & 1 \end{pmatrix} \cdots \begin{pmatrix} 1 & & & \\ & \ddots & & \\ & & 1 & z \\ & & & 1 \end{pmatrix}. \tag{9.7}$$

For numerical stability reasons, it is better to perform a residual correction instead of iterative improvement of the complete complex vector $(q, z)^T$. We use the following notation for the residual correction:

$$x = \widetilde{x} + \Delta_x, \quad \text{that is} \quad \begin{pmatrix} q \\ z \end{pmatrix} = \begin{pmatrix} \widetilde{q} + \Delta_q \\ \widetilde{z} + \Delta_z \end{pmatrix}$$

with $\widetilde{x}$ (resp. $\widetilde{q}$ and $\widetilde{z}$) being approximations of the exact values $x$ (resp. $q$ and $z$). Then the Newton iteration has the form

$$\begin{aligned} x^{(k+1)} - \widetilde{x} &= x^{(k)} - \widetilde{x} - R \cdot f(x^{(k)}), \quad \text{or} \\ \Delta_x^{(k+1)} &= \Delta_x^{(k)} - R \cdot f(\widetilde{x} + \Delta_x^{(k)}). \end{aligned}$$

For our special eigenvalue problem, we get

$$\begin{pmatrix} \Delta_q^{(k+1)} \\ \Delta_z^{(k+1)} \end{pmatrix} = \begin{pmatrix} \Delta_q^{(k)} \\ \Delta_z^{(k)} \end{pmatrix} - R \cdot f(\begin{pmatrix} \widetilde{q} + \Delta_q^{(k)} \\ \widetilde{z} + \Delta_z^{(k)} \end{pmatrix}) =: g(\begin{pmatrix} \Delta_q^{(k)} \\ \Delta_z^{(k)} \end{pmatrix}) = g(\Delta_x^{(k)})$$

with $g : \mathbb{C}^n \to \mathbb{C}^n$, $q, \Delta_q \in \mathbb{C}^{n-1}$, and $z, \Delta_z \in \mathbb{C}$. All further calculations depend only on the residual values $(\Delta_q, \Delta_z)^T$ denoted as $\Delta_x$.

Some transformations of the right-hand side (omitting the iteration index $k$) yield

$$\begin{aligned} g(\Delta_x) &= \Delta_x - R \cdot \Big( A - (\widetilde{z} + \Delta_z) I \Big) \begin{pmatrix} \widetilde{q} + \Delta_q \\ p_n \end{pmatrix} \\ &= \Delta_x - R \cdot \Bigg[ (A - \widetilde{z} I) \begin{pmatrix} \widetilde{q} \\ p_n \end{pmatrix} - \Delta_z \begin{pmatrix} \widetilde{q} \\ p_n \end{pmatrix} \\ &\qquad + (A - \widetilde{z} I) \begin{pmatrix} \Delta_q \\ 0 \end{pmatrix} - \Delta_z \begin{pmatrix} \Delta_q \\ 0 \end{pmatrix} \Bigg] \end{aligned}$$

$$= \Delta_x - R\cdot(A-\tilde{z}I)\begin{pmatrix}\tilde{q}\\ p_n\end{pmatrix} + R\cdot\Delta_z\begin{pmatrix}\Delta_q\\ 0\end{pmatrix}$$

$$-R\cdot\left((A-\tilde{z}I)\begin{pmatrix}1 & & & \\ & \ddots & & \\ & & 1 & \\ & & & 0\end{pmatrix} - \begin{pmatrix}0 & \cdots & 0 & \tilde{q}_0\\ \vdots & & \vdots & \vdots\\ \vdots & & \vdots & \tilde{q}_{n-2}\\ 0 & \cdots & 0 & p_n\end{pmatrix}\right)\cdot\Delta_x\,.$$

We do some additional substitutions for a more algorithmic representation:

$$g(\Delta_x) := -R\cdot d + R\cdot\Delta_z\begin{pmatrix}\Delta_q\\ 0\end{pmatrix} + (I - R\cdot J_f)\Delta_x, \tag{9.8}$$

with

$$d := (A-\tilde{z}I)\begin{pmatrix}\tilde{q}\\ p_n\end{pmatrix}. \tag{9.9}$$

In summary, we have a formulation of the Newton iteration with

$$\Delta_x^{(k+1)} = g(\Delta_x^{(k)}) \quad \text{denoting} \quad \begin{pmatrix}\Delta_q^{(k+1)}\\ \Delta_z^{(k+1)}\end{pmatrix} = g(\begin{pmatrix}\Delta_q^{(k)}\\ \Delta_z^{(k)}\end{pmatrix}) \tag{9.10}$$

to solve the eigenvalue problem (9.3).

### The Approximate Iteration

We must determine good approximations of the exact eigenvector $q^*$ and eigenvalue $z^*$ to avoid inflation effects using the interval version of the Newton iteration (9.10). For this purpose, we first use a non-interval residual iteration algorithm starting with an arbitrary starting approximation $\tilde{z}$ for a root of $p(z)$. The initial eigenvector $\tilde{q}$ corresponding to that eigenvalue $\tilde{z}$ is computed recursively by Horner's evaluation of the polynomial $p$ at the point $\tilde{z}$. We use the iteration function $g$ of (9.8) (in floating-point arithmetic) to improve the corresponding residual vector $\Delta_x = (\Delta_q, \Delta_z)^T$ until sufficient accuracy is achieved.

### The Verification Step

We define the interval function $g_{[]} : I\mathbb{C}^n \to I\mathbb{C}^n$ as an interval extension of (9.8) for the interval version of Newton's iteration algorithm to get

$$[\Delta_x]^{(k+1)} = g([\Delta_x]^{(k)}) \quad \text{denoting} \quad \begin{pmatrix}[\Delta_q]^{(k+1)}\\ [\Delta_z]^{(k+1)}\end{pmatrix} = g_{[]}(\begin{pmatrix}[\Delta_q]^{(k)}\\ [\Delta_z]^{(k)}\end{pmatrix}) \tag{9.11}$$

with $[\Delta_q] \in I\mathbb{C}^{n-1}$, and $[\Delta_z] \in \mathbb{C}$. Omitting the iteration index $k$,

$$g_{[]}([\Delta_x]) := -R\cdot d + R\cdot[\Delta_z]\begin{pmatrix}[\Delta_q]\\ 0\end{pmatrix} + (I - R\cdot J_f)[\Delta_x],$$

with the fixed values $d$ from (9.9), $J_f$ from (9.5) and $R$ from (9.7).

Schauder's fixed-point theorem says: If we have the enclosure

$$[\Delta_x]^{(k+1)} = g_{[]}([\Delta_x]^{(k)}) \overset{\circ}{\subset} [\Delta_x]^{(k)}, \tag{9.12}$$

then there exists a (not necessarily unique) fixed-point of $g_{[]}$, and with this a solution $x^* \in \widetilde{x} + [\Delta_x]^{(k+1)}$ of the eigenvalue problem (9.4).

If we choose $R$ as an exact inverse of $J_f$, i.e. $R \equiv J_f^{-1}$, we get $I - R \cdot J_f = 0$, and a simplified iteration function

$$g_{[]}([\Delta_x]) := -R \cdot d + R \cdot [\Delta_z] \begin{pmatrix} [\Delta_q] \\ 0 \end{pmatrix}.$$

This simplification is also valid if we replace $R$ by an interval matrix $[R]$ that is a verified enclosure of the exact inverse $J_f^{-1}$ (see [10]). Using complex interval arithmetic, we get

$$J_f^{-1} \in [R] = \begin{pmatrix} -[w]_1/[w]_0 & 1 & & \\ \vdots & & \ddots & \\ -[w]_{n-1}/[w]_0 & & & 1 \\ -1/[w]_0 & 0 & \dots & 0 \end{pmatrix} \cdot \begin{pmatrix} 1 & \widetilde{z} & & \\ & 1 & & \\ & & \ddots & \\ & & & 1 \end{pmatrix} \cdots \begin{pmatrix} 1 & & & \\ & \ddots & & \\ & & 1 & \widetilde{z} \\ & & & 1 \end{pmatrix},$$

which allows us to simplify the interval function even on a computer to

$$g_{[]}([\Delta_x]) := -[R] \cdot [d] + [R] \cdot [\Delta_z] \begin{pmatrix} [\Delta_q] \\ 0 \end{pmatrix}. \tag{9.13}$$

Subsequently, we start a new iteration step by evaluating the function $g_{[]}$ for a complex interval vector argument until we achieve enclosure (9.12). For computational reasons (cf. Section 3.6), we start with a slightly inflated approximation. The Schauder fixed-point theorem guarantees that there exists a solution of the fixed-point problem (9.12) in $([\Delta_q]^{(k+1)}, [\Delta_z]^{(k+1)})^T$. If $0 \notin [w_0]$, it also follows that $[R]$ is non-singular. This yields

$$\Big( A - (\widetilde{z} + [\Delta_z])I \Big) \begin{pmatrix} \widetilde{q} + [\Delta_q] \\ p_n \end{pmatrix} \supseteq (A - z^* I) \begin{pmatrix} q^* \\ p_n \end{pmatrix} = 0.$$

That is, $(\widetilde{z} + [\Delta_z])$ is a verified enclosure of an eigenvalue $z^*$, which is a root of the complex polynomial $p$, and $(\widetilde{q} + [\Delta_q])$ is a verified enclosure of a corresponding eigenvector $q^*$, the components of which are the coefficients of the deflated polynomial.

## 9.2 Algorithmic Description

First, we present the algorithm Approximation to improve an initial approximation $\widetilde{z}$ of a root of a complex polynomial $p(z) = \sum_{i=0}^{n} p_i z^i$. Approximation works by transforming problem (9.1) to the equivalent problem of finding an eigenvalue and its corresponding eigenvector for the companion matrix (9.2). It improves the approximations of a root and the coefficients of the corresponding deflated polynomial to avoid overestimations during the floating-point interval calculations. A residual iteration method is used to improve the starting approximation until the accuracy necessary to start the interval algorithm is achieved. No guarantee is claimed for the correctness of the approximation computed by **Approximation**.

In the following algorithm, the vector $\Delta \in \mathbb{C}^n$ denotes the combined vector $\Delta_x = (\Delta_q, \Delta_z)^T$ mentioned in the preceding section. The matrix $R$ is not computed explicitly since the components $w_i$ represent the entire matrix. With this the complexity of the inversion is reduced to $O(n)$. The iteration $\Delta^{(k+1)} := g(\Delta^{(k)})$ is done directly using some loops that are equivalent to the mathematical formulation in (9.8). We use $k_{\max} = 50$ as the maximum number of iterations, and $\varepsilon = 10^{-10}$ as the value for the relative error. If the condition number of the inverse $R$ is extremely large, then the convergence of the residual iteration is slow. To avoid the possibility of an unbounded number of iterations at Step 3, we halt after $k_{\max}$ iterations. It turned out that $k_{\max}$ and $\varepsilon$ are good values for minimizing the effort to get sufficiently accurate approximations $\tilde{z}$ and $\tilde{q}$.

**Algorithm 9.1: Approximation** $(p, \tilde{q}, \tilde{z}, Err)$

1. {Constant for termination criteria}
   $\varepsilon := 10^{-10}$; $k_{\max} := 50$;
2. {Determination of an approximate eigenvector $\tilde{q}$}
   {for the initial polynomial zero $\tilde{z}$ }
   $\tilde{q}_{n-1} := p_n$;
   $\tilde{q}_i := \tilde{q}_{i+1} * \tilde{z} + p_{i+1}$; $(i = n-2, \cdots, 0)$
3. {Floating-point iteration of an eigenvalue $\tilde{z}$ and an eigenvector $\tilde{q}$}
   $k := 0$;
   **repeat**
   (a) {Compute the defect $d = (A - \tilde{z}I)\tilde{q}$ (see (9.9))}
   $d_0 := \square(-\tilde{z} * \tilde{q}_0 - p_0)$;
   $d_i := \square(\tilde{q}_{i-1} - \tilde{z} * \tilde{q}_i - p_i)$; $(i = 1, \cdots, n-1)$
   (b) {Compute the components of $R$,the inverse of the Jacobian $J_f$}
   {(see (9.7)) }
   $w_{n-1} := \tilde{q}_{n-1}$;
   $w_i := \square(\tilde{q}_i + \tilde{z} * w_{i+1})$; $(i = n-2, \cdots, 0)$
   **if** $w_0 = 0$ **then** $Err :=$ "Inversion failed";
   (c) {Compute $\Delta^{(k+1)} = g(\Delta^{(k)})$ (see (9.8)}
   $\Delta_{n-1}^{(k+1)} := d_{n-1}$;
   $\Delta_i^{(k+1)} := \square(d_i + \tilde{z} * \Delta_{i+1}^{(k+1)})$; $(i = n-2, \cdots, 0)$
   $t := \Delta_0^{(k+1)} / w_0$;
   $\Delta_i^{(k+1)} := \square(-\Delta_{i+1}^{(k+1)} + t * w_{i+1})$; $(i = 0, \cdots, n-2)$

$\Delta_{n-1}^{(k+1)} := t;$

(d) {Determine the new iterates $\tilde{q}$ and $\tilde{z}$}

$$\begin{aligned} \tilde{q}_i &:= \tilde{q}_i + \Delta_i^{(k+1)}; \quad (i = 0, \cdots, n-2) \\ \tilde{z} &:= \tilde{z} + \Delta_{n-1}^{(k+1)}; \end{aligned}$$

(e) $k := k + 1;$

**until** $\|\Delta\|_\infty / \max(\|(\tilde{q}_i)_{i=0}^{n-2}\|_\infty, |\tilde{z}|) \leq \varepsilon$ **or** $(k = k_{\max});$

4. **return** $\tilde{q}$, $\tilde{z}$, $Err$;

The algorithm IntervalIteration computes a verified enclosure of a root of a complex polynomial using an interval iteration. Starting with good approximations $\tilde{z}$ for a root of the complex polynomial $p(z)$ and $\tilde{q}$ for the coefficients of the deflated polynomial $q(z) = \frac{p(z)}{z-\tilde{z}}$, a verification strategy based on Schauder's fixed-point theorem is used to determine (if possible) a guaranteed enclosure of a polynomial root $z^*$. In addition, guaranteed enclosures of the coefficients of the deflated polynomial $q^*(z) = \frac{p(z)}{z-z^*}$ are returned as a by-product. We use $k_{\max} = 10$ as the maximum number of iterations, and $\varepsilon = 0.1$ as the value for the epsilon inflation. It turned out that these are good values for minimizing the effort if no verification is possible (see also [10]).

**Algorithm 9.2:** IntervalIteration $(p, \tilde{q}, \tilde{z}, [q], [z], Err)$

1. {Computation of the interval evaluation}

(a) {Compute an enclosure of the defect $[d] = (A - \tilde{z}I)\tilde{q}$ (see 9.9)}

$$\begin{aligned} [d]_0 &:= \Diamond(-\tilde{z} * \tilde{q}_0 - p_0); \\ [d]_i &:= \Diamond(\tilde{q}_{i-1} - \tilde{z} * \tilde{q}_i - p_i); \quad (i = 1, \cdots, n-1) \end{aligned}$$

(b) {Compute an enclosure of $[R]$, the inverse of the Jacobian $J_f$} {(see (9.7)) }

$$\begin{aligned} [w]_{n-1} &:= \tilde{q}_{n-1}; \\ [w]_i &:= \Diamond(\tilde{q}_i + \tilde{z} * [w]_{i+1}); \quad (i = n-2, \cdots, 0) \end{aligned}$$

**if** $0 \in [w]_0$ **then** $Err :=$ "Verified enclosure failed";

(c) {Compute an enclosure $[\Delta]^{(0)} = \begin{pmatrix} [\Delta_q] \\ [\Delta_z] \end{pmatrix} \supset \begin{pmatrix} q^* - \tilde{q} \\ z^* - \tilde{z} \end{pmatrix}$}

$$\begin{aligned} [\Delta]_{n-1}^{(0)} &:= [d]_{n-1}; \\ [\Delta]_i^{(0)} &:= \Diamond([d]_i + \tilde{z} * [\Delta]_{i+1}^{(0)}); \quad (i = n-2, \cdots, 0) \\ [t] &:= [\Delta]_0^{(0)} / [w]_0; \\ [\Delta]_i^{(0)} &:= \Diamond(-[\Delta]_{i+1}^{(0)} + [t] * [w]_{i+1}); \quad (i = 0, \cdots, n-2) \\ [\Delta]_{n-1}^{(0)} &:= [t]; \end{aligned}$$

2. {Interval iteration}

$k := 0;\ \varepsilon := 10^{-1};\ k_{\max} := 10;$

**repeat**

(a) {Slightly enlarge the enclosure interval}
$[\Delta]^{(k)} := [\Delta]^{(k)} \bowtie \varepsilon$;

(b) {Determine a new enclosure interval $[\Delta]^{(k+1)} = g_{[]}([\Delta]^{(k)})$ (see (9.13))}

$$\begin{array}{lll} [v]_{n-1} & := & 0; \\ [v]_i & := & \Diamond([\Delta]^{(k)}_{n-1} * [\Delta]^{(k)}_i + \tilde{z} * [v]_{i+1}); \quad (i = n-2, \cdots, 0) \\ [v]_0 & := & [v]_0 / [w]_0; \\ [\Delta_i]^{(k+1)} & := & \Diamond([\Delta]^{(0)}_i + [v]_{i+1} - [v]_0 * [w]_{i+1}); \quad (i = 0, \cdots, n-2) \\ [\Delta]^{(k+1)}_{n-1} & := & [\Delta]^{(0)}_{n-1} - [v]_0; \end{array}$$

(c) $k := k + 1$;

**until** $([\Delta]^{(k)} \overset{\circ}{\subset} [\Delta]^{(k-1)})$ **or** $(k = k_{\max})$;

3. {Verification of the result (see (9.12))}
**if** $([\Delta]^{(k)} \overset{\circ}{\subset} [\Delta]^{(k-1)})$ **then**

$$\begin{array}{lll} [q]_i & := & \Diamond(\tilde{q}_i + [\Delta]^{(k)}_i); \quad (i = 0, \cdots, n-2) \\ [q]_{n-1} & := & p_n; \\ [z] & := & \Diamond(\tilde{z} + [\Delta]^{(k)}_{n-1}); \end{array}$$

**else** $Err$ := "Inclusion failed";

4. **return** $[q]$, $[z]$, $Err$;

The algorithm CPolyZero uses the algorithms Approximation and IntervalIteration presented above. Complex polynomials $p(z) = \sum_{i=0}^{n} p_i z^i$ of degree one can be solved directly using interval operations. To solve complex polynomials of degree $n > 1$ Algorithms 9.1 and 9.2 are used to determine guaranteed enclosures for a polynomial zero $z^*$ and for the coefficients of the deflated polynomial $q(z) = \frac{p(z)}{z-z^*}$.

**Algorithm 9.3:** CPolyZero $(p, \tilde{z}, [z], [q], Err)$

1. {Initialization and treatment of the special cases}
$Err$ := "No Error";
**if** $(n = 0)$ **or** $((n = 1)$ **and** $(p_1 = 0))$ **then** $Err$ := 'Constant polynomial';
**else if** $(n = 1)$ **then**
$[z] := -(p_0 \oslash p_1)$;
$[q]_0 := p_1$;

2. {Usual case}
**else**
Approximation $(p, \tilde{q}, \tilde{z}, Err)$;
IntervalIteration $(p, \tilde{q}, \tilde{z}, [z], [q], Err)$;

3. **return** $[q]$, $[z]$, $Err$;

## Applicability of the Algorithm

### Simple Zeros

This algorithm can only be used to enclose simple zeros of the given polynomial $p$. If we try to enclose a multiple zero by starting algorithm IntervalIteration with a good approximation of a multiple zero, we will get an error message because computing a verified enclosure of the inverse $R$ of the matrix $J_f$ fails. From (9.6), we get

$$|\det(J_f)| = |w_0| = |\sum_{i=0}^{n-1} q_i z^i| = |p'(z)|, \tag{9.14}$$

where $z$ is an eigenvalue, and $q$ is an eigenvector of the eigenproblem (9.3). If we have a multiple zero $z$ of the polynomial, (9.14) yields

$$|\det(J_f)| = |p'(z)| = 0.$$

Thus, the matrix $J_f$ is singular. The inverse $R$ cannot be enclosed because it does not exist.

### Improvements

If two or more zeros lie very close together, it may happen that the enclosure of the inverse of $J_f$ fails, because $J_f$ is in some numerical sense too close to a singular matrix, i.e. it is too ill conditioned. To overcome this problem in many cases, the enclosure of the inverse has to be computed with double mantissa length using accurate expressions (cf. Chapter 10).

As a further refinement, we can compute the approximate solutions $\tilde{z}$ and $\tilde{q}$ in algorithm Approximation with double length and make use of the double length in algorithm IntervalIteration. Both improvements require the accurate evaluation of dot product expressions (see Section 2.3).

With these improvements, we can separate and verify two zeros differing only in the last digit of a mantissa of single length (cf. [10], [21]).

### Real polynomials

If we have a real polynomial $p(z) = \sum_{i=0}^{n} p_i z^i$ ($p_i \in I\!R$) (imaginary part of each coefficient is zero), and we have a real starting approximation $\tilde{z} \in I\!R$, we will never get enclosures of a pair of conjugate complex zeros of $p(z)$, because all complex arithmetic operations deliver a real result, as the imaginary parts of all operands will stay zero. At any rate, the algorithm described here also works on real polynomials. If we want to find a complex conjugate zero of a real polynomial, we must choose a non-real starting approximation $\tilde{z}$.

# 9.3 Implementation and Examples

## 9.3.1 C++ Program Code

We give the listings of the module *cpzero* for enclosing a root of a complex polynomial. The module requires the data types *CPolynomial* and *CIPolynomial* which are defined in the modules *cpoly* and *cipoly*, respectively. In the listings, interval data are named with double characters, e.g. dd[i] denotes the interval $[d]_i$.

### 9.3.1.1 Module cpoly

The header file of module *cpoly* supplies the definition of the class *CPolynomial* representing a complex polynomial $p(z) = \sum_{i=0}^{n} p_i z^i$, $p_i, z \in C$. Since arithmetic operations for complex polynomials are not required by Algorithm 9.3, they are not provided by module *cpoly*.

```
//----------------------------------------------------------------------------
// File: cpoly (header)
// Purpose: Definition of the class for the representation of a complex
//    polynomial by its coefficients.
// Class CPolynomial:
//    Deg()          : to get the degree of the polynomial
//    CPolynomial(): constructors
//    operator []    : component access
//    operator >>    : input operator for data of type CPolynomial
//    operator <<    : output operator for data of type CPolynomial
//----------------------------------------------------------------------------
#ifndef __CPOLY_HPP
#define __CPOLY_HPP

#include <cvector.hpp>     // Complex vector arithmetic

class CPolynomial {
  private:
    cvector coeff;
  public:
    CPolynomial ( int );
    CPolynomial ( CPolynomial& );
    complex& operator[] ( int i ) { return coeff[i]; }

    friend int Deg ( CPolynomial& );
    friend istream& operator>> ( istream&, CPolynomial& );
    friend ostream& operator<< ( ostream&, CPolynomial );
};
#endif
```

The input operator >> echos x↑0 *, + x↑1 *, + x↑2 *, ... on the standard output device to give an information on the index of the coefficient actually to be entered.

```
//----------------------------------------------------------------------------
// File: cpoly (implementation)
// Purpose: Definition of the class for the representation of a complex
//    polynomial by its coefficients.
// Class CPolynomial:
//    Deg()          : to get the degree of the polynomial
```

```
//    CPolynomial(): constructors
//    operator []  : component access
//    operator >>  : input operator for data of type CPolynomial
//    operator <<  : output operator for data of type CPolynomial
//-----------------------------------------------------------------------------
#include <cpoly.hpp>

int Deg ( CPolynomial& p ) { return Ub(p.coeff); }

CPolynomial::CPolynomial( int n )
{
  Resize(coeff,0,n);
  coeff = 0.0;
}

CPolynomial::CPolynomial( CPolynomial& p )
{
  Resize(coeff,0,Deg(p));
  coeff = p.coeff;
}

istream& operator>> ( istream& in, CPolynomial& p )
{
  cout << "  x↑0 * ";  in >> p[0];
  for (int i = 1; i <= Deg(p); i++)
    { cout << "+ x↑" << i << " * ";  in >> p[i]; }
  return in;
}

ostream& operator<< ( ostream& out, CPolynomial p )
{
  int PolyIsZero, n = Deg(p);

  PolyIsZero = 1;
  for (int i = 0; i <= n; i++) {
    if (p[i] == 0.0) continue;
    if (PolyIsZero)
      out << "  ";
    else
      out << "+ ";
    out << p[i] << " * x↑" << i << endl;
    PolyIsZero = 0;
  }
  if (PolyIsZero) out << "  0 (= zero polynomial)" << endl;
  return out;
}
```

#### 9.3.1.2 Module cipoly

The header file of module *cipoly* supplies the definition of the class *CIPolynomial* representing a complex interval polynomial $p_{[]}([z]) = \sum_{i=0}^{n}[p]_i[z]^i$, $[p]_i, [z] \in IC$. Arithmetic operations for complex interval polynomials are not provided. The friend functions *in()* and *Blow()* realize the contained-in-the-interior relation and the epsilon inflation for complex interval polynomials.

```
//-----------------------------------------------------------------------------
// File: cipoly (header)
// Purpose: Definition of the class for the representation of a complex
//    interval polynomial by its coefficients.
// Class CIPolynomial:
```

```
//    Deg()          : to get the degree of the polynomial
//    CIPolynomial(): constructors
//    operator []    : component access
//    in()           : Contained-in-the-interior relation for two
//                     complex interval polynomials
//    Blow()         : Epsilon inflation
//    operator >>    : input operator for data of type CIPolynomial
//    operator <<    : output operator for data of type CIPolynomial
//----------------------------------------------------------------------------
#ifndef __CIPOLY_HPP
#define __CIPOLY_HPP

#include <civector.hpp>     // Complex interval vector arithmetic

class CIPolynomial {
  private:
    civector coeff;
  public:
    CIPolynomial ( int );
    CIPolynomial ( CIPolynomial& );
    cinterval& operator[] ( int i )  { return coeff[i]; }

    friend int Deg ( CIPolynomial& );
    friend int in ( CIPolynomial&, CIPolynomial& );
    friend CIPolynomial Blow ( CIPolynomial, const real );
    friend istream& operator>> ( istream&, CIPolynomial& );
    friend ostream& operator<< ( ostream&, CIPolynomial );
};
#endif
```

The contained-in-the-interior relation (see Section 3.1) and the epsilon inflation (see Section 3.6) are defined componentwise. As for complex polynomials the input operator **>>** echos **x↑0 *, + x↑1 *, + x↑2 ***, ... on the standard output device to give an information on the index of the coefficient actually to be entered.

```
//----------------------------------------------------------------------------
// File: cipoly (implementation)
// Purpose: Definition of the class for the representation of a complex
//    interval polynomial by its coefficients.
// Class CIPolynomial:
//    Deg()          : to get the degree of the polynomial
//    CIPolynomial(): constructors
//    operator []    : component access
//    in()           : Contained-in-the-interior relation for two
//                     complex interval polynomials
//    Blow()         : Epsilon inflation
//    operator >>    : input operator for data of type CIPolynomial
//    operator <<    : output operator for data of type CIPolynomial
//----------------------------------------------------------------------------
#include <ci_util.hpp>     // Complex interval utility functions
#include <cipoly.hpp>

int Deg ( CIPolynomial& p ) { return Ub(p.coeff); }

CIPolynomial::CIPolynomial( int n )
{
  Resize(coeff,0,n);
  coeff = 0.0;
}

CIPolynomial::CIPolynomial( CIPolynomial& p )
{
```

```
    Resize(coeff,0,Deg(p));
    coeff = p.coeff;
  }

  int in ( CIPolynomial& p, CIPolynomial& q )                // Contained-in-the-
  {                                                          // interior relation
    int i, incl = 1;                                         //------------------

    for (i = 0; (incl == 1) && (i <= Deg(p)); i++)
      incl = in(p[i],q[i]);
    return incl;
  }

  CIPolynomial Blow ( CIPolynomial p, const real eps )       // Epsilon inflation
  {                                                          //------------------
    for (int i = 0; i <= Deg(p); i++)
      p[i] = Blow(p[i],eps);
    return p;
  }

  istream& operator>> ( istream& in, CIPolynomial& p )
  {
    cout << "  x↑0 * ";  in >> p[0];
    for (int i = 1; i <= Deg(p); i++)
      { cout << "+ x↑" << i << " * ";  in >> p[i]; }
    return in;
  }

  ostream& operator<< ( ostream& out, CIPolynomial p )
  {
    int PolyIsZero, n = Deg(p);

    PolyIsZero = 1;
    for (int i = 0; i <= n; i++) {
      if (p[i] == 0.0) continue;
      if (PolyIsZero)
        out << "  ";
      else
        out << "+ ";
      out << p[i] << " * x↑" << i << endl;
      PolyIsZero = 0;
    }
    if (PolyIsZero) out << "  0 (= zero polynomial)" << endl;
    return out;
  }
```

### 9.3.1.3 Module cpzero

The header file of module *cpzero* supplies the interface of the function *CPolyZero()* for computing an enclosure of a root of o complex polynomial and the interface of the function *CPolyZeroErrMsg()* which can be used to get an error message for the error code returned by *CPolyZero()*.

```
//----------------------------------------------------------------------------
// File: cpzero (header)
// Purpose: Determination and enclosure of a root of a complex polynomial,
//    and of the coefficients of the deflated polynomial.
// Global functions:
//    CPolyZero()      : computes an enclosure for a root and for the
//                       deflated complex polynomial
//    CPolyZeroErrMsg(): delivers an error message text
```

```
//----------------------------------------------------------------------------
#ifndef __CPZERO_HPP
#define __CPZERO_HPP

#include <cinterva.hpp>   // Complex interval arithmetic
#include <cpoly.hpp>      // Complex polynomial type
#include <cipoly.hpp>     // Complex interval polynomial type

extern char* CPolyZeroErrMsg ( int );
extern void  CPolyZero ( CPolynomial, complex,
                         CIPolynomial&, cinterval&, int& );
#endif
```

The function *CPolyZero()* is an implementation of Algorithm 9.3. It computes enclosures of a root and of the deflated polynomial for a complex polynomial $p(z) = \sum_{i=0}^{n} p_i z^i$ with $p_i, z \in C$. The static functions *Approximation()* and *IntervalIteration()* are implementations of the Algorithms 9.1 and 9.2.

```
//----------------------------------------------------------------------------
// File: cpzero (implementation)
// Purpose: Determination and enclosure of a root of a complex polynomial,
//    and of the coefficients of the deflated polynomial.
// Method: The root of a complex polynomial matches with the eigenvalue of
//    its companion matrix and the coefficients of the deflated polynomial
//    match to the components of the corresponding eigenvector. The eigen-
//    vector and eigenvalue are determined using the simplified Newton
//    iteration with iterative refinement.
// Global functions:
//    CPolyZero()      : computes an enclosure for a root and for the
//                       deflated complex polynomial
//    CPolyZeroErrMsg(): delivers an error message text
//----------------------------------------------------------------------------
#include <string.h>       // String Handling
#include <r_util.hpp>     // Real utility functions
#include <ci_util.hpp>    // Complex interval utility functions
#include <cpzero.hpp>

//----------------------------------------------------------------------------
// Error codes used in this module. In the comments below p[0],..., p[n] are
// the coefficients of a polynomial p.
//----------------------------------------------------------------------------
const int
  NoError   = 0,   // No Error occurred
  ZeroPoly  = 1,   // Zero polynomial, i.e. n = 0 and p[0] = 0
  ConstPoly = 2,   // Constant polynomial, i.e. n = 0  and p[0] != 0
  InvFailed = 3,   // Inversion of the Jacobian failed
  VerFailed = 4,   // Verified inversion of the Jacobian failed
  IncFailed = 5;   // Inclusion failed

//----------------------------------------------------------------------------
// Error messages depending on the error code.
//----------------------------------------------------------------------------
char* CPolyZeroErrMsg ( int Err )
{
  static char Msg[80] = "";

  if (Err != NoError) {
    char Hlp[60];

    switch (Err) {
      case ZeroPoly:
        sprintf(Hlp,"Zero polynomial occurred");                break;
```

```
      case ConstPoly:
        sprintf(Hlp,"Constant polynomial != 0 occurred");         break;
      case InvFailed:
        sprintf(Hlp,"Inversion of the Jacobian failed");          break;
      case IncFailed:
        sprintf(Hlp,"Inclusion failed");                          break;
      case VerFailed:
        sprintf(Hlp,"Verified inversion of the Jacobian failed"); break;
      default:
        strcpy(Hlp,"Code not defined");
    }
    sprintf(Msg,"Error: %s!",Hlp);
  }
  return(Msg);
} // CPolyZeroErrMsg

//----------------------------------------------------------------------------
// Function used to get the maximum norm of 'p' where 'p' is interpreted as a
// complex vector.
//----------------------------------------------------------------------------
static real MaxNorm ( CPolynomial p )
{
  int  i;
  real max, tmp;

  max = abs(p[0]);
  for (i = 1; i <= Deg(p); i++) {
    tmp = abs(p[i]);
    if (tmp > max)  max = tmp;
  }
  return max;
}

//----------------------------------------------------------------------------
// Purpose: Determination of a floating-point approximation for a root of
//    polynomial p and for the deflated polynomial.
// Parameters:
//    In    : 'p'  : represents a complex polynomial by its coefficients
//    In/Out: 'z'  : starting approximation (in) for a root and is returned
//                   as an improved floating-point approximation
//    Out   : 'q'  : returns an approximation (for the coefficients of) the
//                   deflated polynomial
//            'Err': error flag
// Description: For a given starting approximation 'z', a floating-point
//    Newton iteration is performed to get improved approximations for a root
//    'z' of polynomial 'p', as well as for the coefficients 'q' of the
//    deflated polynomial.
//----------------------------------------------------------------------------
static void Approximation ( CPolynomial  p, complex& z,
                            CPolynomial& q, int&     Err )
{
  const int      kmax  = 50;             // Maximum number of iteration steps
  const real     eps   = 1E-9;           // Relative error of approximation

  int            stop, k, i, n = Deg(p);
  complex        t;
  cdotprecision  accu;
  CPolynomial    delta(n-1), d(n-1), w(n-1), qHlp(n-2);

  Err = NoError;                           // Initialization

  q[n-1] = p[n];                           // Determination of an approximate
  for (i = n-2; i >= 0; i--)               // eigenvector q
    q[i] = q[i+1]*z + p[i+1];              //-------------------------------
```

```
  k = 0;        // Floating-point iteration of z and q
  do {          //------------------------------------
    accu = -p[0];                                   // Computation of the defect
    accumulate(accu,-z,q[0]);                       //      d = (A - z*I)*q
    d[0] = rnd(accu);                               //--------------------------
    for (i = 1; i <= n-1; i++) {
      accu = q[i-1];
      accumulate(accu,-z,q[i]);
      accu -= p[i];
      d[i] = rnd(accu);
    }


    w[n-1] = q[n-1];                                // Computation of R, the inverse
    for (i = n-2; i >= 0; i--) {                    // of the Jacobian J
      accu = q[i];                                  //------------------------------
      accumulate(accu,z,w[i+1]);
      w[i] = rnd(accu);
    }

    if (w[0] == 0.0)
      Err = InvFailed;
    else {
      delta[n-1] = d[n-1];                        // Computation of delta↑(k+1) =
      for (i = n-2; i >= 0; i--)                  //  g(delta↑(k))
        delta[i] = d[i] + z*delta[i+1];           //-----------------------------
      t = delta[0]/w[0];
      for (i = 0; i <= n-2; i++)
        delta[i] = -delta[i+1] + t*w[i+1];
      delta[n-1] = t;

      for (i = 0; i <= n-2; i++) {                // Determine the new iterates of
        q[i] = q[i] + delta[i];                   // q and z and store the first
        qHlp[i] = q[i];                           // n-1 components of q in qHlp
      }                                           //-----------------------------
      z = z + delta[n-1];

      k++;
    } //else

    // Stop iteration if the relative round-off error of the coefficients
    // and of the zero is approximately <= a given epsilon. Stop also if
    // the maximum number of iteration steps is exceeded, or if an error
    // occurred. Some C++ compilers (e.g. HP C++ V2.0) do not support
    // temporaries of class variables in an || expression. Thus, 'stop'
    // is used to avoid this error.
    //-------------------------------------------------------------------
    // stop = in(ddelta,ddelta_old) || (k == kmax);
    // stop = (MaxNorm(delta) / Max(MaxNorm(qHlp),abs(z)) <= eps) ||
    //          (Err != NoError) || (k == kmax);
    stop = (MaxNorm(delta) / Max(MaxNorm(qHlp),abs(z)) <= eps);
    stop = stop || (Err != NoError) || (k == kmax);

  } while ( !stop );

} //Approximation

//----------------------------------------------------------------------------
// Purpose: Determination of enclosures for a root of polynomial p and for
//    the deflated polynomial.
// Parameters:
//    In : 'p'  : represents a complex polynomial by its coefficients
```

```
//         'q'  : approximation of (the coefficients of) the deflated
//                polynomial
//         'z'  : floating-point approximation of a root
//    Out: 'qq' : returns an enclosure for (the coefficients of) the
//                deflated polynomial
//         'zz' : returns an enclosure of the root
//         'Err': error flag
// Description: For starting floating-point approximations 'z' of a root
//    and 'q' of the deflated polynomial an interval residual iteration is
//    performed to get enclosures 'zz' for the root of polynomial 'p', as
//    well as for the coefficients of the deflated polynomial.
//----------------------------------------------------------------------------
static void IntervalIteration ( CPolynomial p,  CPolynomial   q,
                                complex     z,  CIPolynomial& qq,
                                cinterval&  zz, int&    Err       )
{
  const int        kmax = 10;                // Maximum number of iteration steps

  int              stop, i, k, n = Deg(p);
  real             eps;
  cinterval        tt;
  cidotprecision   accu;
  CIPolynomial     dd(n-1), vv(n-1), ww(n-1);
  CIPolynomial     ddelta(n-1), ddelta_0(n-1), ddelta_old(n-1);

  Err = NoError;                              // Initialization

  accu = -p[0];                               // Compute an inclusion of the
  accumulate(accu,-z,q[0]);                   // defect [d] = (A - z*I)*q
  rnd(accu,dd[0]);                            //---------------------------
  for (i = 1; i <= n-1; i++) {
    accu = q[i-1];
    accumulate(accu,-z,q[i]);
    accumulate(accu,-1.0,p[i]);
    dd[i] = rnd(accu);
  }

  ww[n-1] = q[n-1];                           // Compute an inclusion of [R], the
  for (i = n-2; i >=0; i--) {                 // inverse of the Jacobian J
    accu = q[i];                              //--------------------------------
    accumulate(accu,z,ww[i+1]);
    ww[i] = rnd(accu);
  }

  if (_cinterval(0.0) < ww[0])
    Err = VerFailed;
  else {
    ddelta_0[n-1] = dd[n-1];                  // Compute a starting
    for (i = n-2; i >= 0; i--) {              // inclusion [y]↑(0)
      accu = dd[i];                           //-------------------
      accumulate(accu,z,ddelta_0[i+1]);
      ddelta_0[i] = rnd(accu);
    }
    tt = ddelta_0[0]/ww[0];
    for (i = 0; i <= n-2; i++) {
      accu = -ddelta_0[i+1];
      accumulate(accu,tt,ww[i+1]);
      ddelta_0[i] = rnd(accu);
    }
    ddelta_0[n-1] = tt;

    k = 0;  ddelta = ddelta_0;                // Interval iteration
    do {                                      //--------------------
      ddelta_old = ddelta;
```

```
      if (k <= 3)                          // Slightly enlarge the
        eps = 0.125;                       // inclusion interval
      else if (k <= 6)                     //---------------------
        eps = 0.5;
      else
        eps = 5.0;
      ddelta_old = Blow(ddelta,eps);

      vv[n-1] = 0.0;                       // Determine a new
      for (i = n-2; i >= 0; i--) {         // inclusion interval
        accu = 0.0;                        //-------------------
        accumulate(accu,ddelta_old[n-1],ddelta_old[i]);
        accumulate(accu,z,vv[i+1]);
        vv[i] = rnd(accu);
      }
      vv[0] = vv[0] / ww[0];
      for (i = 0; i <= n-2; i++) {
        accu = ddelta_0[i];
        accu += vv[i+1];
        accumulate(accu,-vv[0],ww[i+1]);
        ddelta[i] = rnd(accu);
      }
      ddelta[n-1] = ddelta_0[n-1] - vv[0];

      k++;

      // Some C++ compilers (e.g. HP C++ V2.0) do not support temporaries
      // of class variables in an || expression. Thus, 'stop' is used to
      // avoid this error.
      //-----------------------------------------------------------------
      // stop = in(ddelta,ddelta_old) || (k == kmax);
      stop = in(ddelta,ddelta_old);
      stop = stop || (k == kmax);

    } while ( !stop );

  } //else

  if ( in(ddelta,ddelta_old) ) {           // Verification of the result
    for (i = 0; i <= n-2; i++) {           //---------------------------
      accu = q[i];
      accu += ddelta[i];
      qq[i] = rnd(accu);
    }
    qq[n-1] = p[n];
    accu = z;
    accu += ddelta[n-1];
    zz = rnd(accu);
  }
  else
    Err = IncFailed;

} //IntervalIteration

//----------------------------------------------------------------------
// Purpose: Determination of enclosures for a root of polynomial p and for
//    the deflated polynomial
// Parameters:
//    In : 'p'  : represents a complex polynomial by its coefficients
//         'z'  : floating-point approximation of a root
//    Out: 'qq' : returns an enclosure for (the coefficients of) the
//                deflated polynomial
//         'zz' : returns an enclosure of the root
```

```
//          'Err': error flag
// Description: For starting approximations 'z' of a root, enclosures 'zz'
//    for a root of polynomial 'p', as well as 'qq' for the coefficients of
//    the deflated polynomial are computed. Since the root of a complex
//    polynomial matches with an eigenvalue of its companion matrix and the
//    coefficients of the deflated polynomial match to the components of the
//    corresponding eigenvector, the eigenvector and eigenvalue are determined
//    by iterative refinement. First, good floating-point approximations are
//    computed, and then verified enclosures for a root and for a deflated
//    polynomial are returned.
//----------------------------------------------------------------------------
void CPolyZero ( CPolynomial   p, complex    z,
                 CIPolynomial& qq, cinterval& zz, int& Err )
{
  int  n = Deg(p);

  Err = NoError;                                  // Initialization

  if ( (n == 0) || (n == 1) )
    if (n == 0)                                   // Polynomial of degree n = 0
                                                  //---------------------------
      if (p[0] == 0.0) Err = ZeroPoly;            // Zero polynomial
      else             Err = ConstPoly;           // Constant polynomial

    else                                          // Polynomial of degree n = 1
      if (p[1] == 0.0)                            //---------------------------
        if (p[0] == 0.0) Err = ZeroPoly;          // Zero polynomial
        else             Err = ConstPoly;         // Constant polynomial
      else {
        zz = -_cinterval(p[0]) / p[1];
        qq[0] = p[1];
      }
  else {                                          // Polynomial of degree n > 1
    CPolynomial  q(n-1);                          //---------------------------

    Approximation(p,z,q,Err);
    if (Err == NoError) IntervalIteration(p,q,z,qq,zz,Err);
  }
} //CPolyZero
```

### 9.3.2 Example

We consider the following complex polynomial of $4^{th}$ degree. We want to find and enclose a root for the arbitrary starting approximation $\tilde{z} = (1+i)$ using Algorithm 9.3.

$$p(z) = z^4 - (5+3i)z^3 + (6+7i)z^2 - (54-22i)z + (120+90i)$$

$$\text{exact zeros}: \quad \pm\sqrt{2}(1+2i), \quad 5, \quad \text{and} \quad 3i.$$

$$\sqrt{2} \approx 1.41421356237309\ldots$$

$$\text{starting approximation}: \quad (1+i)$$

The following program calls *CPolyZero()* to calculate an enclosure $[z]$ of an exact root $z^*$. That is, it is guaranteed that $z^* \in [z]$. Guaranteed enclosures $[q]$ of the coefficients of the deflated polynomial $q(z) = \frac{p(z)}{z-[z]}$ are determined, too.

```
//----------------------------------------------------------------------------
// Example: Finding an enclosure for a root of a complex polynomial
//----------------------------------------------------------------------------
#include <cpzero.hpp>     // Roots (zeros) of a complex polynomial

int main ( )
{
  int       Err, n;
  complex   z;
  cinterval zz;

  do {
    cout << "Enter the degree of the polynomial (>=0): ";
    cin  >> n; cout << endl;
  } while (n < 0);

  CPolynomial  p(n);
  CIPolynomial qq(n);

  cout << "Enter the coefficients in increasing order:" << endl;
  cin  >> p; cout << endl;
  cout << "Enter the starting approximation:" << endl;
  cin  >> z; cout << endl;

  CPolyZero(p,z,qq,zz,Err);

  if (!Err) {
    cout << SetPrecision(15,7) << Scientific;    // Output format

    cout << "Polynomial: "
         << endl << p << endl;
    cout << "Zero found in:"
         << endl << zz << endl << endl;
    cout << "The coefficients of the reduced polynomial are:"
         << endl << qq << endl;
  }
  else
    cout << CPolyZeroErrMsg(Err) << endl;

  return 0;
}
```

Our implementation of Algorithm 9.3 produces the following output. The complex rectangular intervals are presented in the form $(re, im)$, where $re$ and $im$ are real intervals represented by their lower and upper bounds, respectively. For clarity, the output format is restricted to 8 significant digits. In fact, the relative accuracy of the computed intervals is about $10^{-15}$. One can change the output format with C–XSC stream manipulator *SetPrecision* (see Section 2.4).

```
Enter the degree of the polynomial (>=0): 4

Enter the coefficients in increasing order:
  x^0 * (120,90)
+ x^1 * (-54,22)
+ x^2 * (6,7)
+ x^3 * (-5,-3)
+ x^4 * (1,0)

Enter the starting approximation: (1,1)

Polynomial:
```

```
  ( 1.2000000E+002, 9.0000000E+001) * x^0
+ (-5.4000000E+001, 2.2000000E+001) * x^1
+ ( 6.0000000E+000, 7.0000000E+000) * x^2
+ (-5.0000000E+000,-3.0000000E+000) * x^3
+ ( 1.0000000E+000, 0.0000000E+000) * x^4

Zero found in:
([ 1.4142135E+000, 1.4142136E+000],[ 2.8284271E+000, 2.8284272E+000])

The coefficients of the reduced polynomial are:
  ([-4.2426407E+001,-4.2426406E+001],[ 2.1213203E+001, 2.1213204E+001]) * x^0
+ ([ 1.4142135E+000, 1.4142136E+000],[-3.3847764E+000,-3.3847763E+000]) * x^1
+ ([-3.5857865E+000,-3.5857864E+000],[-1.7157288E-001,-1.7157287E-001]) * x^2
+ ([ 1.0000000E+000, 1.0000000E+000],[ 0.0000000E+000, 0.0000000E+000]) * x^3
```

### 9.3.3 Restrictions and Hints

#### Close Zeros

If two zeros of the polynomial are so close together that they are identical in their number representation up to the mantissa length and differ only in digits beyond the mantissa, they are called "a numerical multiple zero". Such zeros cannot be verified with our program described above because they cannot be separated by the given number representation. The program handles them just like a true multiple zero (see Section 9.2) and terminates. However, we could implement the algorithm CPolyZero of Section 9.2 using a multi-precision floating-point and floating-point interval arithmetic (see [10], [61]).

#### Clusters of Zeros

If two or more zeros are extremely close together, i.e. they form a cluster, it is not possible to verify a zero of this cluster with the implementation given in Section 9.3.1 because we may enclose a zero of the derivative $p'$ of the polynomial $p$. Hence, an enclosure of the determinant of the Jacobian matrix (see page 172) contains zero, and the verified inversion of the Jacobian matrix fails. We may overcome this limit of the implementation by computing the inverse of the Jacobian matrix with higher accuracy, e.g. with double mantissa length, using the exact scalar product (see [60]).

## 9.4 Exercises

**Exercise 9.1** Use the sample program on page 182 to enclose the other three roots of the polynomial given in Section 9.3.2.

By repeating the deflation of a verified zero from the reduced polynomial *pdeflated* $\in [q]$, the approximation of a new zero in the reduced polynomial, and the verification of the new zero in the original polynomial, we get all simple zeros of the polynomial. The deflated polynomial we can get from the parameter $[q]$ of the algorithm CPolyZero. For approximating a new zero, the deflated polynomial *pdeflated* is used. The verification of the new zero is done in the original polynomial $p$ because the zeros of the approximate deflated polynomial are smeared out because $[q]$ has interval-valued coefficients, while $p$ has point-valued coefficients.

**Algorithm 9.4:** AllCPolyZeros $(p)$

1. $pdeflated := p$;
   **for** $i := 1$ **to** $n - 1$ **do**
   {Verification of a new zero}
   (a) {Approximate a new zero of *pdeflated*}
   (b) {Verify the new zero for $p$}
   (c) {Deflate verified zero from *pdeflated*}

**Exercise 9.2** Implement the above algorithm using the modules of Section 9.3.1. Test the implementation with the example of Section 9.3.2. Note, that algorithm AllCPolyZeros might fail to determine *all* zeros of a polynomial, for example if there exist multiple zeros.

**Exercise 9.3** Determine enclosures for all zeros of the Wilkinson polynomial $p(z) = \prod_{i=1}^{n}(z - i)$ with $n = 5$ using Algorithm 9.4. Start computation with complex and real starting approximations.

**Exercise 9.4** This exercise requires a C++ compiler that supports class templates: Use the module *poly* in our programs as suggested in Exercise 4.3. Replace the complex polynomial types *CPolynomial* and *CIPolynomial* by corresponding template types *Polynomial<complex>* and *Polynomial<cinterval>*.

## 9.5 References and Further Reading

Algorithm 9.3 is an extension of one first proposed by Böhm [10] for polynomials with real coefficients. A commercial implementation was released by IBM with the Fortran subroutine library ACRITH [36]. of IBM [36] The algorithm was generalized for polynomials with complex coefficients by Geörg [21] and Grüner [24]. This modification may be used to enclose simple real and complex zeros of complex polynomials, as well as simple real and complex zeros of real polynomials. Additional comments are given by Krämer in [54].

# Part III

# Multi-Dimensional Problems

# Chapter 10

# Linear Systems of Equations

Finding the solution of a linear system of equations is one of the basic problems in numerical algebra. We will develop a verification algorithm for square systems with full matrix based on a Newton-like method for an equivalent fixed-point problem.

## 10.1 Theoretical Background

### 10.1.1 A Newton-like Method

Let $Ax = b$ be a real system of equations with $A \in I\!R^{n \times n}$ and $b, x \in I\!R^n$. Finding a solution of the system $Ax = b$ is equivalent to finding a zero of $f(x) = Ax - b$. Hence, Newton's method gives the following fixed-point iteration scheme

$$x^{(k+1)} = x^{(k)} - A^{-1}(Ax^{(k)} - b), \quad k = 0, 1, \ldots . \tag{10.1}$$

Here, $x^{(0)}$ is some arbitrary starting value. In general, the inverse of $A$ is not exactly known. Thus instead of (10.1), we use the Newton-like iteration

$$x^{(k+1)} = x^{(k)} - R(Ax^{(k)} - b), \quad k = 0, 1, \ldots, \tag{10.2}$$

where $R \approx A^{-1}$ is an approximate inverse of $A$.

Let us now replace the real iterates $x^{(k)}$ by interval vectors $[x]^{(k)} \in I\,I\!R^n$. If there exists an index $k$ with $[x]^{(k+1)} \subseteq [x]^{(k)}$, then, by Brouwer's fixed-point theorem (see page 51), Equation 10.2 has at least one fixed-point $x \in [x]^{(k)}$. Supposed $R$ is regular. Then this fixed-point is also a solution of $Ax = b$. However, if we consider the diameter of $[x]^{(k+1)}$, we get $d([x]^{(k+1)}) = d([x]^{(k)}) + d(R(A[x]^{(k)} - b)) \geq d([x]^{(k)})$. Thus in general, the subset relation will not be satisfied. For this reason, we modify the right-hand side of (10.2) to

$$x^{(k+1)} = Rb + (I - RA)x^{(k)}, \quad k = 0, 1, \ldots, \tag{10.3}$$

where $I$ denotes the $n \times n$ identity matrix. Rump [82] proved for Equation (10.3) that if there exists an index $k$ with $[x]^{(k+1)} \overset{\circ}{\subset} [x]^{(k)}$, then the matrices $R$ and $A$ are regular, and there is a unique solution $x$ of the system $Ax = b$ with $x \in [x]^{(k+1)}$. Moreover this result is valid for any matrix $R$. Note that the $\overset{\circ}{\subset}$ operator denotes the inclusion in the interior as defined in Section 3.1.

### 10.1.2 The Residual Iteration Scheme

It is a well-know numerical principle that an approximate solution $\widetilde{x}$ of $Ax = b$ may be improved by solving the system $Ay = d$, where $d = b - A\widetilde{x}$ is the residual of $A\widetilde{x}$. Since $y = A^{-1}(b - A\widetilde{x}) = x - \widetilde{x}$, the exact solution of $Ax = b$ is given by $x = \widetilde{x} + y$. Applying Equation (10.3) to the residual system leads to

$$y^{(k+1)} = \underbrace{R(b - A\widetilde{x})}_{=:\,z} + \underbrace{(I - RA)}_{=:\,C}\, y^{(k)}, \quad k = 0, 1, \dots . \tag{10.4}$$

According to Rump's results mentioned above, the residual equation $Ay = d$ has a unique solution $y \in [y]^{(k+1)}$ if we succeed in finding an index $k$ satisfying $[y]^{(k+1)} \overset{\circ}{\subset} [y]^{(k)}$ for the corresponding interval iteration scheme. Moreover, since $y = x - \widetilde{x} \in [y]^{(k+1)}$, we then have a verified enclosure of the unique solution of $Ax = b$ given by $\widetilde{x} + [y]^{(k+1)}$.

These results remain valid if we replace the exact expressions for $z$ and $C$ in (10.4) by interval extensions. However, to avoid overestimation effects, it is highly recommended to evaluate $b - A\widetilde{x}$ and $I - RA$ without any intermediate rounding.

### 10.1.3 How to Compute the Approximate Inverse

We recall that the results summarized above are valid even for an arbitrary matrix $R$. However, it is a non-surprising empirical fact that the better $R$ approximates the inverse of $A$, the faster the contained-in-the-interior relation for two successive iterates will be satisfied. We favor Crout's algorithm (cf. [89]) with partial pivoting for matrix inversion.

Let $A' = PA$ be the matrix which comes from $A$ by row interchanges so that a factorization $A' = LU$ with a lower-triangular matrix $L$ and an upper-triangular matrix $U$ exists. If we normalize the diagonal entries of $L$ to unity, and if we save the indices for the pivotal rows in a vector $p$, initialized by $p = (1, 2, \dots, n)^{\mathrm{T}}$, the decomposition procedure is as follows:

$$\left.\begin{array}{l} v_k = a_{ki} - \displaystyle\sum_{j=1}^{i-1} l_{kj} a_{ji}, \quad k = i, \dots, n \\ |v_j| = \displaystyle\max_{i \le k \le n} |v_k| \\ \text{if } j \ne i\text{:} \quad p_i \leftrightarrow p_j, v_i \leftrightarrow v_j, a_i \leftrightarrow a_j, l_i \leftrightarrow l_j, u_i \leftrightarrow u_j \\ u_{ii} = v_i \\ \left.\begin{array}{l} u_{ik} = a_{ik} - \displaystyle\sum_{j=1}^{i-1} l_{ij} u_{jk} \\ l_{ki} = \dfrac{v_k}{v_i} \end{array}\right\} k = i+1, \dots, n \end{array}\right\} i = 1, \dots, n. \tag{10.5}$$

Here, the double arrow ($\leftrightarrow$) indicates that the values of the specified elements or rows are swapped.

Once the LU-decomposition is available, the approximate inverse $R$ is computed column by column using simple forward/backward substitution. Let $e^{(k)}$ denote the $k$-th unit vector. We successively solve the systems $Ly = Pe^{(k)}$ and $Ux = y$ by

$$y_i = e^{(k)}_{p_i} - \sum_{j=1}^{i-1} l_{ij} y_j, \quad i = 1, \ldots, n, \text{ and} \tag{10.6}$$

$$x_i = \frac{1}{u_{ii}} \left( y_i - \sum_{j=i+1}^{n} u_{ij} x_j \right), \quad i = n, n-1, \ldots, 1. \tag{10.7}$$

The vector $x$ is an approximation for the $k$-th column of $R$.

We define the condition number of a regular square matrix $A$ by

$$\text{cond}(A) := \|A\|_\infty \cdot \|A^{-1}\|_\infty.$$

For solving a linear system of equations with conventional floating-point arithmetic, it is a rule of thumb that $\log(\text{cond}(A))$ is a rough indicator for the number of decimals of lost accuracy in the solution. For instance, a condition number of $10^5$ signals a loss of about five decimals of accuracy in the solution. Some of the algorithms described below return the quantity $\|A\|_\infty \cdot \|R\|_\infty$ as an estimate for the condition number of the matrix $A$.

## 10.2 Algorithmic Description

First, we give the algorithm for computing an approximate inverse. Except for the special case of $2 \times 2$ matrices, which is handled by applying well-known explicit rules, Crout's method is used as described in Section 10.1.3. To save memory, the matrices $L$ and $U$ are not explicitly allocated. Actually, their elements are stored by overwriting the input matrix $A$. Thus, any appearance of components of $L$ and $U$ in Equations (10.5)–(10.7) is replaced by the corresponding elements of $A$. The inversion procedure is stopped if a pivotal element is less than *Tiny*. Otherwise, we would risk an overflow exception at runtime when executing the division by the pivotal element. The value of *Tiny* depends on the exponent range of the floating-point format being used. For the IEEE binary floating-point format (see Section 3.7) with a minimum positive element of about $10^{-324}$, a value of $10^{-200}$ is a good choice. The forward substitution stage of the algorithm takes advantage of the leading zero elements in the unit vector $e^{(k)}$.

**Algorithm 10.1:** MatInv ($A, R, Err$)

1. {Initialization}
   $Tiny := 10^{-200}$;  $p := (1, 2, \ldots, n)^T$;  $Err :=$ "No error";
2. {Special case: $n = 2$}
   **if** ($n = 2$) **then**
   $Det := \Box(a_{11} \cdot a_{22} - a_{21} \cdot a_{12})$;

**if** $(|Det| < Tiny)$ **then**
  **return** $Err$ := "Matrix is probably singular!";
$r_{11} := \ a_{22}/Det; \quad r_{12} := -a_{12}/Det;$
$r_{21} := -a_{21}/Det; \quad r_{22} := \ a_{11}/Det;$

3. {Usual case: $n \neq 2$}
**else**
  {LU-Factorization, see (10.5)}
  **for** $i = 1$ to $n$ **do**

$$v_k := \square\left(a_{ki} - \sum_{j=1}^{i-1} a_{kj} \cdot a_{ji}\right); \quad (k = i, \ldots, n)$$

$|v_j| := \max_{i \leq k \leq n} |v_k|;$
  **if** $(|v_j| < Tiny)$ **then**
    **return** $Err$ := "Matrix is probably singular!";
  **if** $(j \neq i)$ **then**
    swap $(p_i, p_j)$; swap $(v_i, v_j)$; swap $(a_i, a_j)$;
  $a_{ii} := v_i;$
  **for** $k = i + 1$ to $n$ **do**

$$a_{ik} := \square\left(a_{ik} - \sum_{j=1}^{i-1} a_{ij} \cdot a_{jk}\right);$$

$a_{ki} := v_k / v_i;$

4. {Computation of the inverse column by column}
**for** $k = 1$ to $n$ **do**
  (a) {Forward substitution, see (10.6)}
  Find index $l$ with $p_l = k$;
  $y_i := 0; \quad (i = 1, \ldots, l-1)$
  $y_l := 1;$

$$y_i := -\square\left(\sum_{j=1}^{i-1} a_{ij} \cdot y_j\right); \quad (i = l+1, \ldots, n);$$

  (b) {Backward substitution, see (10.7)}

$$x_i := \square\left(y_i - \sum_{j=i+1}^{n} a_{ij} \cdot x_j\right) / a_{ii}; \quad (i = n, n-1, \ldots, 1)$$

  $r_{*,k} := x; \quad$ {$k$-th column of $R$}

5. {Return: Approximate inverse $R$ and error code $Err$}
**return** $R$, $Err$;

The procedure for solving the linear system of equations $Ax = b$ has two principal steps: We compute an approximate solution $\tilde{x}$, and then we try to find an enclosure for the error of this approximation. That is, the method is an *a posteriori* method as described in Section 3.8. To get a good approximation, we use (10.2) for a simple real residual iteration (see Step 2 of Algorithm 10.5). Since this pre-iteration will be followed by a verification step, we may apply some heuristic considerations to improve the value it computes. We try to predict whether some of the components of the exact solution might vanish. Let $x$ and $y$ be two successive iterates. Our heuristic is that any component of $y$ which has diminished by more than $n$ orders of magnitude is a good candidate for a zero entry in the exact solution. In Algorithm 10.2, those components of the new iterate $y$ are changed to zero. It turns out that $n = 5$ is a good value for practical use.

**Algorithm 10.2:** CheckForZeros $(x, y)$

1. {Initialization}
   *Factor* := $10^5$;
2. {Updating zero-suspicious entries of $y$, i.e. those entries with $|y_i/x_i| \le 10^{-5}$}
   **if** ($|y_i| \cdot \mathit{Factor} < |x_i|$) **then** $y_i := 0$; $(i = 1, \ldots, n)$
3. {Return the updated $y$}
   **return** $y$;

In general, the real iteration will be stopped if the relative error of any component of two successive iterates is less than $\delta$. Otherwise, we halt the iteration after $k_{\max}$ steps. Here, we use $\delta = 10^{-12}$ because the IEEE binary floating-point format corresponds to about 16 significant decimal digits (see Section 3.7). That is, we try to compute about 12 correct digits of the mantissa of the approximate solution. Algorithm 10.3 describes the stopping criterion for the iteration. Here, we make use of another heuristic. If the components of two successive iterates differ in sign, or if one of them vanishes, we take this as an indicator for a zero entry in the exact solution. Thus, we implicitly set its relative error to zero.

**Algorithm 10.3:** Accurate $(x, y)$

1. {Initialization: Desired relative accuracy}
   $\delta := 10^{-12}$;
2. {Check if the relative error of $y$ with respect to $x$ is less than or equal to $\delta$}
   $i := 1$;
   **repeat**
     **if** ($x_i \cdot y_i \le 0$) **then** $ok$ := *True* {Relative error implicitly set to 0}
     **else**
       $ok := (|y_i - x_i| \le \delta \cdot |y_i|)$;
     i := i+1;
   **until** (**not** $ok$) **or** ($i > n$);
3. {Return value of the flag $ok$}
   **return** Accurate := $ok$;

For the verification step, refer to Section 10.1.2. To accelerate the procedure, the iterates are inflated at the beginning of each iteration loop. For a general remark about the $\varepsilon$-inflation, refer to Section 3.8. As recommended in [52], we use a constant value of $\varepsilon = 1000$ for the $\varepsilon$-inflation. Refer to Falcó Korn [17] for a more sophisticated inflation strategy. It is an empirical fact that the inner inclusion is satisfied nearly always after a few steps or never. Thus in Algorithm 10.4, the iteration is stopped after $p_{\max}$ steps.

**Algorithm 10.4:** VerificationStep ($[x], [z], [C], \mathit{IsVerified}$)

1. {Initialization}
   $\varepsilon := 1000;\quad p_{\max} := 3;\quad p := 0;\quad [x]^{(0)} := [z];$
2. **repeat**
   $[x]^{(p)} := [x]^{(p)} \bowtie \varepsilon;\qquad \{\varepsilon\text{-Inflation}\}$
   $[x]^{(p+1)} := \Diamond([z] + [C] \cdot [x]^{(p)});$
   $\mathit{IsVerified} := ([x]^{(p+1)} \overset{\circ}{\subset} [x]^{(p)});$
   $p := p + 1;$
   **until** *IsVerified* **or** $(p \geq p_{\max})$;
   $[x] := [x]^{(p)};$
3. {Return enclosure $[x]$ and flag *IsVerified*}
   **return** $[x]$, *IsVerified*;

We now give the complete algorithm based on the Algorithms 10.2–10.4 for computing a verified enclosure of the solution of a linear system of equations. The procedure fails if the computation of an approximate inverse $R$ fails or if the inclusion in the interior cannot be established. A condition number estimate is returned by the variable *Cond*. We need a narrow enclosure $[z]$ of $z = R(b - A\tilde{x})$ to start the verification step. Thus in Step 3, we first compute the residual $d = \Box(b - A\tilde{x})$. If the computation was afflicted with roundoff errors, an enclosure of these errors is given by $[d] = \Diamond(b - A\tilde{x} - d)$. So, we have $b - A\tilde{x} \in d + [d]$ and $z \in \Diamond(Rd + R[d])$. In Step 4, we remark that $\tilde{x}$ is an exact, but not necessarily unique, solution of the system if $[z] = 0$.

**Algorithm 10.5:** LinSolve ($A, b, [x], \mathit{Cond}, \mathit{Err}$)

1. {Computation of an approximate solution}
   MatInv ($A, R, \mathit{Err}$);
   **if** ($\mathit{Err} \neq$ "No error") **then**
   **return** $\mathit{Err} :=$ "Matrix is probably singular!"
   $\mathit{Cond} := \|A\|_\infty \cdot \|R\|_\infty;$
2. {Real residual iteration for an approximate solution}
   $k_{\max} := 10;\quad k := 0;\quad \tilde{x}^{(0)} := R \cdot b;$
   **repeat**
   $d := \Box(b - A \cdot \tilde{x}^{(k)});$
   $\tilde{x}^{(k+1)} := \Box(\tilde{x}^{(k)} + R \cdot d);$
   CheckForZeros ($\tilde{x}^{(k)}, \tilde{x}^{(k+1)}$);
   $\mathit{Success} :=$ Accurate ($\tilde{x}^{(k)}, \tilde{x}^{(k+1)}$);
   $k := k + 1;$

**until** *Success* **or** $(k \geq k_{\max})$;
$\tilde{x} := \tilde{x}^{(k)}$;

3. {Computation of enclosures $[C]$ and $[z]$ for $C = I - RA$ and $z = R(b - A\tilde{x})$}
$[C] := \Diamond(I - R \cdot A)$;
$d := \Box(b - A \cdot \tilde{x})$;
$[d] := \Diamond(b - A \cdot \tilde{x} - d)$;
$[z] := \Diamond(R \cdot d + R \cdot [d])$;
4. {Verification step}
**if** $([z] = 0)$ **then**
$[x] := \tilde{x}$; {Exact solution}
**else**
VerificationStep ($[x], [z], [C]$, *IsVerified*);
**if** (**not** *IsVerified*) **then**
*Err* := "Verification failed, the system is probably ill-conditioned!";
**return** *Cond*, *Err*;
**else**
$[x] := \tilde{x} + [x]$; {Approximation plus correction interval}
5. {Return: Enclosure, condition number estimate, and error code}
**return** $[x]$, *Cond*, *Err*;

# 10.3 Implementation and Examples

## 10.3.1 C++ Program Code

We give the listings of the modules *matinv* for matrix inversion and *linsys* for solving a linear system of equations. We emphasize that those parts of the algorithms of the previous section which are marked by $\Box(\ldots)$ and $\Diamond(\ldots)$ are implemented using accurate evaluation of dot product expressions (see Section 2.3).

### 10.3.1.1 Module matinv

The header file of module *matinv* provides the interface of the function *MatInv()*, for computing an approximate inverse of a real matrix and the interface of the function *MatInvErrMsg()* which can be used to get an error message for the error code returned by *MatInv()*.

```
//----------------------------------------------------------------------------
// File: matinv (header)
// Purpose: Computation of an approximate inverse of a real square matrix.
// Global functions:
//   MatInv()      : matrix inversion
//   MatInvErrMsg(): to get an error message text
//----------------------------------------------------------------------------
#ifndef __MATINV_HPP
#define __MATINV_HPP
```

```
#include <rmatrix.hpp>      // Real matrix/vector arithmetic

extern char* MatInvErrMsg ( int );
extern void  MatInv ( rmatrix, rmatrix&, int& );
#endif
```

The function *MatInv()* is an implementation of Algorithm 10.1. It takes into account that the index range of the input matrix $A$ may start with lower index bounds different from one. The output matrix $R$ assumes its index range from the input matrix $A$. Thus, input and output matrix will have the same shape after calling *MatInv()*. The function checks whether the system is square and if it is supposed to be singular.

```
//----------------------------------------------------------------------------
// File: matinv (implementation)
// Purpose: Computation of an approximate inverse of a real square matrix.
// Method: LU decomposition applying Crout's algorithm.
// Global functions:
//   MatInv()      : matrix inversion
//   MatInvErrMsg(): to get an error message text
//----------------------------------------------------------------------------
// In the comments below the notations #*(...) and ##(...) are used to indi-
// cate the evaluation of the expression specified in round brackets using
// the exact scalar product. I.e. each component of the result is computed
// with only one final rounding. The symbol #* holds for rounding to nearest
// whereas ## holds for rounding to the smallest enclosing interval. An exact
// scalar product may be implemented using dotprecision accumulators.
//----------------------------------------------------------------------------
#include <string.h>       // String handling
#include <matinv.hpp>

//----------------------------------------------------------------------------
// Error codes used in this module.
//----------------------------------------------------------------------------
const int
  NoError      = 0,      // No error occurred
  NotSquare    = 1,      // Matrix to be inverted is not square
  Singular     = 2;      // Matrix to be inverted is probably singular

//----------------------------------------------------------------------------
// Error messages depending on the error code.
//----------------------------------------------------------------------------
char* MatInvErrMsg ( int Err )
{
  static char Msg[80] = "";

  if (Err != NoError) {
    char Hlp[60];

    switch (Err) {
      case NotSquare:
        sprintf(Hlp,"Matrix to be inverted is not square");           break;
      case Singular:
        sprintf(Hlp,"Inversion failed, matrix is probably singular"); break;
      default:
        strcpy(Hlp,"Code not defined");
    }
    sprintf(Msg,"Error: %s!",Hlp);
  }
  return(Msg);
} // MatInvErrMsg
```

```
//----------------------------------------------------------------------------
// Purpose: The function 'MatInv()' may be used for the computation of an
//    approximate inverse of a real square matrix.
// Parameters:
//    In : 'A'  : matrix to be inverted
//    Out: 'R'  : approximate inverse
//         'Err': error code
// Description:
//    Inversion of a regular matrix A stored in 'A' using LU decomposition.
//    For LU decomposition, formally a permutation matrix P is determined so
//    that the product P*A may be decomposed into a lower-triangular matrix L
//    and an upper-triangular matrix U with P*A = L*U. The diagonal elements
//    of L are set to 1. Using Crout's algorithm, the elements of both matri-
//    ces L and U are stored by temporary overwriting a copy of the input
//    matrix 'A'. The permutation matrix P is not explicitly generated. The
//    indices of row interchanges are stored in the index vector 'p' instead.
//    The i-th element of P*b may be accessed indirectly using the p[i]-th
//    entry of 'b'. Accurate expressions are used to avoid cancellation
//    errors. The k-th column of the inverse R of P*A is computed by
//    forward/backward substitution with the k-th unit vector e_k as the
//    right-hand side of the system: U*y = P*e_k ==> y, L*x = y ==> x. For
//    error codes, see above.
//----------------------------------------------------------------------------
void MatInv ( rmatrix A, rmatrix& R, int& Err )
{
  const real
    Tiny = 1E-200;          // A divisor less than 'Tiny' is handled like zero

  dotprecision Accu;                          // Long accumulator
  real     Max, Temp;                         // Help variables
  int      p1 = Lb(A,1), q1 = Lb(A,2);        // Lower bounds of 'A'.
  int      pn = Ub(A,1), qm = Ub(A,2);        // Upper bounds of 'A'.
  int      n = pn-p1+1;                       // Length of the rows of 'A'
  int      m = qm-q1+1;                       // Length of the columns of 'A'
  int      i, j, k, l, kk;                    // For loops

  Err = NoError;                              // Initialization

  if (n != m)                                 // Error: 'A' is not square
    { Err = NotSquare; return; }

  // Normalize index range of 'A' to standard range (1..n,1..n)
  //-----------------------------------------------------------
  SetLb(A,ROW,1);  SetLb(A,COL,1);
  Resize(R,p1,pn,q1,qm);                      // Resize 'R' to same shape as 'A'

  if (n == 2) {                               // Special case: (2,2)-matrix
    Accu = 0.0;                               // Compute the determinant of 'A'
    accumulate(Accu, A[1][1],A[2][2]);
    accumulate(Accu,-A[2][1],A[1][2]);
    Temp = rnd(Accu);

    if (abs(Temp) < Tiny)                     // Error: 'A' is probably singular
      { Err = Singular; return; }

    R[p1][q1] =  A[2][2] / Temp;  R[p1][qm] = -A[1][2] / Temp;
    R[pn][q1] = -A[2][1] / Temp;  R[pn][qm] =  A[1][1] / Temp;
    return;
  }

  // Start usual case: Dimension of 'A' > 2
  //---------------------------------------
  rvector v(n), x(n);           // Help vectors
```

```
int*    p = new int [n+1];   // Dynamic array of type integer
                             // Note: p[0] not used !

for (i = 1; i <= n; i++)  p[i] = i; // Initializing permutation vector

// Start LU factorization
//-----------------------
i = 1;
while ( (Err == NoError) && (i <= n) ) {
  // Compute the numerators of those elements of the i-th column
  // of L which are not updated so far and store them in 'v'.
  //-------------------------------------------------------------
  for (k = i; k <= n; k++) {   // v_k = #*(A_ki - sum(j=1(1)i-1,A_kj*A_ji))
    Accu = A[k][i];            //-----------------------------------------
    for (j = 1; j < i; j++) accumulate(Accu,-A[k][j],A[j][i]);
    v[k] = rnd(Accu);
  }

  // Look for the column pivot
  //--------------------------
  j = i;  Max = abs(v[i]);
  for (k = i+1; k <= n; k++)
    if ( (Temp = abs(v[k])) > Max ) { j = k; Max = Temp; }

  // Swap rows of 'A' and 'v', store the permutation in 'p'
  //-------------------------------------------------------
  if (j != i) {
    x = A[i];      A[i] = A[j];  A[j] = x;
    k = p[i];      p[i] = p[j];  p[j] = k;
    Temp = v[i];   v[i] = v[j];  v[j] = Temp;
  }

  if (Max < Tiny)                    // Pivot element < 'Tiny', inversion
    { Err = Singular; return; }      // failed, matrix 'A' assumed to be
                                     // singular
  Temp = v[i];
  A[i][i] = Temp;    // Update the diagonal element of U

  // Update U's row and L's column elements
  //---------------------------------------
  for (k = i+1; k <= n; k++) { // A_ik = #*(A_ik - sum(j=1(1)i-1,A_ij*A_jk))
    Accu = A[i][k];            //-----------------------------------------
    for (j = 1; j < i; j++) accumulate(Accu,-A[i][j],A[j][k]);
    A[i][k] = rnd(Accu);
    A[k][i] = v[k] / Temp;
  }
  i++;
} // while

// Now 'A' is overwritten with the subdiagonal elements of L in its
// lower left triangle and with the elements of U in its diagonal and
// its upper right triangle. The elements of the inverse matrix are
// computed column by column using forward/backward substitution.
//-------------------------------------------------------------------
for (k = 1; k <= n; k++) {
  // Forward substitution: L*x = P*e_k, where e_k is the k-th unit
  // vector. Note: If P*e_k has m leading zeros, this results in
  // x_i = 0 for 1,..,l-1 and x_l = 1. Thus, forward substitution
  // starts at index l+1.
  //-------------------------------------------------------------
  l = 1;
  while (p[l] != k) { x[l] = 0.0; l++; }
  x[l] = 1.0;
  for (i = l+1; i <= n; i++) {    // x_i = - #*(sum(j=1(1)i-1,A_ij*x_j))
```

```
      Accu = 0.0;                    //-----------------------------------
      for (j = 1; j < i; j++) accumulate(Accu,A[i][j],x[j]);
      x[i] = -rnd(Accu);
    }

    // Backward substitution: U * x = x, where the right-hand side is
    // the result of the forward substitution. It will be overwritten
    // by the solution of the system, the k-th column of the inverse
    // matrix.
    //-------------------------------------------------------------------
    kk = q1+k-1;                     // Index offset for the k-th column of R
    for (i = n; i >= 1; i--) {       // x_i = #*(x_i-sum(j=i+1(1)n,A_ij*x_j)/A_ii)
      Accu = x[i];                   //-----------------------------------------
      for (j = i+1; j <= n; j++) accumulate(Accu,-A[i][j],x[j]);
      x[i] = rnd(Accu) / A[i][i];
      R[p1+i-1][kk] = x[i];          // Remember index offset !
    }
  } // for (k = 1,...

  delete [] p;    // Free dynamically allocated memory
} // MatInv
```

#### 10.3.1.2 Module linsys

The header file of module *linsys* provides the interfaces of two versions of function *LinSolve()*, the second one with an additional parameter of type *real*. The function *LinSolveErrMsg()* can be used to get an explicit error message for the integer error code returned by any of these two functions.

```
//----------------------------------------------------------------------------
// File: linsys (header)
// Purpose: Computation of a verified solution of a square linear system of
//    equations A*x = b with full real matrix A and real right-hand side b.
// Global functions:
//    LinSolve()       : to get a verified enclosure of the solution (two
//                       versions)
//    LinSolveErrMsg(): to get an error message text
//----------------------------------------------------------------------------
#ifndef __LINSYS_HPP
#define __LINSYS_HPP

#include <rmatrix.hpp>     // Real matrix/vector arithmetic
#include <ivector.hpp>     // Interval vector arithmetic

extern char* LinSolveErrMsg ( int );
extern void  LinSolve ( rmatrix&, rvector&, ivector&, int& );
extern void  LinSolve ( rmatrix&, rvector&, ivector&, real&, int& );
#endif
```

The local function *LinSolveMain()* is an implementation of Algorithm 10.5. It takes into account that the index range of its parameters may start with a lower bound different from one. The function checks if the system is square. The local functions *CheckForZeros()*, *Accurate()*, and *VerificationStep()* are implementations of the Algorithms 10.2–10.4. The module provides two global functions *LinSolve()* one with and the other without computing a condition number estimate for the input matrix. Both functions call *LinSolveMain()* with an appropriate list of parameters.

```
//----------------------------------------------------------------------------
// File: linsys (implementation)
// Purpose: Computation of a verified solution of a square linear system of
//    equations A*x = b with full real matrix A and real right-hand side b.
// Method: Transformation of A*x = b to fixed-point form and applying an
//    interval residual iteration.
// Global functions:
//    LinSolve()      : to get a verified enclosure of the solution (two
//                      versions)
//    LinSolveErrMsg(): to get an error message text
//----------------------------------------------------------------------------
// In the comments below the notations #*(...) and ##(...) are used to indi-
// cate the evaluation of the expression specified in round brackets using
// the exact scalar product. I.e. each component of the result is computed
// with only one final rounding. The symbol #* holds for rounding to nearest
// whereas ## holds for rounding to the smallest enclosing interval. An exact
// scalar product may be implemented using dotprecision accumulators.
//----------------------------------------------------------------------------
#include <string.h>         // String handling
#include <imatrix.hpp>      // Interval matrix/vector arithmetic
#include <mvi_util.hpp>     // Interval matrix/vector utility functions
#include <matinv.hpp>       // Matrix inversion
#include <linsys.hpp>

//----------------------------------------------------------------------------
// Error codes used in this module.
//----------------------------------------------------------------------------
const int
  NoError      = 0,   // No error occurred
  NotSquare    = 1,   // System to be solved is not square
  DimensionErr = 2,   // Dimensions of A and b are not compatible
  InvFailed    = 3,   // System is probably singular
  VerivFailed  = 4;   // Verification failed, system is probably
                      // ill-conditioned

//----------------------------------------------------------------------------
// Error messages depending on the error code.
//----------------------------------------------------------------------------
char* LinSolveErrMsg ( int Err )
{
  static char Msg[80] = "";

  if (Err != NoError) {
    char Hlp[60];

    switch (Err) {
      case NotSquare:
        strcpy(Hlp,"System to be solved is not square"); break;
      case DimensionErr:
        strcpy(Hlp,"Dimensions of A and b are not compatible"); break;
      case InvFailed:
        strcpy(Hlp,"System is probably singular"); break;
      case VerivFailed:
        strcpy(Hlp,"Verification failed, system is probably ill-conditioned");
        break;
      default:
        strcpy(Hlp,"Code not defined");
    }
    sprintf(Msg,"Error: %s!",Hlp);
  }
  return(Msg);
} // LinSolveErrMsg

//----------------------------------------------------------------------------
```

```
// Computes an upper bound for the maximum norm of a real matrix 'M'.
//----------------------------------------------------------------------------
static real MaxNorm ( rmatrix& M )
{
  int          i, j;
  real         Max, Tmp;
  dotprecision Accu;

  Max = 0.0;
  for (i = Lb(M,1); i <= Ub(M,1); i++) {
    // Compute Tmp = #>( sum(j=1(1)n,|M_ij|) )
    //--------------------------------------
    Accu = 0.0;
    for (j = Lb(M,2); j <= Ub(M,2); j++)
      Accu += abs(M[i][j]);
    Tmp = rnd(Accu,RND_UP);
    if (Tmp > Max) Max = Tmp;
  }
  return Max;
} // MaxNorm

//----------------------------------------------------------------------------
// The vectors x and y are successive approximations for the solution of a
// linear system of equations computed by iterative refinement. If a compo-
// nent of y is diminished by more than 'Factor', it is a good candidate for
// a zero entry. Thus, it is set to zero.
//----------------------------------------------------------------------------
static void CheckForZeros ( rvector& x, rvector& y )
{
  const real Factor = 1E+5;
  int        i;

  for (i = Lb(y); i <= Ub(y); i++)
    if ( abs(y[i])*Factor < abs(x[i]) ) y[i] = 0.0;
}

//----------------------------------------------------------------------------
// The vectors x and y are successive iterates. The function returns true if
// the relative error of all components x_i and y_i is <= 10↑(-12), i.e. y_i
// has about 12 correct decimals. If x_i or y_i vanishes, the relative error
// is implicitly set to zero.
//----------------------------------------------------------------------------
static int Accurate ( rvector& x, rvector& y )
{
  const real Delta = 1E-12;   // Relative error bound
  int        i, n, ok;
  real       abs_yi;

  i = Lb(y); n = Ub(y);
  do {
    if (sign(x[i])*sign(y[i]) <= 0)       // Relative error set to 0
      ok = 1;
    else {
     abs_yi = abs(y[i]);                  // Relative error > Delta ?
     ok = (abs(y[i] - x[i]) <= Delta * abs_yi );
    }
    i++;
  } while (ok && (i <= n));
  return ok;
} // Accurate

//----------------------------------------------------------------------------
// This function 'VerificationStep()' performs the iteration
// [y] = Blow([x],Eps), [x] = [z] + [C]*[y] for k = 1,2,... until the new
```

```
// iterate [x] lies in the interior of [y] or until the maximum number of
// iterations is exceeded. The flag 'IsVerified' is set if an inclusion in
// the interior could be established.
//----------------------------------------------------------------------------
static void VerificationStep ( ivector& xx, ivector& zz,
                               imatrix& C,  int& IsVerified )
{
  const int  MaxIter = 3;          // Maximal number of iteration steps
  const real Epsilon = 1000.0;     // Factor for the epsilon inflation

  int     p;
  ivector yy(Lb(xx),Ub(xx));

  xx = zz; p = 0;                    // Initialize:  [x] = [z]
  do {
    yy = Blow(xx,Epsilon);           // Epsilon inflation
    xx = zz + C*yy;                  // New iterate: [x] = [z]+[C]*[y]
    IsVerified = in(xx,yy);          // Inclusion in the interior?
    p++;
  } while (!IsVerified && (p < MaxIter));
}

//----------------------------------------------------------------------------
// Purpose: The function 'LinSolveMain()' computes a verified solution of a
//    square linear system of equations A*x=b.
// Parameters:
//    In : 'A'          : matrix of the system, passed as reference
//                        parameter to save time for copying it
//         'b'          : right-hand side of the system, passed as
//                        reference parameter to save time for copying it
//         'ComputeCond': flag signalling whether a condition number
//                        estimate is to be computed
//    Out: 'xx'         : enclosure of the unique solution, resized to
//                        standard index range with lower index bound 1
//         'Cond'       : condition number estimate
//         'Err'        : error code
// Description: An approximate inverse 'R' of 'A' is computed by calling
//   function 'MatInv()'. Then an approximate solution 'x' is computed
//   applying a conventional real residual iteration. For the final verifica-
//   tion, an interval residual iteration is performed. An enclosure of the
//   unique solution is returned in the interval vector 'xx'. The function
//   also returns a condition number estimate 'Cond' if the flag 'ComputeCond'
//   is set. 'Cond' is initialized by -1. A negative value for 'Cond' signals
//   that an estimate could not be computed.
//----------------------------------------------------------------------------
static void LinSolveMain ( rmatrix&  A,
                           rvector&  b,
                           ivector&  xx,
                           real&     Cond,
                           int       ComputeCond,
                           int&      Err )
{
  const int
    MaxResCorr = 10;     // Maximal number of real residual corrections

  int     n = ColLen(A);                // Length of the columns of 'A'
  int     m = RowLen(A);                // Length of the rows of 'A'
  int     IsVerified, i, j, k;
  int     bLow, ARowLow, AColLow;       // Lower index bounds od 'A' and 'b'
  rmatrix R;                            // To store the inverse of 'A'

  Cond = -1.0;                          // Initial value for the condition

  if (n != m)                           // Error: 'A' is not square
```

```
    { Err = NotSquare; return; }

  if (n != VecLen(b))                    // Error: Dimensions of 'A' and 'b'
    { Err = DimensionErr; return; }      //        are not compatible

  MatInv(A,R,Err);
  if (Err != NoError)                    // Error: Inversion failed
    { Err = InvFailed; return; }

  // Start algorithm
  //----------------
  rvector       x(n), y(n), d(n);        // Allocate dynamic arrays
  ivector       yy(n), zz(n);
  imatrix       C(n,n);
  dotprecision  Accu;                    // Allocate accumulators
  idotprecision IAccu;

  if (ComputeCond)                       // Compute condition number
    Cond = mulu(MaxNorm(A),MaxNorm(R)); //------------------------

  // Normalize index range of 'A', 'R' and 'b' to standard range 1..n
  // and resize 'xx' to length 'n'. Save lower bounds of 'b' and 'A'
  // to restore them before leaving 'LinSolveMain()'.
  //-----------------------------------------------------------------
  Resize(xx,n);
  bLow = Lb(b); SetLb(b,1);
  ARowLow = Lb(A,ROW); SetLb(A,ROW,1);
  AColLow = Lb(A,COL); SetLb(A,COL,1);
  SetLb(R,ROW,1); SetLb(R,COL,1);

  x = R*b; k = 0;    // Real residual iteration
  do {               //------------------------
    y = x;
    for (i = 1; i <= n; i++) {   // Compute: d = #*(b-A*y)
      Accu = b[i];               //-----------------------
      accumulate(Accu,-A[i],y);
      d[i] = rnd(Accu);
    }
    for (i = 1; i <= n; i++) {   // Compute: x = #*(y+R*d)
      Accu = y[i];               //-----------------------
      accumulate(Accu,R[i],d);
      x[i] = rnd(Accu);
    }
    CheckForZeros(y,x);
    k++;
  } while (!Accurate(y,x) && (k < MaxResCorr));

  // Prepare verification step, i.e. compute enclosures [C]
  // and [z] of C = (I - R*A) and z = R*(b - A*x).
  //-------------------------------------------------------
  for (i = 1; i <= n; i++)
    for (j = 1; j <= n; j++) {          // Compute [C] = ##(I-R*A)
      Accu = (i == j) ? 1.0 : 0.0;      //------------------------
      accumulate(Accu,-R[i],A[Col(j)]);
      rnd(Accu,C[i][j]);
    }
  for (i = 1; i <= n; i++) {            // Compute: d = #*(b-A*x)
    Accu = b[i];                        //-----------------------
    accumulate(Accu,-A[i],x);
    d[i] = rnd(Accu);
  }
  for (i = 1; i <= n; i++) {            // Compute: [y] = ##(b-A*x-d)
    Accu = b[i];                        //---------------------------
    accumulate(Accu,-A[i],x);
```

```
    Accu -= d[i];
    rnd(Accu,yy[i]);
  }

  // Now b-A*x lies in d+[y]. Thus R*(b-A*x) lies in [z] = ##(R*d+R*[y])
  //----------------------------------------------------------------------
  for (i = 1; i <= n; i++) {
    IAccu = 0.0;
    accumulate(IAccu,R[i],d);
    accumulate(IAccu,R[i],yy);
    zz[i] = rnd(IAccu);
  }

  // If R*(b-A*x) = 0 then x is an exact solution else try to find a
  // verified enclosure [x] for the absolute error of x.
  //------------------------------------------------------------------
  if (Zero(zz))
    xx = x;
  else {
    VerificationStep(xx,zz,C,IsVerified);   // Attempt to compute [x]
    if ( !IsVerified )
      Err = VerivFailed;
    else
      xx = x + xx;                          // The exact solution lies x+[x]
  }

  // Restore lower index bounds of 'A' and 'b'
  //-----------------------------------------
  SetLb(b,bLow);
  SetLb(A,ROW,ARowLow); SetLb(A,COL,AColLow);
} //LinSolveMain

//----------------------------------------------------------------------------
// Purpose: The function 'LinSolve()' computes a verified solution of a
//    square linear system of equations A*x=b without returning a condition
//    number estimate.
// Parameters:
//    In : 'A'   : matrix of the system, passed as reference
//                 parameter to save time for copying it
//         'b'   : right-hand side of the system, passed as
//                 reference parameter to save time for copying it
//    Out: 'xx'  : enclosure of the unique solution
//         'Err': error code
// Description: Calls 'LinSolveMain()' for solving the linear system with the
//    flag 'ComputeCond' not set.
//----------------------------------------------------------------------------
void LinSolve ( rmatrix& A, rvector& b, ivector& xx, int& Err )
{
  real DummyCond;          // Dummy parameter for call of 'LinSolveMain()'

  LinSolveMain(A,b,xx,DummyCond,0,Err);
}

//----------------------------------------------------------------------------
// Purpose: The function 'LinSolve()' computes a verified solution of a
//    square linear system of equations A*x=b and returns a condition
//    number estimate.
// Parameters:
//    In : 'A'    : matrix of the system, passed as reference
//                  parameter to save time for copying it
//         'b'    : right-hand side of the system, passed as
//                  reference parameter to save time for copying it
//    Out: 'xx'   : enclosure of the unique solution
//         'Cond': condition number estimate
```

```
//          'Err' : error code
// Description: Calls 'LinSolveMain()' for solving the linear system with the
//    flag 'ComputeCond' set.
//----------------------------------------------------------------------------
void LinSolve ( rmatrix& A, rvector& b, ivector& xx, real& Cond, int& Err )
{
  LinSolveMain(A,b,xx,Cond,1,Err);
}
```

### 10.3.2 Example

In our sample program, we demonstrate how to use the modules defined in the previous section. In function *main()* the matrix of the system, its right-hand side, and the vector for the enclosure of the solution are allocated dynamically depending on the dimension $n$ of the system. After entering the system, the program tries to enclose a solution. If the algorithm succeeds, an enclosure of the solution and a naive floating-point approximation are printed. Otherwise, an error message is displayed. Finally, a condition number estimate for the input matrix is printed.

```
//----------------------------------------------------------------------------
// Example: Linear systems of equations
//----------------------------------------------------------------------------
#include <matinv.hpp>     // Matrix inversion
#include <linsys.hpp>     // Linear system solver

int main ( )
{
  int  Err, n;
  real Cond;

  cout << SetPrecision(23,15) << Scientific;    // Output format

  do {
    cout << "Enter the dimension of the system: ";
    cin  >> n;  cout << endl;
  } while (n <= 0);

  rmatrix A(n,n), R(n,n);       // Dynamic allocation
  rvector b(n);
  ivector x(n);

  cout << "Enter matrix A:" << endl;
  cin  >> A;  cout << endl;
  cout << "Enter vector b:" << endl;
  cin  >> b;  cout << endl;

  LinSolve(A,b,x,Cond,Err);

  if (!Err) {
    // Compare the result to naive floating-point approximation
    MatInv(A,R,Err);
    cout << "Naive floating-point approximation:" << endl
         << R*b << endl;
    cout << "Verified solution found in:" << endl
         << x << endl;
  }
  else
    cout << LinSolveErrMsg(Err) << endl;
```

```
  if (!Err || Err == 4)
    cout << "Condition estimate: " << SetPrecision(9,1)
         << Cond << endl;

  return 0;
}
```

As sample input data, we use a Boothroyd/Dekker system (see [97]) whose elements are integers defined by

$$a_{ij} = \binom{n+i-1}{i-1} \cdot \binom{n-1}{n-j} \cdot \frac{n}{i+j-1}, \quad i,j = 1,\ldots,n.$$

With the right-hand side $b = (1, 2, \ldots, n)$, the exact solution of the system is given by $x = (0, 1, \ldots, n-1)$. For $n = 10$, our sample program leads to the following output:

```
Enter the dimension of the system: 10

Enter matrix A:
   10     45     120     210     252     210     120      45     10     1
   55    330     990    1848    2310    1980    1155     440     99    10
  220   1485    4752    9240   11880   10395    6160    2376    540    55
  715   5148   17160   34320   45045   40040   24024    9360   2145   220
 2002  15015   51480  105105  140140  126126   76440   30030   6930   715
 5005  38610  135135  280280  378378  343980  210210   83160  19305  2002
11440  90090  320320  672672  917280  840840  517440  205920  48048  5005
24310 194480  700128 1485120 2042040 1884960 1166880  466752 109395 11440
48620 393822 1432080 3063060 4241160 3938220 2450448  984555 231660 24310
92378 755820 2771340 5969040 8314020 7759752 4849845 1956240 461890 48620

Enter vector b:
1 2 3 4 5 6 7 8 9 10

Naive floating-point approximation:
 3.769740075654227E-010
 9.999999949022040E-001
-1.999999975192608E+000
 2.999999904595484E+000
-3.999999691503490E+000
 4.999999132005541E+000
-5.999997815384631E+000
 6.999994975405571E+000
-7.999989266972989E+000
 8.999978443156579E+000

Verified solution found in:
[ 0.000000000000000E+000, 0.000000000000000E+000]
[ 1.000000000000000E+000, 1.000000000000000E+000]
[-2.000000000000000E+000,-2.000000000000000E+000]
[ 3.000000000000000E+000, 3.000000000000000E+000]
[-4.000000000000000E+000,-4.000000000000000E+000]
[ 5.000000000000000E+000, 5.000000000000000E+000]
[-6.000000000000000E+000,-6.000000000000000E+000]
[ 7.000000000000000E+000, 7.000000000000000E+000]
[-8.000000000000000E+000,-8.000000000000000E+000]
[ 9.000000000000000E+000, 9.000000000000000E+000]

Condition estimate:  1.1E+015
```

Note that the resulting enclosure of the solution is a thin interval vector. This is a consequence of our strategy for updating zero-suspicious entries in Algorithm 10.2. After the real residual iteration of Algorithm 10.5, the approximate solution is already exact! The verification step is used only to get a mathematical proof of this fact. In general, the enclosure of the solution is not a thin interval vector.

### 10.3.3 Restrictions and Improvements

The method described in this chapter demands exactly representable input data. Neither the matrix $A$ nor the right-hand side $b$ of the system may be disturbed by conversion errors during input. Actually, this is no restriction if $A$ and $b$ result from previous computations. To see why this is important, let us consider the $2 \times 2$ system with

$$A = \begin{pmatrix} 1 & 1 \\ \varepsilon & 0 \end{pmatrix}, \quad b = \begin{pmatrix} 1 \\ 1 \end{pmatrix}, \quad \varepsilon \neq 0.$$

The exact solution of the system is $x = (1/\varepsilon, 1 - 1/\varepsilon)^{\mathrm{T}}$. If we solve the system for $\varepsilon = 10^{-20}$ using the above program, we get the following result

```
Verified solution found in:
[  1.000000000000000E+020,  1.000000000000001E+020 ]
[ -1.000000000000001E+020, -1.000000000000000E+020 ]
```

which does not contain the exact solution $(10^{20}, 1 - 10^{20})^{\mathrm{T}}$. The reason for this incorrect result is that $\varepsilon = 10^{-20}$ is not exactly representable on a binary number screen (see Section 3.7). Thus after data conversion, we solved a perturbed system with a $\Box(\varepsilon)$ entry for the lower left element of $A$.

To get rid of the problem of input conversion, we might enclose all input data in small intervals. We would have to modify our algorithm so that it works with interval data. The modifications of Algorithm 10.5 to accept interval input as well as complex and complex interval data are subject to a subsequent volume of this book. This future volume is also planned to deal about the treatment of rectangular linear systems of equations.

## 10.4 Exercises

**Exercise 10.1** How large a Hilbert matrix can the function *LinSolve()* handle? Hilbert matrices are very ill-conditioned for large dimensions. The elements of a Hilbert matrix $H$ are defined by

$$h_{ij} = \frac{1}{i + j - 1} \qquad i, j = 1, \ldots, n.$$

Multiply $H$ by the common denominator of all its elements to get an exactly representable input matrix. Use an arbitrary vector as right-hand side of the system.

**Exercise 10.2** Algorithm 10.5 fails for very ill-conditioned systems. For instance, we are not able to solve a Boothroyd/Dekker system of dimension 15 using function *LinSolve()*. As mentioned earlier, the better $R$ approximates the inverse of $A$, the better the algorithm works. Therefore, it is a good idea to improve the behavior of our algorithm by refining the approximate inverse.

If $R$ is an already computed approximation of $A^{-1}$, we can use a little trick to improve its value. By computing

$$\begin{aligned} R_1 &= (RA)^{-1}R, \\ R_2 &= \Box((RA)^{-1}R - R_1) \end{aligned}$$

we get $A^{-1} = R_1 + R_2$. $R_1$ is a new approximation to $A^{-1}$, and $R_2$ is its refinement. This fact can be used to modify Algorithm 10.5. If the verification step has failed, it has to be restarted after computing an improved double-length approximate inverse. To accomplish this goal, we do not want Step 4 to return with an error message. An improved version of Algorithm 10.5 has the following form:

**Algorithm 10.6:** NewLinSolve ($A, b, [x], Err$)

1. – 3. see Algorithm 10.5
4. {Verification step}
   **if** ($[z] = 0$) **then**
     **return** $[x] := \tilde{x}$, *Cond*, *Err*; {Exact solution}
   **else**
     VerificationStep ($[x], [z], [C]$, *IsVerified*);
     **if** (*IsVerified*) **then**
       $[x] := \tilde{x} + [x]$; {Approximation plus correction interval}
       **return** $[x]$, *Cond*, *Err*;
5. {Computing a double-length approximate inverse stored in $R_1$ and $R_2$}
   MatInv ($R \cdot A, R_2, Err$);
   **if** ($Err \neq$ "No error") **then**
     **return** $Err :=$ "Verification failed, system is probably ill-conditioned!";
   $R_1 := R_2 \cdot R$;
   $R_2 := \Box(R_2 \cdot R - R_1)$;
6. {Computation of enclosures $[C]$ and $[z]$}
   $[C] := \Diamond(I - R_1 \cdot A - R_2 \cdot A)$;
   $d := \Box(b - A \cdot \tilde{x})$;
   $[d] := \Diamond(b - A \cdot \tilde{x} - d)$;
   $[z] := \Diamond(R_1 \cdot d + R_2 \cdot d + R_1 \cdot [d] + R_2 \cdot [d])$;
7. – 8. see 4. – 5. of Algorithm 10.5

Supplement the module *linsys* by a function *NewLinSolve()* according to the modifications described in Algorithm 10.6. Test the new function for a Boothroyd/Dekker system of dimension 15. The improved version should succeed in enclosing the solution.

**Exercise 10.3** Another possible extension of Algorithm 10.5 is to use the staggered type arithmetic from Chapter 8 to store and to compute $\widetilde{x}$ and $R$. Make these modifications and compare the accuracy, set of problems for which it works, and speed with those of the function *LinSolve()*.

## 10.5 References and Further Reading

The basic principles of a self-verifying algorithm for linear systems of equations are due to Rump [82]. A straight-forward extension of the algorithms to interval systems and complex systems is given in Rump [84]. A compact summary of the mathematical theory is found in Rump [85]. Comprehensive discussions on interval linear systems of equations including preconditioning strategies are given in the books of Hansen [31] and Neumaier [70]. There are also commercial implementations for the verified computation of linear systems of equations which come with the subroutine libraries ACRITH of IBM and ARITHMOS of SIEMENS (see [36], [86]).

# Chapter 11

# Linear Optimization

A linear programming problem consists of a linear function to be maximized (or minimized) subject to linear equality and inequality constraints. Any linear program (LP) can be put by well-known transformations into standard form

$$(\text{LP}) \qquad \left( \begin{array}{ccccc} z & = & c^{\mathrm{T}}x & = & \max! \\ & & Ax & = & b \\ & & x & \geq & 0 \end{array} \right) \tag{11.1}$$

$$\Leftrightarrow$$

$$\max\{c^{\mathrm{T}}x \mid x \in X\}, \quad X := \{x \in I\!R^n \mid Ax = b, \ x \geq 0\},$$

where $A$ is a real $m \times n$ matrix, $b \in I\!R^m$, $c \in I\!R^n$, and $m < n$. The input data of (11.1) are given by the triple $P = (A, b, c) \in I\!R^{m \cdot n + m + n}$.

Linear programming problems occur frequently in many economic, engineering, and scientific applications, so many software libraries include routines to estimate optimal solutions. Of course, such routines for computing approximate solutions are subject to rounding and cancellation errors which may result in "answers" that are not truly optimal or are not even feasible. We describe an algorithm for the self-validated solution of linear programming problems. The algorithm we present computes an enclosure of the true optimal solution set. The self-validated nature of our algorithm is especially important as the applications of linear programming become larger and more complex, and often deeply embedded as sub-problems in even more complex models.

## 11.1 Theoretical Background

### 11.1.1 Description of the Problem

The function $c^{\mathrm{T}}x$ is called the *objective function*, and $X$ is called the *set of feasible solutions* of (11.1). $X$ is a convex polyhedron which is the intersection of the affine variety $\{x \in I\!R^n \mid Ax = b\}$ with the positive orthant in $I\!R^n$.

A point $x \in I\!R^n$ is called *feasible* if $x \in X$. It is called *optimal* if, in addition, $c^{\mathrm{T}}x \geq c^{\mathrm{T}}x'$ for all $x' \in X$. If such a point exists, the value $z_{\text{opt}} := \max\{c^{\mathrm{T}}x \mid x \in X\}$ is called the *optimal value*. We seek an enclosure of the set of all optimal points.

A main theorem in linear programming asserts that if there exists an optimal solution, it has at most $m$ components different from zero. Furthermore, there is only a finite number of such solutions. The *simplex method* invented by Dantzig [14] is an algorithm to compute an approximation of such an optimal solution efficiently (see [92]).

Solutions of $Ax = b$ where $n - m$ components of $x$ are zero can be described by partitioning the rectangular matrix $A$ consisting of the columns $(a_{*,1}, \ldots, a_{*,n})$ into two submatrices $A_B$ and $A_N$, where $A_B = (a_{*,\beta_1}, \ldots, a_{*,\beta_m})$ is a nonsingular $m \times m$ submatrix with the column vectors $a_{*,\beta_i}$ of $A$, and $A_N = (a_{*,\nu_1}, \ldots, a_{*,\nu_{n-m}})$ is the $m \times (n-m)$ submatrix with the remaining column vectors of $A$. The set of indices $\mathcal{B} = \{\beta_1, \ldots, \beta_m\} \subseteq \{1, \ldots, n\}$ which determines the nonsingular quadratic matrix $A_B$ is called a *basic index set* of (11.1), and $\mathcal{N} := \{\nu_1, \ldots, \nu_{n-m}\} := \{1, \ldots, n\} \setminus \mathcal{B}$ is the set of *nonbasic indices*.

For a given partitioning $\mathcal{B}$ and $\mathcal{N}$, the linear equation $Ax = b$ can be expressed in the form

$$A_B \cdot x_B + A_N \cdot x_N = b \tag{11.2}$$

where $x = (x_B, x_N)^{\mathrm{T}}$ is partitioned analogously. Multiplying Equation (11.2) by the inverse $A_B^{-1}$ yields

$$x_B = A_B^{-1} \cdot b - H \cdot x_N \quad \text{with} \quad H := A_B^{-1} \cdot A_N.$$

Partitioning $c = (c_B, c_N)^{\mathrm{T}}$ similarly, the objective function can be expressed in the form

$$\begin{aligned} c^{\mathrm{T}} x = c_B^{\mathrm{T}} \cdot x_B + c_N^{\mathrm{T}} \cdot x_N &= c_B^{\mathrm{T}} \cdot (A_B^{-1} \cdot b - H \cdot x_N) + c_N^{\mathrm{T}} \cdot x_N \\ &= c_B^{\mathrm{T}} \cdot A_B^{-1} \cdot b - (c_B^{\mathrm{T}} \cdot H - c_N^{\mathrm{T}}) \cdot x_N. \end{aligned}$$

Therefore,

$$c^{\mathrm{T}} x = c_B^{\mathrm{T}} \cdot A_B^{-1} b - d^{\mathrm{T}} \cdot x_N \quad \text{with} \quad d := H^{\mathrm{T}} \cdot c_B - c_N. \tag{11.3}$$

Thus, both the $m$ variables $x_B$ and the value of the objective function are determined by the $n - m$ components of $x_N$.

A *basic solution* (corresponding to the basic index set $\mathcal{B}$) is a solution for which the $n - m$ components of $x_N$ are zero:

$$x := (x_B, x_N)^{\mathrm{T}} \quad \text{with} \quad x_B := A_B^{-1} \cdot b \quad \text{and} \quad x_N := 0.$$

In addition, if $x \geq 0$ (this is the case if $x_B := A_B^{-1} b \geq 0$), then $x \in X$ is called a *basic feasible solution*. The value of the objective function is $z = c^{\mathrm{T}} x = c_B^{\mathrm{T}} A_B^{-1} b$.

Geometrically, the basic feasible solutions correspond to the vertices of the convex polyhedron $X$. The vector $d$ in (11.3) is called the *vector of reduced costs* since it indicates how the objective function depends on $x_N$.

### 11.1.2 Verification

The "results" of interest for the linear programming problem (11.1) are:

1. The *optimal value* $z_{\mathrm{opt}}$,

2. the *optimal basic index set* $\mathcal{B}$, and
3. the *optimal basic feasible solution* $x$.

The simplex method efficiently determines an optimal value $z_{opt}$ and one optimal basic index set $\mathcal{B}$ with the corresponding optimal feasible solution $x$. Implementations of the simplex algorithm in software libraries are usually quite reliable, but the accumulation of rounding and cancellation errors may cause them to return a non-optimal or infeasible solution, or to fail to find all optimal solutions in case the algorithm is unable to recognize that more than one vertex of $X$ is optimal. Therefore, we compute enclosures for the *optimal value* $z_{\text{opt}}$, the *set of optimal basic index sets* $\mathcal{V}_{\text{opt}}$, and the *set of optimal basic feasible solutions* $\mathcal{X}_{\text{opt}}$ for a linear programming problem $P$. Algorithm 11.7 calculates an interval $[z]$, a *set of index sets* $\mathcal{V} = \{\mathcal{B}^{(1)}, \ldots, \mathcal{B}^{(s)}\}$, and a set of interval vectors $\mathcal{X} = \{[x]^{(1)}, \ldots, [x]^{(s)}\}$, $[x]^{(i)} \in I\mathbb{R}^n$ such that the following three conditions are satisfied:

1. $z_{opt} \in [z]$
2. $\mathcal{B} \in \mathcal{V}_{opt} \Rightarrow \mathcal{B} \in \mathcal{V}$
3. $x \in \mathcal{X}_{opt} \Rightarrow$ there exists a $\mathcal{B}^{(i)} \in \mathcal{V}$ with $x \in [x]^{(i)} \in \mathcal{X}$

The following *a posteriori* method either computes such enclosures $[z]$, $\mathcal{V}$, and $\mathcal{X}$ satisfying Conditions 1, 2, and 3 and guarantees that the real problem $P$ admits optimal solutions, or else the method gives a warning that no enclosures could be calculated. The latter is the case if the set of feasible solutions $X$ is empty or if the objective function is unbounded on $X$.

Let $\mathcal{B} \in \mathcal{V}$ be an index set $\{\beta_1, \ldots, \beta_m\} \subseteq \{1, \ldots, n\}$. Let $[x_B]$, $[y]$, $[h]_{*,\nu}$ with $\nu \in \mathcal{N} := \{\nu_1, \ldots, \nu_{n-m}\} := \{1, \ldots, n\} \setminus \mathcal{B}$ be vectors of inclusion for the solutions of the systems of linear equations

$$\begin{aligned} A_B \cdot x_B &= b \\ A_B^{\mathrm{T}} \cdot y &= c_B \\ A_B \cdot h_{*,\nu} &= a_{*,\nu}. \end{aligned} \tag{11.4}$$

These vector enclosures are calculated by the self-validating Algorithm 10.5 for solving square linear systems. If the algorithm succeeds, it is verified that the real matrix $A_B$ is nonsingular.

We define

$$\begin{aligned} [d] &:= A_N^{\mathrm{T}} \cdot [y] - c_N \\ [x] &:= ([x_B], [x_N])^{\mathrm{T}}, \quad [x_N] := 0 \\ [H] &:= ([h]_{*,\nu_1}, \ldots, [h]_{*,\nu_{m-n}}) \\ [z] &:= c_B^{\mathrm{T}} \cdot [x_B] \cap b^{\mathrm{T}} \cdot [y]. \end{aligned}$$

The interval tableau $[T]$ corresponding to the basic index set $\mathcal{B} \in \mathcal{V}$ has the form

$$[T] := \left( \begin{array}{c|c} [z] & [d]^{\mathrm{T}} \\ \hline [x_B] & [H] \end{array} \right) = \left( \begin{array}{c|ccc} [z] & [d]_{\nu_1} & \cdots & [d]_{\nu_{n-m}} \\ \hline [x]_{\beta_1} & [h]_{\beta_1\nu_1} & \cdots & [h]_{\beta_1\nu_{n-m}} \\ \vdots & \vdots & & \vdots \\ [x]_{\beta_m} & [h]_{\beta_m\nu_1} & \cdots & [h]_{\beta_m\nu_{n-m}} \end{array} \right). \tag{11.5}$$

The self-validating algorithm for solving linear programming problems with input data $P = (A, b, c)$ is based on the following two theorems whose proofs appear in Krawczyk [57] and Jansson [38].

**Theorem 11.1** *Let $\mathcal{B} \in \mathcal{V}$ be a basic index set of $P$ and let $[T]$ be the corresponding interval tableau.*

1. *If the conditions*
$$\underline{x}_B > 0 \quad \text{and} \quad \underline{d} > 0 \tag{11.6}$$
*are satisfied, then the linear programming problem $P$ has a unique optimal solution contained in $[x]$. Moreover $\mathcal{V}_{\text{opt}} = \mathcal{B}$, and $z_{\text{opt}} \in [z]$.*
2. *If $\mathcal{B} \in \mathcal{V}_{\text{opt}}$, then*
$$\overline{x}_B \geq 0 \quad \text{and} \quad \overline{d} \geq 0. \tag{11.7}$$

If the Condition (11.6) is satisfied, then $P$ is called *basis stable*, i.e. the linear programming problem $P$ has a unique optimal solution. Inequalities (11.7) give a necessary condition that $\mathcal{B}$ is an optimal basic index set for the problem $P$. Such basic index sets are only of interest if $P$ is not basis stable.

The second theorem shows how to compute an enclosure for a *set of neighboring basic index sets* $\mathcal{L}$ of $\mathcal{B}$, where $\mathcal{B} \in \mathcal{V}_{\text{opt}}$ and all $\mathcal{B}' \in \mathcal{L}$ differ in exactly one index from $\mathcal{B}$. Each neighboring basic index set denotes a different neighboring vertex of the set of feasible solutions $X$. With Theorem 11.2 we have a sufficient criterion to determine a list of candidates that is a true superset of all neighboring basic index sets that may represent an optimal solution.

**Theorem 11.2** *Let $\mathcal{B}$ be a basic index set of $P$ with $\overline{x}_B \geq 0$ and $\overline{d} \geq 0$. Let $[T]$ be the corresponding interval tableau. Let $\widetilde{\mathcal{L}}$ be defined as the set of all index sets $\mathcal{B}' = (\mathcal{B} \setminus \{\beta\}) \cup \{\nu\}$ with $\beta \in \mathcal{B}$ and $\nu \in \mathcal{N}$ that satisfy one of the following conditions:*

$$0 \in [d]_\nu,\ \overline{h}_{\beta_\nu} > 0 \quad \text{and} \quad \frac{\underline{x}_\beta}{\overline{h}_{\beta\nu}} \leq \min\left\{ \frac{\overline{x}_\beta}{\underline{h}_{\beta\nu}} \mid \underline{h}_{\beta\nu} > 0,\ \beta \in \mathcal{B} \right\} \tag{11.8}$$

$$0 \in [x]_\beta,\ \underline{h}_{\beta_\nu} < 0 \quad \text{and} \quad \frac{\overline{d}_\nu}{\underline{h}_{\beta\nu}} \geq \max\left\{ \frac{\underline{d}_\nu}{\overline{h}_{\beta\nu}} \mid \overline{h}_{\beta\nu} < 0,\ \nu \in \mathcal{N} \right\}. \tag{11.9}$$

*If $\mathcal{B} \in \mathcal{V}_{\text{opt}}$, then $\widetilde{\mathcal{L}} \supseteq \mathcal{L}$.*

## 11.2 Algorithmic Description

With Algorithm 11.1 we compute the interval tableau $[T]$ of (11.5) using the algorithm LinSolve from Chapter 10 for solving the square linear systems of equations. If LinSolve fails to solve one of the linear systems, an error code is returned.

**Algorithm 11.1:** ComputeTableau $(A, b, c, \mathcal{B}, \mathcal{N}, [T], Err)$

1. {Initialization}
   $Err$ := "No error";
2. {Solve linear systems of equations (11.4)}
   LinSolve $(A_B, b, [x_B], Err)$; $[x_N] := 0$;
   LinSolve $(A_B^T, c_B, [y], Err)$;
   **for** $\nu \in \mathcal{N}$ **do**
      LinSolve $(A_B, a_{*,\nu}, [h]_{*,\nu}, Err)$;
   **if** $Err \neq$ "No error" {for any LinSolve call} **then**
      **return** $Err$ := "Submatrix $A_B$ probably singular";
3. {Compute components of interval tableau $[T]$}
   $[x] := ([x_B], [x_N])^T$;
   $[d] := A_N^T \cdot [y] - c_N$;
   $[H] := ([h]_{\nu_1}, \ldots, [h]_{\nu_{m-n}})$;
   $[z] := c_B^T \cdot [x_B] \cap b^T \cdot [y]$;
4. **return** $[T]$, $Err$;

Algorithm 11.2 determines whether the interval tableau $[T]$ corresponding to the actual basic index set $\mathcal{B}$ represents a basis stable solution. According to Condition (11.6) the existence of a basis stable solution indicates a unique optimal solution for the linear programming problem $P$, i.e. $\{\mathcal{B}\} = \mathcal{V} \supseteq \mathcal{V}_{opt}$, $\{[x]\} = \mathcal{X} \supseteq \mathcal{X}_{opt}$ and $[z] \supseteq z_{opt}$. Thus, no further calculations are necessary.

**Algorithm 11.2:** BasisStable $([T])$

1. {Check whether $\underline{x}_B > 0$ (cf. (11.6))}
   **for** $\beta \in \mathcal{B}$ **do**
      **if** $\underline{x}_\beta \leq 0$ **then return** *BasisStable* := *false*;
2. {Check whether $\underline{d} > 0$ (cf. (11.6))}
   **for** $\nu \in \mathcal{N}$ **do**
      **if** $\underline{d}_\nu \leq 0$ **then return** *BasisStable* := *false*;
3. **return** *BasisStable* := *true*;

We use Algorithm 11.3 to decide whether the interval tableau $[T]$ corresponding to the actual basic index set $\mathcal{B}$ represents a possibly optimal solution for the linear programming problem $P$, i.e. $\mathcal{B} \in \{\mathcal{B}^{(1)}, \ldots, \mathcal{B}^{(s)}\} = \mathcal{V} \supseteq \mathcal{V}_{opt}$, $[x] \in \mathcal{X} = \{[x]^{(1)}, \ldots, [x]^{(s)}\} \supseteq \mathcal{X}_{opt}$, and $[z] \supseteq z_{opt}$. If (11.7) is satisfied, then $\mathcal{B}$, $[x]$, and $[z]$ are stored to the list of optimal solutions.

**Algorithm 11.3:** PossiblyOptimalSolution $([T])$

1. {Check whether $\overline{x}_B >= 0$ (cf. (11.7))}
   **for** $\beta \in \mathcal{B}$ **do**
      **if** $\overline{x}_\beta < 0$ **then return** *PossiblyOptimalSolution* := *false*;

2. {Check whether $\overline{d} >= 0$ (cf. (11.7))}
   **for** $\nu \in \mathcal{N}$ **do**
   **if** $\overline{d}_\nu < 0$ **then return** *PossiblyOptimalSolution* := *false*;
3. **return** *PossiblyOptimalSolution* := *true*;

Algorithms 11.4 and 11.5 check whether the set of feasible solutions $X$ is empty or if the objective function $c^T x$ is unbounded on $X$. This is done by an examination of the interval tableau $[T]$. $X$ is probably (i.e. numerically) empty if there does not exist a $\nu \in \mathcal{N}$ with $[h]_{\beta\nu} < 0$ for all $\beta \in \mathcal{B}$ with $0 \in [x]_\beta$. The objective function is probably (i.e. numerically) unbounded if there does not exist a $\beta \in \mathcal{B}$ with $[h]_{\beta\nu} > 0$ for all $\nu \in \mathcal{N}$ with $0 \in [d]_\nu$.

**Algorithm 11.4:** EmptySolutionSet ($[T]$)

1. {Initialization}
   *IsEmpty* := *false*;
2. {Check whether for all $\beta \in \mathcal{B}$ with $0 \in [x]_\beta$ there}
   {exists a $\nu \in \mathcal{N}$ with $\overline{h}_{\beta\nu} < 0$ }
   **for** $\beta \in \mathcal{B}$ **with** $0 \in [x]_\beta$ **do**
   **while** (**not** *IsEmpty*) **do**
   *IsEmpty* := *true*;
   **for** $\nu \in \mathcal{N}$ **do**
   **if** $\overline{h}_{\beta\nu} < 0$ **then**
   *IsEmpty* := *false*; **exit**$_{\text{for-loop}}$;
3. **return** *EmptySolutionSet* := *IsEmpty*;

**Algorithm 11.5:** Unbounded ($[T]$)

1. {Initialization}
   *IsUnbounded* := *false*;
2. {Check whether for all $\nu \in \mathcal{N}$ with $0 \in [d]_\nu$ there}
   {exists a $\beta \in \mathcal{B}$ with $\underline{h}_{\beta\nu} > 0$ }
   **for** $\nu \in \mathcal{N}$ **with** $0 \in [d]_\nu$ **do**
   **while** (**not** *IsUnbounded*) **do**
   *IsUnbounded* := *true*;
   **for** $\beta \in \mathcal{B}$ **do**
   **if** $\underline{h}_{\beta\nu} > 0$ **then**
   *IsUnbounded* := *false*; **exit**$_{\text{for-loop}}$;
3. **return** *Unbounded* := *IsUnbounded*;

For the actual basic index set $\mathcal{B}$, Algorithm 11.6 determines a set of neighboring basic index sets $\mathcal{L}$ that are candidates for being optimal basic index sets. A true superset of all relevant candidates is determined according to Theorem 11.2.

**Algorithm 11.6:** NeighboringList ($[T], \mathcal{B}, \mathcal{N}$)

1. {Initialization}
   $\mathcal{L} := \emptyset$;

2. {Determine a list of neighboring basic index sets that are}
   {candidates for optimal solutions (see Theorem 11.2) }

   (a) {Search for candidates $\nu \in \mathcal{N}$ and $\beta \in \mathcal{B}$ by determination}
   {of the minimum of quotients $[x]_\beta/[h]_{\beta\nu}$ (cf. (11.8)) }
   **for** all $\nu \in \mathcal{N}$ **with** $0 \in d_\nu$ **do**
     {Determine minimum of $\overline{x}_\beta/\underline{h}_{\beta\nu}$ for column $\nu$}
     *colmin* $:= \min(\overline{x}_\beta/\underline{h}_{\beta\nu})$;
     {Determine candidates $\beta \in \mathcal{B}$ for exchange of indices}
     **if** $\underline{x}_\beta/\overline{h}_{\beta\nu} \leq$ *colmin* **then**
       {Determine new basic index set}
       $\mathcal{B}' := (\mathcal{B} \setminus \{\beta\}) \cup \{\nu\}$;
       $\mathcal{L} := \mathcal{L} + \mathcal{B}'$;

   (b) {Search for candidates $\beta \in \mathcal{B}$ and $\nu \in \mathcal{N}$ by determination}
   {of the maximum of quotients $[d]_\nu/[h]_{\beta\nu}$ (cf. (11.9)) }
   **for** all $\beta \in \mathcal{B}$ **with** $0 \in x_\beta$ **do**
     {Determine maximum of $\underline{d}_\nu/\overline{h}_{\beta\nu}$ for row $\beta$}
     *rowmax* $:= \max(\underline{d}_\nu/\overline{h}_{\beta\nu})$;
     {Determine candidates $\nu \in \mathcal{N}$ for exchange of indices}
     **if** $\overline{d}_\nu/\underline{h}_{\beta\nu} \geq$ *rowmax* **then**
       {Determine new basic index set}
       $\mathcal{B}' := (\mathcal{B} \setminus \{\beta\}) \cup \{\nu\}$;
       $\mathcal{L} := \mathcal{L} + \mathcal{B}'$;

3. **return** *NeighboringList* $:= \mathcal{L}$;

We now present the algorithm LinOpt to determine enclosures for the optimal value $z_{opt}$, the complete set of optimal basic index sets $\mathcal{V}_{opt}$, and the set of optimal basic feasible solutions $\mathcal{X}_{opt}$ for a linear programming problem $P$. The problem is given by the triple $P = (A, b, c)$. As stated in the previous section, we also need an initial optimal basic index set $\mathcal{B}_{start}$ to start the algorithm.

During the iteration in Step 2(a) of Algorithm 11.7, we select a basic index set $\mathcal{B}$ out of the list of candidates for optimal basic index sets $\mathcal{C}$ and update the list of examined basic index sets $\mathcal{E}$. In Step 2(b) we compute the interval tableau (11.5) using Algorithm 11.1. Next, we check whether the actual basic index set $\mathcal{B}$ represents a basis stable solution using Algorithm 11.2. If not, we use in Step 2(d) Algorithms 11.4 and 11.5 to decide if the linear programming problem is irregular, i.e. $X$ is empty or $c^T x$ is unbounded on $X$. In Step 2(e), we validate the optimality of the actual solution using Algorithm 11.3. If this is the case, we store it to the list of optimal solutions and then determine new candidates for optimal basic index sets, the list of neighboring basic index sets $\mathcal{L}$, using Algorithm 11.6.

**Algorithm 11.7:** LinOpt $(A, b, c, \mathcal{B}_{start}, [z], \mathcal{V}, \mathcal{X}, No, Err)$

1. {Initialization}
   $Err :=$ "No error"; $k_{max} := 100$;
   $k := 0$; $\mathcal{C} := \mathcal{B}_{start}$; $\mathcal{E} := \emptyset$; $\mathcal{V} := \emptyset$; $\mathcal{X} := \emptyset$; $No := 0$;

2. {Iteration}
   **repeat**

   (a) {Determine index sets $\mathcal{B}$ and $\mathcal{N}$}
   select $\mathcal{B} \in \mathcal{C}$; $\mathcal{C} := \mathcal{C} \setminus \mathcal{B}$; $\mathcal{E} := \mathcal{E} \cup \mathcal{B}$;
   $\mathcal{N} := \{1, \ldots, n\} \setminus \mathcal{B}$;

   (b) {Compute interval tableau $[T]$}
   ComputeTableau $(A, b, c, \mathcal{B}, \mathcal{N}, [T], Err)$;
   **if** $Err \neq$ "No error" **then exit**$_{\text{repeat-loop}}$

   (c) {Check whether the tableau is basis stable}
   **if** BasisStable $([T])$ **then**
   {Store optimal value, optimal basic index set and optimal solution}
   $[z] := c_B^T \cdot [x_B]$; $\mathcal{X} := ([x_B], [x_N])^T$; $\mathcal{V} := \mathcal{B}$; $No := No + 1$;
   **return** $[z]$, $\mathcal{V}$, $\mathcal{X}$;

   (d) {Check whether the set of feasible solutions is empty or}
   {if the objective function is unbounded }
   **if** EmptySolutionSet $([T])$ **then**
   $Err :=$ "Set of feasible solutions is empty";
   **exit**$_{\text{repeat-loop}}$
   **if** Unbounded $([T])$ **then**
   $Err :=$ "Objective function is unbounded";
   **exit**$_{\text{repeat-loop}}$

   (e) {Check whether the actual solution is possibly optimal}
   **if** PossiblyOptimalSolution $([T])$ **then**
   {Store optimal value, optimal basic index set, and optimal solution}
   $[z] := [z] \cup c_B^T \cdot [x_B]$; $\mathcal{X} := \mathcal{X} \cup ([x_B], [x_N])^T$;
   $\mathcal{V} := \mathcal{V} \cup \mathcal{B}$; $No := No + 1$;
   {Determine the list of neighboring basic index sets $\mathcal{L}$}
   $\mathcal{L} :=$ NeighboringList $([T], \mathcal{B}, \mathcal{N})$;
   $\mathcal{L} := (\mathcal{L} \setminus \mathcal{E}) \setminus \mathcal{C}$; $\mathcal{C} := \mathcal{C} \cup \mathcal{L}$;

   (f) $k := k + 1$;

   **until** $(\mathcal{C} = \emptyset)$ **or** $(k = k_{max})$;

3. {Check for errors}
   **if** $(k = k_{max})$ **then** $Err :=$ "Maximum number of iterations exceeded";
   **if** $(No = 0)$ **then** $Err :=$ "Initial basic index set is not optimal";

4. **return** $[z]$, $\mathcal{V}$, $\mathcal{X}$, $No$, $Err$;

## Applicability of the Algorithm

### Initial Index Set

A basic index set $\mathcal{B}_{start} \in \mathcal{V}$ can be determined by computing an approximation of the optimal solution and the corresponding basic index set $\mathcal{B}$ for the linear programming problem $P = (A, b, c)$ using the standard simplex algorithm. Calculate vectors of inclusion $[x_B]$ and $[y]$ for the systems of linear equations

$$A_B \cdot x_B = b, \quad A_B^{\mathrm{T}} \cdot y = c_B.$$

If $\overline{x}_B \geq 0$ and $\overline{d} \geq 0$ with $[d] := A_N^{\mathrm{T}} \cdot [y] - c_N$, then it is proved by Theorem 11.1 that $\mathcal{B}$ is a possibly optimal basic index set of the linear programming problem $P = (A, b, c)$. Therefore, $\mathcal{B}_{start} := \mathcal{B} \in \mathcal{V}$.

### Stopping Criteria

Algorithm 11.7 stops in Step 2(b) or 2(d) and returns an error if either $[T]$ is not computable or one of the conditions in Step 2(d) is fulfilled.

If $[T]$ is not computable, Algorithm 11.1 has failed to solve the systems of linear equations (11.4). This failure indicates that the actual submatrix $A_B$ is singular or nearly singular.

If in Step 2(d) the first or second condition is satisfied, this indicates that the set of feasible solutions $X$ is probably empty or the objective function is unbounded on $X$. In this case, it is not possible to compute enclosures for $z_{opt}$, $\mathcal{V}_{opt}$, and $\mathcal{X}_{opt}$.

### Improvements

The main work of the Algorithm 11.7 is to compute the interval tableau $[T]$, i.e. to solve the systems of linear equations (11.4). The only purpose of the interval tableaus is to compute an enclosure of the set of basic index sets $\mathcal{L}$ by Theorem 11.2. For the computation of $\mathcal{L}$, it is enough to get enclosures $[x_B]$ and $[d]$ and enclosures for the columns of $H$ with $0 \in [d]_\nu$ and the rows of $[H]$ with $0 \in [x]_\beta$ (cf. (11.8) and (11.9)). In practical applications, $[d]_\nu > 0$ and $[x]_\beta > 0$ holds for most indices $\nu \in \mathcal{N}$ and $\beta \in \mathcal{B}$. Therefore, it is much less effort to compute only the necessary columns and rows of $[H]$.

The accuracy of Algorithm 11.7 is reflected in the sharpness of the computed bounds for the solution of the linear programming problem $P$. The tightness of these bounds only depends on the algorithm LinSolve in Step 2 of Algorithm 11.1 for calculating the bounds of the linear systems (11.4).

In practice the input data of a linear programming problem often are afflicted with tolerances (e.g. measurement errors) and even more often the coefficients of the problem are not exactly representable on the machine. We can perform sensitivity and error analysis as well as achieve correct results if we specify the input data by a triple

$$[P] = ([A], [b], [c])$$

where $[A] \in I\mathbb{R}^{m\times n}$, $[b] \in I\mathbb{R}^{m}$, and $[c] \in I\mathbb{R}^{n}$. The exact values for all input data are not necessarily known, but their lower and upper bounds are given. The triple $[P]$ defines a whole set of linear programming problems $P \in [P]$ with real input data. This can be viewed as a linear parametric optimization problem, where each coefficient varies independently between the given lower and upper interval bounds. This approach overcomes the problem of conversion errors when reading decimal input data using a binary machine arithmetic (see 3.7). We can transform the Algorithms 11.1 to 11.7 to *full interval versions* by changing the input data $A$, $b$, and $c$ as well as of the submatrices and subvectors $A_B$, $A_N$, $c_B$, and $c_N$ from real to interval types. For this purpose, we also must adapt the function LinSolve used in Algorithm 11.1 to solve a interval linear system of equations (cf. Section 10.5). Since all the theory of this chapter holds for the interval case (cf. [38]), the *full interval algorithm* determines all optimal solutions for the entire set of linear programming problems $P \in [P]$.

## 11.3 Implementation and Examples

### 11.3.1 C++ Program Code

Here we list the C++ code of the modules needed for solving a linear programming problem determined by the triple $P = (A, b, c)$. All algorithms of the previous section are implemented in module *lop*. Their implementation requires some basic set operations for sets of integers. These operations are provided by module *set_ari*. Module *lop_ari* supplies an implementation of a linearly linked list which is used to handle the list of basic index sets that are candidates for being optimal basic index sets. Module *rev_simp* supplies an implementation of the revised simplex algorithm and is used to compute an initial optimal basic index set.

#### 11.3.1.1 Module set_ari

The header file of module *set_ari* supplies the definition of the class *IndexSet* used for the representation of a set of integers. The operators + and − are used to form the union and the difference of two sets, respectively.

```
//----------------------------------------------------------------------------
// File: set_ari (header)
// Purpose: Implementation of basic operations for sets of integers (also
//    called indices).
// Class IndexSet:
//    IndexSet()    : constructors
//    ~IndexSet()   : destructor
//    operator =    : assignment operator for sets
//    operator +    : unification of two sets or of a set and an index
//    operator -    : difference of two sets or of a set and an index
//    operator []   : returns an element of the set
//    operator ==   : test for equality of two sets
// Global functions:
//    Size()        : returns the actual number of elements in the set
//    Complement() : returns the complementary set
```

```
//    SetToVector(): assignment of an index set to an integer vector
//----------------------------------------------------------------------------
#ifndef __SET_ARI_HPP
#define __SET_ARI_HPP

#include <intvecto.hpp>     // Integer vector type

class IndexSet {
  int   MaxSize;
  char* Index;

  public:
    ~IndexSet ( );
    IndexSet  ( int = 0, char = '\0' );
    IndexSet  ( const IndexSet& );
    IndexSet& operator= ( const IndexSet& );
    IndexSet  operator+ ( const IndexSet& ) const;
    IndexSet  operator- ( const IndexSet& ) const;
    IndexSet  operator+ ( int ) const;
    IndexSet  operator- ( int ) const;
    int       operator[] ( int ) const;
    int       operator== ( const IndexSet& );

  friend int      Size        ( const IndexSet& );
  friend IndexSet Complement  ( const IndexSet& );
  friend void     SetToVector ( const IndexSet&, intvector& );
};
#endif
```

In our simple implementation a set is represented by an array of type *char*. The maximum length of that array is set to *MaxSize* elements when an object of type *IndexSet* is allocated. An integer $n$ is element of the set if the $n$-th component of the array does not vanish. Thus, the elements of the set are restricted to the range $1,\ldots,MaxSize$. For example, the array $(1,0,0,0,1,1,0,0,0,1,0,0)$ corresponds to the set $\{1,5,6,10\}$.

```
//----------------------------------------------------------------------------
// File: set_ari (implementation)
// Purpose: Implementation of basic operations for sets of integers (also
//    called indices).
// Method : A set of integers is represented by an array of characters INDEX.
//    An index N is element of the set if INDEX[N] != '\0'. For example
//
//       INDEX =  ('\0','1','1','\0','1','\0','1')
//
//    represents the set {2,3,5,7}. The maximum number of set elements is
//    restricted to MAXSIZE elements.
// Class IndexSet:
//    IndexSet()    : constructors
//    ~IndexSet()   : destructor
//    operator =    : assignment operator for sets
//    operator +    : unification of two sets or of a set and an index
//    operator -    : difference of two sets or of a set and an index
//    operator []   : returns an element of the set
//    operator ==   : test for equality of two sets
// Global functions:
//    Size()        : returns the actual number of elements in the set
//    Complement()  : returns the complementary set
//    SetToVector(): assignment of an index set to an integer vector
//----------------------------------------------------------------------------
#include <stdlib.h>          // For error handling with 'exit()'
```

```
#include <string.h>          // For memory copying with 'memcpy()'
#include <iostream.h>        // I/O handling with 'cerr'
#include <mv_util.hpp>       // Integer and real matrix/vector utility functions
#include <set_ari.hpp>

//----------------------------------------------------------------------------
// Fatal errors for set handling
//----------------------------------------------------------------------------
const int
  IllSetRange = 1,  // Illegal set range
  DiffSetSize = 2;  // Different maximum sizes of two sets

static void FatalError ( int Err )
{
  switch (Err) {
    case IllSetRange : cerr << "Illegal set range, index can not be stored!";
                       break;
    case DiffSetSize : cerr << "Sets have different maximum size!"; break;
  }
  cerr << endl;
  exit(-1);
}

//----------------------------------------------------------------------------
// Destructor and constructors
//----------------------------------------------------------------------------
IndexSet::~IndexSet ( )                                           // Destructor
  { delete[] Index; }                                             //-----------

//----------------------------------------------------------------------------
// By default all components of the index array are initialized by '\0'.
// Thus, the default initialization for a set is the empty set {}.
//----------------------------------------------------------------------------
IndexSet::IndexSet ( int NewSize, char FillChar )
{
  MaxSize = NewSize;
  if (MaxSize == 0)
    { Index = NULL; return; }
  Index = new char[MaxSize];
  for (int i = 0; i < MaxSize; i++) Index[i] = FillChar;
}

IndexSet::IndexSet ( const IndexSet& s )                    // Copy constructor
{                                                           //-----------------
  MaxSize = s.MaxSize;
  if (MaxSize == 0)
    { Index = NULL; return; }
  Index = new char[MaxSize];
  memcpy(Index, s.Index, MaxSize*sizeof(char));
}

//----------------------------------------------------------------------------
// Assignment operator
//----------------------------------------------------------------------------
IndexSet& IndexSet::operator= ( const IndexSet& s )
{
  MaxSize = s.MaxSize;
  Index = new char[MaxSize];
  memcpy(Index, s.Index, MaxSize*sizeof(char));
  return *this;
}

//----------------------------------------------------------------------------
// Unification and difference of two sets. Both sets must have the same
```

```
// maximum size.
//--------------------------------------------------------------------------------
IndexSet IndexSet::operator+ ( const IndexSet& s ) const
{
  // Sets to be unified must have the same maximum size!
  //--------------------------------------------------
  if (MaxSize != s.MaxSize) FatalError(DiffSetSize);

  IndexSet res(MaxSize);
  for (int i = 0; i < MaxSize; i++) res.Index[i] = Index[i] | s.Index[i];
  return res;
}

IndexSet IndexSet::operator- ( const IndexSet& s ) const
{
  // Sets to be subtracted must have the same maximum size!
  //-------------------------------------------------------
  if (MaxSize != s.MaxSize) FatalError(DiffSetSize);

  IndexSet res(MaxSize);
  for (int i = 0; i < MaxSize; i++)
    res.Index[i] = (s.Index[i]) ? '\0' : Index[i];
  return res;
}

//--------------------------------------------------------------------------------
// Union and difference of a set and a single element. The integer value must
// lie within the range of the set.
//--------------------------------------------------------------------------------
IndexSet IndexSet::operator+ ( int i ) const
{
  // 'i' must lie in the range of the set!
  //--------------------------------------
  if (i < 1 || MaxSize < i) FatalError(IllSetRange);

  IndexSet res = *this;
  res.Index[i-1] = '\1';
  return res;
}

IndexSet IndexSet::operator- ( int i ) const
{
  // 'i' must lie in the range of the set!
  //--------------------------------------
  if (i < 1 || MaxSize < i) FatalError(IllSetRange);

  IndexSet res = *this;
  res.Index[i-1] = '\0';
  return res;
}

//--------------------------------------------------------------------------------
// Returns the n-th element of the set. The elements of the set are
// assumed to be sorted in increasing order.
//--------------------------------------------------------------------------------
int IndexSet::operator[] ( int n ) const
{
  int i = 0, cnt = 0;

  while (i < MaxSize && cnt != n) {
    if (Index[i] != '\0') cnt++;
    i++;
  }
  return i;
```

```
}

int IndexSet::operator== ( const IndexSet& s )              // Test for equality
{                                                           //------------------
  // Sets to be compared must have the same size!
  //---------------------------------------------
  if (MaxSize != s.MaxSize) FatalError(DiffSetSize);

  int res = 1, i = 0;

  while (res == 1 && i < MaxSize) {
    res = (Index[i] == s.Index[i]);
    i++;
  }
  return res;
}

int Size ( const IndexSet& s )          // Returns the number of elements in 's'
{                                       //--------------------------------------
  int i, cnt = 0;

  for (i = 0; i < s.MaxSize; i++)
    if (s.Index[i] != '\0') cnt++;
  return cnt;
}

//----------------------------------------------------------------------------
// Returns the complementary set of 's', i.e. {1,2,..,MaxSize} - s
//----------------------------------------------------------------------------
IndexSet Complement ( const IndexSet& s )
{
  IndexSet t(s.MaxSize,'\1');
  return t - s;
}

void SetToVector ( const IndexSet& B, intvector& x )      // Assign index set to
{                                                         // integer vector
                                                          //--------------------
  // Lower index bound 1 assumed!
  //----------------------------
  for (int i = 1; i <= Ub(x); i++) x[i] = B[i];
}
```

### 11.3.1.2 Module lop_ari

The header file of module *lop_ari* supplies the type *Baselist* and functions for managing lists of basic index sets. The constant pointer *EmptyList* denotes an empty list.

```
//----------------------------------------------------------------------------
// File: lop_ari (header)
// Purpose: Definition of a linearly linked list as an abstract data type for
//    the representation of a list of integer sets.
// Global type:
//    BaseList : representation of a linearly linked list of index sets
// Global constant:
//    EmptyList: the empty list (NULL pointer)
// Global functions:
//    FreeAll(): free complete list
//    extract(): extracts submatrix/subvector depending on the
//               actual index set
```

```
//    in()     : returns TRUE, if index set is in list
//    select() : selects first element of a list
//    insert() : inserts index set at the head of a list
//    del()    : deletes index set from list
//    append() : appends 2nd list to end of 1st list
//    remove() : removes elements of 2nd list from 1st list
//----------------------------------------------------------------------------
#ifndef __LOP_ARI_HPP
#define __LOP_ARI_HPP

#include <set_ari.hpp>      // Index set handling
#include <rmatrix.hpp>      // Real matrix/vector arithmetic

#define EmptyList NULL             // The empty list
                                   //---------------

struct                             // Structure used for a list of sets
  BaseListElement;                 //---------------------------------

typedef                            // Pointer to a list of integer sets
  BaseListElement* BaseList;       //---------------------------------

extern void     FreeAll ( BaseList& );
extern rmatrix  extract ( rmatrix&, const IndexSet& );
extern rvector  extract ( rvector&, const IndexSet& );
extern int      in      ( const IndexSet&, BaseList );
extern IndexSet select  ( BaseList );
extern void     insert  ( BaseList&, const IndexSet& );
extern void     del     ( BaseList&, const IndexSet& );
extern void     append  ( BaseList&, BaseList& );
extern void     remove  ( BaseList&, BaseList );
#endif
```

A *BaseList* is realized as a linearly linked list. The static variable *FreeList* is used to hold the list of freed elements. The static functions *NewBL()*, *Free()*, and the global function *FreeAll()* work on that list to prevent memory garbage. The global functions *insert()* and *del()* are used to insert an index set at the head of the list and to remove an index set from the list, respectively. Function *append()* concatenates two lists and function *remove()* removes all elements of a list from another list.

```
//----------------------------------------------------------------------------
// File: lop_ari (implementation)
// Purpose: Definition of a linearly linked list as an abstract data type for
//    the representation of a list of integer sets.
// Global type:
//    BaseList : representation of a linearly linked list of index sets
// Global constant:
//    EmptyList: the empty list (NULL pointer)
// Global functions:
//    FreeAll(): free complete list
//    extract(): extracts submatrix/subvector depending on the
//               actual index set
//    in()     : returns TRUE, if index set is in list
//    select() : selects first element of a list
//    insert() : inserts index set at the head of a list
//    del()    : deletes index set from list
//    append() : appends 2nd list to end of 1st list
//    remove() : removes elements of 2nd list from 1st list
//----------------------------------------------------------------------------
#include <lop_ari.hpp>
```

```
struct BaseListElement {              // Type for a list of integer sets
  IndexSet Member;                    //--------------------------------
  BaseList Next;
};

static BaseList                       // Local variable storing list of free
  FreeList = EmptyList;               // elements (automatic garbage recycling)
                                      //---------------------------------------

//----------------------------------------------------------------------------
// Functions for generating and freeing of list elements
//----------------------------------------------------------------------------
static void NewBL ( BaseList& P )       // 'NewBL' generates a new list element
{                                       // or gets one from 'FreeList'.
  if (FreeList == EmptyList)            //------------------------------------
    { P = new BaseListElement;  P->Next = NULL; }
  else
    { P = FreeList;  FreeList = FreeList->Next;  P->Next= NULL; }
}

static void Free ( BaseList& P )              // 'Free' enters one element of a
{                                             // list in the 'FreeList'.
  if (P != NULL)                              //------------------------------
    { P->Next = FreeList;  FreeList = P;  P = NULL; }
}

void FreeAll ( BaseList& List )                // 'FreeAll' enters all elements
{                                              // of 'List' in the 'FreeList'.
  BaseList H;                                  //-----------------------------

  if (List != EmptyList) {
    H = List;
    while (H->Next != NULL) H = H->Next;
    H->Next = FreeList;  FreeList = List;  List = NULL;
  }
}

rmatrix extract ( rmatrix& A, const IndexSet& B )     // Extract submatrix of
{                                                     // of columns A[*,Index]
  int     k, n = Size(B);                             // with (Index in B)
  rmatrix res(Ub(A,1),n);                             //---------------------

  for (k = 1; k <= n; k++) Col(res,k) = Col(A,B[k]);
  return res;
}

rvector extract ( rvector& x, const IndexSet& B )      // Extract subvector
{                                                      // x[Index] with (Index
  int     k, n = Size(B);                              // in B)
  rvector res(n);                                      //--------------------

  for (k = 1; k <= n; k++) res[k] = x[B[k]];
  return res;
}

//----------------------------------------------------------------------------
// Global functions for lists of index sets
//----------------------------------------------------------------------------
int in ( const IndexSet& B, BaseList L )                     // Check if B in L
{                                                            //----------------
  while (L != EmptyList)
    if (L->Member == B)
      return 1;
    else
```

```
      L = L->Next;
  return 0;
}

IndexSet select ( BaseList L )               // Select first index set from L
  { return L->Member; }                      //------------------------------

void insert ( BaseList& L, const IndexSet& B )    // Insert index set B to the
{                                                 // head of the list of index
  BaseList P;                                     // sets L
                                                  //-------------------------
  if ( !in(B,L) ) {
    P = L;
    NewBL(L);  L->Member = B;  L->Next = P;
  }
}

void del ( BaseList& L, const IndexSet& B )    // Delete index set B from list
{                                              // of index sets L
  BaseList  P, D;                              //-----------------------------

  if (L != EmptyList)
    if (L->Member == B)                        // B is 1st element
      { D = L; L = L->Next; Free(D); }
    else {                                     // B is not 1st element
      P = L;
      while (P->Next != NULL)
        if (P->Next->Member == B)
          { D = P->Next;  P->Next = P->Next->Next;  Free(D); }
        else
          P = P->Next;
    }
}

void append ( BaseList& Res, BaseList& Add )  // Append list of index sets Add
{                                             // to list Res
  BaseList P;                                 //------------------------------

  if (Res == EmptyList)
    Res = Add;
  else {
    P = Res;
    while (P->Next != NULL) P = P->Next;
    P->Next = Add;
  }
  Add = EmptyList;
}

void remove ( BaseList& Res, BaseList Sub )   // Remove list of index sets Sub
{                                             // from Res
  IndexSet B;                                 //------------------------------

  while (Sub != EmptyList) {
    B = Sub->Member;
    del(Res,B);
    Sub = Sub->Next;
  }
}
```

#### 11.3.1.3 Module rev_simp

The header file of module *rev_simp* provides the interface of the void function *RevSimplex()* which is an implementation of the *revised simplex method.* The function *RevSimplexErrMsg()* can be used to get an error message for the error code returned by *RevSimplex()*.

```
//----------------------------------------------------------------------------
// File: rev_simp (header)
// Purpose: Determine an optimal value 'z', an optimal basic index set 'v',
//     and an optimal solution vector 'x' for a linear programming problem
//     P = (A,b,c) given in the standard form:
//                     ( z = c↑t * x = max! )
//          (LP)       (     A * x = b      )
//                     (       x >= 0       ).
// Global function:
//     RevSimplex()       : determines the values z, v, x for the linear
//                          programming problem P = (A,b,c)
//     RevSimplexErrMsg(): to get an error message text
//----------------------------------------------------------------------------
#ifndef __REV_SIMP_HPP
#define __REV_SIMP_HPP

#include <rmatrix.hpp>       // Real matrix/vector arithmetic
#include <intvecto.hpp>      // Integer vector type
#include <mv_util.hpp>       // Matrix/vector utility functions
                             // (needed for output of integer vectors)

extern char* RevSimplexErrMsg( int );
extern void  RevSimplex      ( rmatrix, rvector, rvector,
                               rvector&, intvector&, real&, int& );
#endif
```

*RevSimplex()* solves a linear programming problem in standard form (11.1). The implementation closely follows the one published in Syslo, Deo, and Kowalic [92, p. 14 ff]. Beside the correction of an index calculation error (marked by `/*!!!*/`), the source code has just slightly been changed for our special needs. This variant returns an optimal value $z_{\text{opt}}$, an optimal basic feasible solution $x_{\text{opt}}$, and an optimal basic index set $v_{\text{opt}} = \mathcal{B}_{start}$ as required to start Algorithm 11.7.

```
//----------------------------------------------------------------------------
// File: rev_simp (implementation)
// Purpose: Determine an optimal value 'z', an optimal basic index set 'v',
//     and an optimal solution vector 'x' for a linear programming problem
//     P = (A,b,c) given in the standard form:
//                     ( z = c↑t * x = max! )
//          (LP)       (     A * x = b      )
//                     (       x >= 0       ).
// Method: An initial basic index set is determined using the 2-phases method.
//     Then the revised simplex method described by Syslo, Deo, and Kowalic in
//     'Discrete Optimization Algorithms with Pascal' [p.14], Prentice Hall,
//     New Jersey (1983) is used.
// Global function:
//     RevSimplex()       : determines the values z, v, x for the linear
//                          programming problem P = (A,b,c)
//     RevSimplexErrMsg(): to get an error message text
//----------------------------------------------------------------------------
#include <rev_simp.hpp>

//----------------------------------------------------------------------------
```

```
// Error codes used in this module.
//----------------------------------------------------------------------------
const int
  NoError            = 0,   // No error occurred
  WrongDimension     = 1,   // Wrong dimension of problem
  NoOptimalSolution = 2,    // No initial optimal solution found
  NoInitialVertex    = 3;   // No initial vertex found

//----------------------------------------------------------------------------
// Error messages depending on the error code.
//----------------------------------------------------------------------------
char* RevSimplexErrMsg ( int Err )
{
  static char Msg[80] = "";

  if (Err != NoError) {
    char Hlp[60];

    switch (Err) {
      case WrongDimension:
        strcpy(Hlp,"Wrong dimension of problem (e.g. m >= n)"); break;
      case NoOptimalSolution:
        strcpy(Hlp,"No initial optimal solution found"); break;
      case NoInitialVertex:
        strcpy(Hlp,"No initial vertex found"); break;
      default:
        strcpy(Hlp,"Code not defined");
    }
    sprintf(Msg,"Error: %s!",Hlp);
  }
  return(Msg);
} // RevSimplexErrMsg

//----------------------------------------------------------------------------
// Purpose: Determination of the solutions of a linear optimization problem.
// Parameters:
//    In : 'A'   : matrix of constraints
//         'b'   : right-hand side of constraints
//         'c'   : objective function by its coefficients
//    Out: 'x'   : optimal solution vector
//         'v'   : optimal basic index set
//         'z'   : optimal value
//         'Err' : error flag
// Description: For a detailed description of the revised simplex method see
//    Syslo, Deo, and Kowalic.
//    The result vectors 'x' and 'v' are automatically resized to standard
//    index range with lower index bounds 1.
//----------------------------------------------------------------------------
void RevSimplex ( rmatrix A, rvector b, rvector c,
                  rvector& x, intvector& v, real& z, int& Err )
{
  // Check for correct dimensions of the input data
  //-----------------------------------------------
  int  m = ColLen(A), n = RowLen(A);

  if ( m >= n || VecLen(b) != m || VecLen(c) != n )
    { Err = WrongDimension;  return; }

  SetLb(A,1,1); SetLb(A,2,1);              // Resize input data to standard index
  SetLb(b,1); SetLb(c,1);                  // range with lower index bound 1
                                           //------------------------------------
  Resize(x,n);                                    // Resize the result vectors to
  Resize(v,m);                                    // appropriate length
                                                  //-----------------------------
```

```
// Start algorithm
//----------------
const real
  eps = 1E-30;     // if (|a| < eps), it is handled as (a == 0)

real        min, s;
rmatrix     T(m+2,n+2), U(m+2,m+2);
rvector     w(n+2), y(n+2);
int         ex, SolutionFound;
int         i, j, k, l, p, q, phase;

Err = NoError; phase = 1;                                    // Initialization
p = m + 2; q = m + 2; k = m + 1;                             //---------------
SolutionFound = 0;

T(m,n) = A;                                                  // Compute tableau
                                                             //----------------
for (j = 1; j <= n; j++) {
  T[k][j] = -c[j];
  s = 0.0;
  for (i = 1; i <= m; i++) s -= T[i][j];
  T[p][j] = s;
}
U = Id(U);

s = 0.0; w = 0.0;                      // Determine initial basic index set v,
for (i = 1; i <= m; i++) {             // initial optimal value s, and initial
  v[i] = n - m + i;          /*!!!*/   // optimal solution x
  w[i] = b[i];                         //------------------------------------
  s -= b[i];
}
w[k] = 0.0; w[p] = s;

do {
  if (w[p] >= -eps*10.0 && phase == 1)          // Only phase 1:
    { phase = 2; q = m+1; }                     // Solution for phase 1 found
                                                //---------------------------
  min = 0.0;                               // Phase 1 and phase 2:
  for (j = 1; j <= n; j++) {               // Determine index k that minimizes
    s = Row(U,q) * Col(T,j);               // U[q][*] * T[*][j] (j == 1,..,n)
    if (s < min ) { min = s; k = j; }      //--------------------------------
  }

  if (-eps < min) {                        // Vector of reduced costs vanishes
    if (phase == 1)                        //--------------------------------
      { Err = NoInitialVertex; return; }
    SolutionFound = 1;                                       // phase == 2
    z = w[q];                                      // Optimal solution found
  }                                                //-----------------------
  else {                                        // Determine candidate l for
    for (i = 1; i <= q; i++)                    // exchange of indices
      y[i] = Row(U,i) * Col(T,k);               //--------------------------
    ex = 1;
    for (i = 1; i <= m; i++)
      if (y[i] >= eps) {
        s = w[i]/y[i];
        if (ex || s < min) { min = s; l = i; }
        ex = 0;
      }

    if (ex) { Err = NoOptimalSolution; return; }       // No candidate found
                                                       //-------------------
    v[l] = k;                                 // Determine new basic index set
    s = 1.0 / y[l];                           // and compute new tableau
```

```
      for (j = 1; j <= m; j++) U[l][j] *= s;  //-----------------------------
      i = (l == 1) ? 2 : 1;

      do {
        s = y[i];
        w[i] = w[i] - min * s;
        for (j = 1; j <= m; j++) U[i][j] -= U[l][j] * s;
        i = (i == l-1) ? i+2 : i+1;
      } while (i <= q);

      w[l] = min;
    } // Determine candidate l ...

  } while (!SolutionFound);

  x = 0.0;                                   // Return optimal value z, optimal
  for (i = 1; i <= m; i++)                   // basic index set v, and optimal
    x[v[i]] = w[i];                          // solution vector x
                                             //--------------------------------
} // RevSimplex
```

#### 11.3.1.4 Module lop

The header file of module *lop* supplies the interfaces of the function *LinOpt()* for solving a linear programming problem in standard form (11.1) and the function *LinOptErrMsg()* which can be used to get an error message for the error code returned by *LinOpt()*.

```
//----------------------------------------------------------------------------
// File: lop (header)
// Purpose: Determine enclosures for the optimal value 'z_opt', for the set
//    of optimal basic index sets 'V_opt', and for the set of solution
//    vectors 'X_opt' for a linear programming problem P = (A,b,c) given in
//    the standard form:
//                ( z = c↑t * x = max! )
//          (LP)  (     A * x = b      )
//                (       x >= 0       )
//    with an initial optimal basic index set.
// Global functions:
//    LinOpt()        : determines the enclosures of z_opt, V_opt, and X_opt
//                      for LP P = (A,b,c)
//    LinOptErrMsg(): delivers an error message text
//----------------------------------------------------------------------------
#ifndef __LOP_HPP
#define __LOP_HPP

#include <intmatri.hpp>      // Integer matrix/vector types
#include <rmatrix.hpp>       // Real matrix/vector arithmetic
#include <imatrix.hpp>       // Interval matrix/vector arithmetic
#include <mv_util.hpp>       // Real matrix/vector utility functions
#include <mvi_util.hpp>      // Interval matrix/vector utility functions

extern char* LinOptErrMsg ( int );
extern void  LinOpt ( rmatrix, rvector, rvector, intvector,
                      interval&, intmatrix&, imatrix&, int&, int& );
#endif
```

The function *LinOpt()* is an implementation of Algorithm 11.7. Its input data are the the linear optimization problem $P = (A, b, c)$ and an initial optimal basic index set $\mathcal{B}_{start}$. It returns the enclosures of $z_{opt}$, $\mathcal{V}_{opt}$, $\mathcal{X}_{opt}$, the number of optimal solutions, and an error code. The static functions *ComputeTableau()*, *BasisStable()*, *PossiblyOptimalSolution()*, *EmptySolutionSet()*, *Unbounded()*, and *NeighboringList()* are implementations of the Algorithms 11.1 through 11.6.

```
//----------------------------------------------------------------------------
// File: lop (implementation)
// Purpose: Determine enclosures for the optimal value 'z_opt', for the set
//    of optimal basic index sets 'V_opt', and for the set of solution
//    vectors 'X_opt' for a linear programming problem P = (A,b,c) given in
//    the standard form:
//              ( z = c↑t * x = max! )
//         (LP) (     A * x = b      )
//              (       x >= 0       )
//    with an initial optimal basic index set.
// Method: Starting from an initial basic index set, all neighboring index
//    sets are examined to determine all optimal solutions. To validate the
//    optimality of a solution, all linear systems of equations are solved
//    using a verifying linear system solver.
// Global functions:
//    LinOpt()      : determines the enclosures of z_opt, V_opt, and X_opt
//                    for LP P = (A,b,c)
//    LinOptErrMsg(): delivers an error message text
//----------------------------------------------------------------------------
#include <string.h>        // String handling
#include <set_ari.hpp>     // Integer set arithmetic
#include <linsys.hpp>      // Verifying linear system solver
#include <lop_ari.hpp>     // Linear linked list for index sets
#include <lop.hpp>

const int
  kmax = 100;              // Maximum number of iterations

//----------------------------------------------------------------------------
// Error codes used in this module.
//----------------------------------------------------------------------------
const int
  NoError                 = 0,  // No error occurred
  WrongDimension          = 1,  // Wrong dimension of problem
  SolutionSetIsEmpty      = 2,  // Set of feasible solutions is empty
  FunctionUnbounded       = 3,  // Objective function is unbounded
  SubmatrixSingular       = 4,  // Submatrix A_B is singular
  StartIndexSetNotOptimal = 5,  // Initial basic index set not optimal
  IterationError          = 6;  // Maximum number of iteration exceeded

//----------------------------------------------------------------------------
// Error messages depending on the error code.
//----------------------------------------------------------------------------
char* LinOptErrMsg ( int Err )
{
  static char Msg[100] = "";

  if (Err != NoError) {
    char Hlp[80];

    switch (Err) {
      case WrongDimension:
        strcpy(Hlp,"Wrong dimension of problem (e.g. m >= n)");       break;
      case SolutionSetIsEmpty:
        strcpy(Hlp,"Set of feasible solutions is probably empty");    break;
      case FunctionUnbounded:
        strcpy(Hlp,"Objective function is probably unbounded");       break;
      case SubmatrixSingular:
        strcpy(Hlp,"Submatrix A_B in [A_B] is probably singular");    break;
```

```
      case StartIndexSetNotOptimal:
        strcpy(Hlp,"Initial basic index set not optimal (i.e. ");
        strcat(Hlp,"B_start not in V_opt)");                          break;
      case IterationError:
        sprintf(Hlp,"Maximum number of iterations (=%1d) exceeded",
                kmax);                                                 break;
      default:
        strcpy(Hlp,"Code not defined");
    }
    sprintf(Msg,"Error: %s!",Hlp);
  }
  return(Msg);
} // LinOptErrMsg

static ivector Solution ( imatrix  TT,              // Determine solution vector
                          IndexSet Base,            // corresponding to Base
                          IndexSet NonBase )        //-------------------------
{
  int      Index, i, sB = Size(Base);
  ivector  res(sB+Size(NonBase));
  interval Rplus = _interval(0.0,MaxReal);

  res = 0.0;
  for (i = 1; i <= sB; i++) {
    Index = Base[i];
    res[Index] = TT[i+1][1] & Rplus;
  }
  return res;
}

//----------------------------------------------------------------------------
// Compute the interval tableau
//                             ([z]    | [d]↑t     )
//                     [T] = (------+----------)
//                             ([x]_B | [H]        ).
//----------------------------------------------------------------------------
static void ComputeTableau ( rmatrix   A,
                             rvector   b,
                             rvector   c,
                             IndexSet  Base,
                             IndexSet  NonBase,
                             imatrix&  TT,
                             int&      Err )
{
  const int
    m = Ub(A,1), n = Ub(A,2), sB = Size(Base), sN = Size(NonBase);

  rmatrix   A_B(m,sB), A_N(m,sN);
  rvector   c_B(sB), c_N(sN);
  ivector   xx_B(sB), yy(sB), dd(sN);
  imatrix   HH(sB,sN);
  interval  zz;
  int       i, j, Index, local_Error;

  // Determine submatrices and subvectors according to index sets Base and
  // NonBase
  A_B = extract(A, Base); A_N = extract(A, NonBase);  // Determine submatrices
  c_B = extract(c, Base); c_N = extract(c, NonBase);  // and subvectors accor-
                                                      // ding to B and N
  // Solve linear systems of equations                //---------------------
  LinSolve(A_B, b, xx_B, local_Error);

  if (local_Error == NoError)
    LinSolve(transp(A_B), c_B, yy, local_Error);
```

```
  if (local_Error == NoError)
    for (i = 1; i <= n-m; i++) {
      Index = NonBase[i];
      if (local_Error == NoError) {
        // Temporary variables colHH and colA needed since automatic
        // resizing of Col(HH,i) and Col(A,Index) is not supported by C++
        //----------------------------------------------------------------
        ivector colHH = Col(HH,i);
        rvector colA = Col(A,Index);
        LinSolve(A_B, colA, colHH, local_Error);
        Col(HH,i) = colHH;
      }
    }

  if (local_Error == NoError) {
    dd = transp(A_N) * yy - c_N;              // Compute components of interval
    zz = (c_B * xx_B) & (b * yy);             // tableau
    TT[1][1] = zz;                            //------------------------------
    for (j = 2; j <= n-m+1; j++) TT[1][j] = dd[j-1];
    for (i = 2; i <= m+1; i++) TT[i][1] = xx_B[i-1];
    for (i = 2; i <= m+1; i++)
      for (j = 2; j <= n-m+1; j++)
        TT[i][j] = HH[i-1][j-1];
  }
  else
    Err = SubmatrixSingular;
} // ComputeTableau

//----------------------------------------------------------------------------
// Determine whether the interval tableau [T] represents a basis stable
// solution, i.e. (Inf([x]_B) > 0) and (Inf([d]) > 0).
//----------------------------------------------------------------------------
static int BasisStable ( imatrix& TT )
{
  int IsStable = 1, i = 1, j = 1;

  while (i < Ub(TT,1) && IsStable) {
    i++;                                     // [x]_B = [T_i][1] (i = 2 .. m+1)
    if (Inf(TT[i][1]) <= 0.0) IsStable = 0;
  }
  while (j < Ub(TT,2) && IsStable) {
    j++;                                     // [d] = [T_1][j] (j = 2 .. n-m+1)
    if (Inf(TT[1][j]) <= 0.0) IsStable = 0;
  }
  return IsStable;
}

//----------------------------------------------------------------------------
// Determine whether the interval tableau [T] represents a possibly optimal
// solution, i.e (Sup([x]_B) >= 0) and (Sup([d]) >= 0).
//----------------------------------------------------------------------------
static int PossiblyOptimalSolution ( imatrix& TT )
{
  int IsOptimal = 1, i = 1, j = 1;

  while (i < Ub(TT,1) && IsOptimal) {
    i++;                                     // [x]_B = [T_i][1] (i = 2 .. m+1)
    if (Sup(TT[i][1]) < 0.0) IsOptimal = 0;
  }
  while (j < Ub(TT,2) && IsOptimal) {
    j++;                                     // [d] = [T_1][j] (j = 2 .. n-m+1)
    if (Sup(TT[1][j]) < 0.0) IsOptimal = 0;
  }
```

```
 return IsOptimal;
}

//----------------------------------------------------------------------------
// Check whether the set of feasible solutions is empty, i.e. not for all(beta
// in B) with (0 in [x]_beta) there exists a (nu in N) with ([H]_beta,nu < 0)
//----------------------------------------------------------------------------
static int EmptySolutionSet ( imatrix& TT )
{
  int IsEmpty = 0, i = 1, j;

  while (!IsEmpty && i < Ub(TT,1)) {
    i++;
    if ( in(0.0,TT[i][1]) ) {
      IsEmpty = 1;
      j = 1;
      while (IsEmpty && j < Ub(TT,2)) {
        j++;
        if (Sup(TT[i][j]) < 0.0) IsEmpty = 0;
      }
    }
  }
  return IsEmpty;
}

//----------------------------------------------------------------------------
// Check whether the objective function is unbounded, i.e. not for all (nu
// in N) with (0 in [d]_nu) there exists a (beta in B) with ([H]_beta,nu > 0)
//----------------------------------------------------------------------------
static int Unbounded ( imatrix& TT )
{
  int IsUnbounded = 0, i, j = 1;

  while (!IsUnbounded && j < Ub(TT,2)) {
    j++;
    if ( in(0.0,TT[1][j]) ) {
      IsUnbounded = 1;
      i = 1;
      while (IsUnbounded && i < Ub(TT,1)) {
        i++;
        if (Inf(TT[i][j]) > 0.0) IsUnbounded = 0;
      }
    }
  }
  return IsUnbounded;
}

//----------------------------------------------------------------------------
// Determine list of neighboring basic index sets L for index set Base that
// are good candidates for being optimal basic index sets.
//----------------------------------------------------------------------------
static BaseList NeighboringList ( imatrix&   TT,
                                  IndexSet&  Base,
                                  IndexSet&  NonBase )
{
  BaseList  L = NULL;
  IndexSet  NewBase;
  int       i, j, m = Size(Base),n = m+Size(NonBase);
  int       beta, nu;
  real      colmin, rowmax;
  interval  xx, dd, HH;

  colmin = MaxReal;                                 // Search for candidates (nu
  // for all (nu in N)                              // in N) and (beta in B) by
```

```
  for (j = 1; j <= n-m; j++) {                  // determining of the minimum
    dd = TT[1][j+1];                            // of the quotients
    if ( in(0.0,dd) ) {                         // [x]_beta / [H]_beta,nu
      // for all (beta in B)                    //---------------------------
      for (i = 1; i <= m; i++) {                    //  Determine minimum of
        xx = TT[i+1][1];                            // [x]_beta / [H]_beta,nu
        HH = TT[i+1][j+1];                          // for column nu of [T]
                                                    //-----------------------
        if (Inf(HH) > 0.0 && divd(Sup(xx),Inf(HH)) < colmin)
          colmin = divd(Sup(xx),Inf(HH));
      } // (beta in B)

      // for all (beta in B)
      for (i = 1; i <= m; i++) {                 // Determine candidate (beta
        xx = TT[i+1][1];                         // in B) for exchange of
        HH = TT[i+1][j+1];                       // indices
                                                 //-------------------------
        if (Sup(HH) > 0.0 && divd(Inf(xx),Sup(HH)) <= colmin) {
          beta = Base[i];                      // Determine new index set and
          nu = NonBase[j];                     // add it to L
          NewBase = Base - beta + nu;          //----------------------------
          insert(L, NewBase);
        }
      } // (beta in B)
    }
  } // (nu in N)

  rowmax = -MaxReal;                            // Search for candidates beta
  // for all (beta in B)                        // in B) and (nu in N) by
  for (i = 1; i <= m; i++) {                    // determining the maximum
    xx = TT[i+1][1];                            // of the quotients
    if ( in(0.0,xx) ) {                         // [d]_nu / [H]_beta,nu
      // for all (nu in N)                      //---------------------------
      for (j = 1;j <= n-m; j++) {                     // Determine maximum of
        dd = TT[1][j+1];                              // [d]_nu/[H]_beta,nu
        HH = TT[i+1][j+1];                            // for row beta of [T]
                                                      //---------------------
        if (Sup(HH) < 0.0 && divu(Inf(dd),Sup(HH)) > rowmax)
          rowmax = divu(Sup(dd),Sup(HH));
      } // (nu in N)

      // for all (nu in N)
      for (j = 1;j <= n-m; j++) {                   // Determine candidate
        dd = TT[1][j+1];                            // (nu in N) for exchange
        HH = TT[i+1][j+1];                          // of indices
                                                    //-----------------------
        if (Inf(HH) < 0.0 && divu(Inf(dd),Inf(HH)) >= rowmax) {
          beta = Base[i];                      // Determine new index set and
          nu = NonBase[j];                     // add it to L
          NewBase = Base - beta + nu;          //----------------------------
          insert(L, NewBase);
        }
      } // (nu in N)
    }
  } // (beta in B)

  return L;
} // NeighboringList

//----------------------------------------------------------------------------
// Purpose: Determination of enclosures for the solutions of a linear opti-
//    mization problem:      ( z = c↑t * x = max! )
//                      (LP) (     A * x = b      )
//                           (      x >= 0        )
```

```
//    with an initial optimal basic index set.
// Parameters:
//    In : 'A'               : matrix of constraints
//         'b'               : right-hand side of constraints
//         'c'               : represents the objective function by its
//                             coefficients
//         'B_Start_Vector': initial basic index set
//    Out: 'z'               : enclosure of the optimal value 'z_opt'
//         'V'               : superset of all optimal basic index sets 'V_opt'
//         'X'               : superset of all optimal solution vectors 'X_opt'
//         'No'              : number of solutions computed
//         'Err'             : error code
// Description: Starting from an initial basic index set 'B_start', a true
//    superset of all neighboring index sets 'L' that are candidates for being
//    optimal index sets are determined and stored in a list of candidates
//    'CList'. As long as there are candidates left, they are examined using
//    a verifying linear system solver. The linear system solver returns
//    enclosures for the solutions of the square linear systems A_B [x] = b,
//    where A_B denotes a submatrix of the constraint matrix A according to a
//    basic index set B. The result arrays 'V' and 'X' are automatically
//    resized to standard index range with lower index bounds 1.
//----------------------------------------------------------------------------
void LinOpt ( rmatrix      A,
              rvector      b,
              rvector      c,
              intvector    B_Start_Vector,
              interval&    z,
              intmatrix&   V,
              imatrix&     X,
              int&         No,
              int&         Err )
{
  const int
    MinSize = 10;   // Minimum size (number of rows) of the result matrices

  // Check for correct dimensions of the input data
  //-----------------------------------------------
  int  m = ColLen(A), n = RowLen(A);

  if ( m >= n || VecLen(b) != m
              || VecLen(c) != n
              || VecLen(B_Start_Vector) != m )
    { Err = WrongDimension;  return; }

  SetLb(A,1,1); SetLb(A,2,1);           // Resize input data to standard index
  SetLb(b,1); SetLb(c,1);               // range with lower index bound 1
  SetLb(B_Start_Vector,1);              //------------------------------------

  Resize(V,MinSize,m);                        // Resize the result matrices to
  Resize(X,MinSize,n);                        // 'MinSize' rows
                                              //------------------------------

  IndexSet  B_Start(n), Base(n,'\0'), NonBase(n,'\0');
  BaseList  L, CList, E;
  imatrix   TT(m+1,n-m+1);
  int       k, i, TempSize = MinSize;


  for (i = 1; i <= m; i++) B_Start = B_Start + B_Start_Vector[i];

  V = 0; X = 0.0; k = 0; No = 0; Err = NoError;
  E = NULL; CList = NULL; insert(CList, B_Start);

  do { // while (CList != EmptyList ... )
```

```
  Base = select(CList);                               // Select Base out of
  NonBase = Complement(Base);                         // CList and update
  del(CList, Base); insert(E, Base);                  // examined list E
                                                      //-------------------
  ComputeTableau(A, b, c, Base, NonBase, TT, Err);

  if (Err == NoError)
    if (BasisStable(TT)) {
      No = 1;                              // Store unique optimal solution
      z = TT[1][1];                        //-----------------------------

      Row(X,No) = Solution(TT, Base, NonBase);
      SetToVector(Base,Row(V,No));
    }
    else { // not basis stable
      if (EmptySolutionSet(TT))
        Err = SolutionSetIsEmpty;
      else if (Unbounded(TT))
        Err = FunctionUnbounded;
      else if (PossiblyOptimalSolution(TT) && Err == NoError) {
        No++;                                        // Store optimal solution
        if (No == 1)                                 //-----------------------
          z = TT[1][1];
        else
          z = TT[1][1] | z;

        if (No > TempSize)                 // Double size of result matrices
          { DoubleSize(V); DoubleSize(X); TempSize *= 2; }

        Row(X,No) = Solution(TT, Base, NonBase);
        SetToVector(Base,Row(V,No));

        L = NeighboringList(TT,Base,NonBase);        // Determine list of
        remove(L, E); remove(L, CList);              // neighboring basic
                                                     // index sets
                                                     //------------------
        append(CList, L);                // Compute new list of candidates
      }                                  //------------------------------
    } // not basis stable

  k++;

} while (CList != EmptyList && Err == NoError && k < kmax);

if (Err == NoError)                                  // Determine error code
  if (CList != EmptyList && k == kmax)               //---------------------
    Err = IterationError;
  else if (No == 0)
    Err = StartIndexSetNotOptimal;

// Resize the result matrices to appropriate number of rows
// if no error has occurred. Otherwise, resize them to minimum
// dimensions with undefined value.
//-------------------------------------------------------------
if (Err == NoError) {
  Resize(V,No,m);
  Resize(X,No,n);
}
else {
  Resize(V,1,1);     // Values undefined
  Resize(X,1,1);     //-----------------
}

FreeAll(CList); FreeAll(E);
```

```
} // LinOpt
```

### 11.3.2 Examples

**Example 11.1** We consider the following linear programming problem:

maximize $50x_1 + 9x_2$ subject to the linear constraints

$$\begin{array}{rcrcl} x_1 & & & \leq & 50 \\ & & x_2 & \leq & 200 \\ 100\ x_1 & + & 18\ x_2 & \leq & 5000 \\ & & x_1, x_2 & \geq & 0. \end{array} \tag{11.10}$$

This two-dimensional optimization problem is illustrated by Figure 11.1, where $X$ (shaded area) denotes the set of feasible solutions.

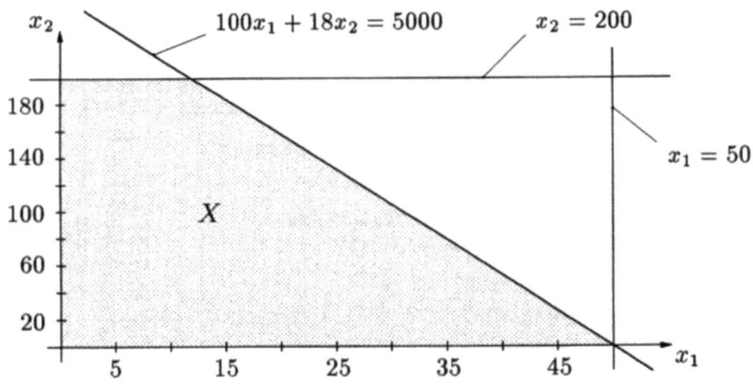


Figure 11.1: Two-dimensional optimization problem

We introduce the slack variables $x_3$, $x_4$, and $x_5$ and transform LP (11.10) into standard form (11.1).

maximize $c^T \cdot x$ subject to the linear constraints

$$\begin{array}{rcl} A \cdot x & = & b \\ x & \geq & 0, \end{array}$$

where

$$A = \begin{pmatrix} 1 & 0 & 1 & 0 & 0 \\ 0 & 1 & 0 & 1 & 0 \\ 100 & 18 & 0 & 0 & 1 \end{pmatrix}, \quad b = \begin{pmatrix} 50 \\ 200 \\ 5000 \end{pmatrix}, \quad c = \begin{pmatrix} 50 \\ 9 \\ 0 \\ 0 \\ 0 \end{pmatrix}.$$

The following sample program determines enclosures for the optimal value $z_{opt}$, the set of optimal basic index sets $\mathcal{V}_{opt}$, and the set of optimal basic feasible solutions $\mathcal{X}_{opt}$ for Example 11.1. The work arrays needed for computation are allocated dynamically depending on the dimension $n \times m$ of the optimization problem. Function *RevSimplex()* computes an approximate solution and returns an initial basic index set $\mathcal{B}_{start}$ which is used as input for the function *LinOpt()*.

```
//----------------------------------------------------------------------------
// Example for the use of module 'lop' to solve the linear programming problem
// (A,b,c) with:
//                                                             ( 50 )
//         (  1    0 1 0 0 )        (   50 )                   (  9 )
//    A =  (  0    1 0 1 0 ),   b = (  200 ),  and  c =        (  0 )
//         ( 100  18 0 0 1 )        ( 5000 )                   (  0 )
//                                                             (  0 )
//----------------------------------------------------------------------------
#include <rev_simp.hpp>        // Revised Simplex algorithm
#include <lop.hpp>             // Linear optimization

int main ( )
{
  int m, n;

  cout << "Enter dimensions of linear "                     // Read dimension of
       << "optimization problem (m and n): " << endl;  // optimization problem
  cin >> m >> n;

  rmatrix   A(m,n);
  rvector   c(n), b(m), x_start;
  intvector B_start;
  intmatrix V;
  imatrix   X;
  real      z_start;
  interval  z;
  int       Error, i, NoOfSolutions;

  // Read optimization problem P = (A,b,c)
  //--------------------------------------
  cout << endl << "Enter objective function (c[1]..c[n]): " << endl;
  cin  >> c;
  cout << endl << "Enter matrix of constraints (A[1,1]..A[m,n]): " << endl;
  cin  >> A;
  cout << endl << "Enter right-hand side (b[1]..b[m]): " << endl;
  cin  >> b;
  cout << endl;

  // Call revised simplex function
  //------------------------------
  RevSimplex(A, b, c, x_start, B_start, z_start, Error);

  if (Error)
    { cout << RevSimplexErrMsg(Error) << endl; return -1; }

  // Display results of calculation
  //-------------------------------
  cout << SetPrecision(23,15) << Scientific;  // output format

  cout << "Results of approximation:" << endl
       << "Optimal value :  " << z_start << endl << endl
       << "Initial basic index set : " << B_start << endl
       << "Initial optimal solution : " << endl << x_start << endl;
```

```
  // Call verifying linear optimization solver
  //-----------------------------------------
  LinOpt(A, b, c, B_start, z, V, X, NoOfSolutions, Error);

  if (Error)
    cout << LinOptErrMsg(Error) << endl;
  else {
    // Display results of calculation
    //-------------------------------
    cout << "Results of verification:" << endl;
    cout << "Optimal value interval :  " << z << endl << endl
         << "List of optimal basic index sets : " << endl;
    for (i = 1; i <= NoOfSolutions; i++)
      cout << "B" << i << " = " << V[i];
    cout << endl
         << "List of optimal basic feasible solutions : " << endl;
    for (i = 1; i <= NoOfSolutions; i++)
      cout << "x of B" << i << " = " << endl << X[i];
    cout << endl;
  }
  return 0;
}
```

The function *LinOpt()* calculates three optimal basic index sets and the corresponding three optimal solution vectors as listed below.

```
Enter dimensions of linear optimization problem (m and n):
3 5

Enter objective function (c[1]..c[n]):
50 9 0 0 0

Enter matrix of constraints (A[1,1]..A[m,n]):
  1  0 1 0 0
  0  1 0 1 0
100 18 0 0 1

Enter right-hand side (b[1]..b[m]):
50 200 5000

Results of approximation:
Optimal value :   2.500000000000000E+003

Initial basic index set : 1 3 2

Initial optimal solution :
 1.400000000000000E+001
 2.000000000000000E+002
 3.600000000000000E+001
 0.000000000000000E+000
 0.000000000000000E+000

Results of verification:
Optimal value interval :  [ 2.500000000000000E+003, 2.500000000000000E+003]

List of optimal basic index sets :
B1 = 1 2 3
B2 = 1 2 4
B3 = 1 3 4

List of optimal basic feasible solutions :
x of B1 =
[ 1.400000000000000E+001, 1.400000000000000E+001]
```

```
[ 2.000000000000000E+002, 2.000000000000000E+002]
[ 3.600000000000000E+001, 3.600000000000000E+001]
[ 0.000000000000000E+000, 0.000000000000000E+000]
[ 0.000000000000000E+000, 0.000000000000000E+000]
x of B2 =
[ 5.000000000000000E+001, 5.000000000000000E+001]
[ 0.000000000000000E+000, 0.000000000000000E+000]
[ 0.000000000000000E+000, 0.000000000000000E+000]
[ 2.000000000000000E+002, 2.000000000000000E+002]
[ 0.000000000000000E+000, 0.000000000000000E+000]
x of B3 =
[ 5.000000000000000E+001, 5.000000000000000E+001]
[ 0.000000000000000E+000, 0.000000000000000E+000]
[ 0.000000000000000E+000, 0.000000000000000E+000]
[ 2.000000000000000E+002, 2.000000000000000E+002]
[ 0.000000000000000E+000, 0.000000000000000E+000]
```

The standard simplex algorithm is neither able to recognize whether there exist more than one optimal solution nor if the computed basic index set and the corresponding solution are truly optimal. Example 11.1 has three optimal basic index sets representing two different vertices of the set of feasible solutions $X$.

**Example 11.2** If we modify the objective function of Example 11.1 to

$$\text{maximize } 50x_1 + 9x_2 + 10^{-40}x_3$$

subject to the linear constraints (11.10), the standard algorithm still returns an optimal solution even though the objective function is unbounded on the set of feasible solutions:

```
Enter dimensions of linear optimization problem (m and n):
3 6

Enter objective function (c[1]..c[n]):
50 9 10e-40 0 0 0

Enter matrix of constraints (A[1,1]..A[m,n]):
  1  0 0 1 0 0
  0  1 0 0 1 0
100 18 0 0 0 1

Enter right-hand side (b[1]..b[m]):
50 200 5000

Results of approximation:
Optimal value :   2.500000000000000E+003

Initial basic index set : 1 4 2

Initial optimal solution :
 1.400000000000000E+001
 2.000000000000000E+002
 0.000000000000000E+000
 3.600000000000000E+001
 0.000000000000000E+000
 0.000000000000000E+000

Error: Initial basic index set not optimal (i.e. B_start not in V_opt)!
```

### 11.3.3 Restrictions and Hints

#### Problem Size

This implementation does not take advantage of any special characteristic of the constraint matrix $A$ in (11.1). It is always assumed to be dense so we can use module *linsys* from Chapter 10. If the problem turns out to be *basis stable* (the most common case, cf. Theorem 11.1), or if it has only a few optimal basic index sets, the size of the problem is bounded by the memory of the host computer necessary to allocate the constraint matrix $A$ and not by the calculation time.

#### Starting Problem

The determination of the initial basic index set $\mathcal{B}_{start}$ using the implementation of the revised simplex algorithm published by Syslo, Deo, and Kowalic [92] may be replaced by any other implementation. We require only that the approximate algorithm returns a set of basic indices of an optimal solution of the linear programming problem $P$.

#### Input Data

The function *LinOpt()* assumes that all input data $P = (A, b, c)$ are exact, i.e. they are not afflicted with tolerances, and they are representable on the machine without conversion errors (see Section 3.7). Only in this case verification of the calculated results is possible. See the improvements paragraph on page 218 for a way around these difficulties by solving a parametric linear programming problem.

## 11.4 Exercises

**Exercise 11.1** Determine the optimal value, the set of optimal solutions, and the set of optimal basic index sets for the linear programming problem

$$\begin{array}{rcrcr} x_1 & + & 2\ x_2 & \leq & 80 \\ & & x_2 & \leq & 30 \\ 2\ x_1 & + & x_2 & \leq & 100 \\ & & x_1, x_2 & \geq & 0, \end{array}$$

using the sample program on page 239.

**Exercise 11.2** Consider the unbalanced transportation problem with three suppliers and three markets:

$$\text{minimize } f(x) = \sum_{i=1}^{3} \sum_{j=1}^{3} c_{ij} x_{ij} \text{ subject to the linear constraints}$$

$$\begin{aligned}\sum_{j=1}^{3} x_{ij} + s_i &= a_i \quad (i = 1,2,3)\\ \sum_{i=1}^{3} x_{ij} - d_j &= b_j \quad (j = 1,2,3)\\ x_{ij},\, s_i,\, d_j &\geq 0 \quad (i,j = 1,2,3)\end{aligned}$$

Let the transportation costs be $c_{ij} = 1$, $(i,j = 1,2,3)$. The supplies are $a_1 = 2$, $a_2 = 6$ and $a_3 = 5$, while all demands are equal to 3, that is $b_j = 3$, $(j = 1,2,3)$.

Transform the linear optimization problem into standard form. Since function *RevSimplex()* from module *rev_simp* will fail to return an appropriate initial basic index set $\mathcal{B}_{start}$, you have to extend the input section of the sample program by the possibility of entering the initial basic index set directly. Change the given implementation by adding an input facility for $\mathcal{B}_{start}$ and remove the call of function *RevSimplex()*.

Use $\mathcal{B}_{start} = (1,5,7,8,9,11)$ and determine the optimal value, the set of optimal solutions, and the set of optimal basic index sets of the unbalanced transportation problem.

## 11.5 References and Further Reading

The standard simplex algorithm was first introduced by Dantzig [14]. A first verifying algorithm for linear programming problems was published by Jansson [38]. There are recent developments inspired by the work of Karmarkar [41] in the field of nonlinear methods for solving linear programming problems. These polynomial-time algorithms that go back to Karmarkar [41] and Khachiyan [48] have been implemented and investigated by a great number of researchers. Especially the projective approach of Karmarkar has been tested successfully solving large-scale optimization problems. A self-validating algorithm using the primal-dual interior point method that is based on Karmarkar's algorithm has been implemented by Hocks [34]. This implementation also involves the verification strategy presented in this chapter.

# Chapter 12

# Automatic Differentiation for Gradients, Jacobians, and Hessians

In Chapter 5, we considered automatic differentiation in one variable, but there are also many applications of numerical and scientific computing where it is necessary to compute derivatives of multi-dimensional functions. In this chapter, we extend the concept of automatic differentiation to the multi-dimensional case as given by Rall [72] and many others. We apply well-known differentiation rules for gradients, Jacobians, or Hessians with the computation of numerical values, combining the advantages of symbolic and numerical differentiation. Only the algorithm or formula for the function is required. No explicit formulas for the gradient, Jacobian, or Hessian have to be derived and coded.

In this chapter, we deal with automatic differentiation based on interval operations to get guaranteed enclosures for the function value and the values of the gradients, Jacobians, or Hessians. We will use gradients, Jacobians and Hessians generated by automatic differentiation in our algorithms for solving nonlinear systems (Chapter 13) and for global multi-dimensional optimization (Chapter 14). The techniques of automatic differentiation are also frequently applied in solving ordinary differential equations, sensitivity analysis, bifurcation studies, and continuation methods. Rall [72] and Griewank and Corliss [23] contain many applications of automatic differentiation.

## 12.1 Theoretical Background

Let $f : \mathbb{R}^n \to \mathbb{R}$ be a scalar-valued and twice continuously differentiable function. We are interested in computing the gradient

$$\nabla f(x) = \begin{pmatrix} \frac{\partial f}{\partial x_1}(x) \\ \frac{\partial f}{\partial x_2}(x) \\ \vdots \\ \frac{\partial f}{\partial x_n}(x) \end{pmatrix}$$

or the Hessian matrix

$$\nabla^2 f(x) = \begin{pmatrix} \frac{\partial^2 f}{\partial x_1^2}(x) & \frac{\partial^2 f}{\partial x_1 \partial x_2}(x) & \cdots & \frac{\partial^2 f}{\partial x_1 \partial x_n}(x) \\ \frac{\partial^2 f}{\partial x_2 \partial x_1}(x) & \frac{\partial^2 f}{\partial x_2^2}(x) & \cdots & \frac{\partial^2 f}{\partial x_2 \partial x_n}(x) \\ \vdots & \vdots & \vdots & \vdots \\ \frac{\partial^2 f}{\partial x_n \partial x_1}(x) & \frac{\partial^2 f}{\partial x_n \partial x_2}(x) & \cdots & \frac{\partial^2 f}{\partial x_n^2}(x) \end{pmatrix}.$$

In the case of functions of a single variable described in Chapter 5, we parsed the function to be differentiated into a code list (see [72]) consisting of a finite sequence of arithmetic operations $+$, $-$, $\cdot$, and $/$ and elementary functions sqrt, exp, sin, etc. Then the code list is interpreted in a differentiation arithmetic. We do exactly the same for multivariable functions in this chapter. We consider only first- and second-order derivative objects, although the same techniques have been used to generate much higher order partial derivatives [9].

*Gradient and Hessian arithmetic* is an arithmetic for ordered triples of the form

$$U = (u_f, u_g, u_h), \quad \text{with } u_f \in I\!R,\ u_g \in I\!R^n,\ u_h \in I\!R^{n \times n},$$

where the scalar $u_f$ denotes the function value $u(x)$ of a twice differentiable function $u : I\!R^n \to I\!R$. The vector $u_g$ and the matrix $u_h$ denote the value of the gradient $\nabla u(x)$ and the value of the Hessian matrix $\nabla^2 u(x)$, respectively, each at a fixed point $x \in I\!R^n$. Our treatment is very similar to that of Rall (see [73]).

For the constant function $u(x) = c$, we set $U = (u_f, u_g, u_h) = (c, 0, 0)$. For the function $u(x) = x_k$ with $k \in \{1, 2, \ldots, n\}$, we define $U = (u_f, u_g, u_h) = (x_k, e^{(k)}, 0)$, where $e^{(k)} \in I\!R^n$ denotes the $k$-th unit vector, and 0 denotes the zero vector or zero matrix, respectively. The rules for the arithmetic are

$$W = U + V \Rightarrow \begin{cases} w_f = u_f + v_f, \\ w_g = u_g + v_g, \\ w_h = u_h + v_h, \end{cases} \tag{12.1}$$

$$W = U - V \Rightarrow \begin{cases} w_f = u_f - v_f, \\ w_g = u_g - v_g, \\ w_h = u_h - v_h, \end{cases} \tag{12.2}$$

$$W = U \cdot V \Rightarrow \begin{cases} w_f = u_f \cdot v_f, \\ w_g = u_f \cdot v_g + v_f \cdot u_g, \\ w_h = v_f \cdot u_h + u_g \cdot v_g^{\mathrm{T}} + v_g \cdot u_g^{\mathrm{T}} + u_f \cdot v_h, \end{cases} \tag{12.3}$$

$$W = U \,/\, V \Rightarrow \begin{cases} w_f = u_f / v_f, \\ w_g = (u_g - w_f \cdot v_g)/v_f, \\ w_h = (u_h - w_g \cdot v_g^{\mathrm{T}} - v_g \cdot w_g^{\mathrm{T}} - w_f \cdot v_h)/v_f, \end{cases} \tag{12.4}$$

where the familiar rules of calculus have to be used in the second and third components, and $v_f \neq 0$ is assumed in the case of division. The operations for $w_f$, $w_g$, and $w_h$ in these definitions are operations on real numbers, vectors, and matrices. If the independent variables $x_i$ of a formula for a function $f : I\!R^n \to I\!R$ and $x \mapsto f(x)$

are replaced by $X_i = (x_i, e^{(k)}, 0)$, and if all constants $c$ are replaced by their $(c, 0, 0)$ representation, then evaluation of $f$ using the rules of differentiation arithmetic gives

$$f(X) = f\left(\begin{pmatrix} X_1 \\ X_2 \\ \vdots \\ X_n \end{pmatrix}\right) = f\left(\begin{pmatrix} (x_1, e^{(1)}, 0) \\ (x_2, e^{(2)}, 0) \\ \vdots \\ (x_n, e^{(n)}, 0) \end{pmatrix}\right) = (f(x), \nabla f(x), \nabla^2 f(x)).$$

**Example 12.1** We want to compute the function, gradient, and Hessian values of the function $f(x) = x_1 \cdot (4 + x_2)$ at the point $x = (1,2)^{\mathrm{T}}$. Differentiation arithmetic gives

$$\begin{aligned} f(X) = (f_f, f_g, f_h) &= (x_1, \begin{pmatrix} 1 \\ 0 \end{pmatrix}, \begin{pmatrix} 0 & 0 \\ 0 & 0 \end{pmatrix}) \cdot ((4,0,0) + (x_2, \begin{pmatrix} 0 \\ 1 \end{pmatrix}, \begin{pmatrix} 0 & 0 \\ 0 & 0 \end{pmatrix})) \\ &= (1, \begin{pmatrix} 1 \\ 0 \end{pmatrix}, \begin{pmatrix} 0 & 0 \\ 0 & 0 \end{pmatrix}) \cdot ((4,0,0) + (2, \begin{pmatrix} 0 \\ 1 \end{pmatrix}, \begin{pmatrix} 0 & 0 \\ 0 & 0 \end{pmatrix})) \\ &= (1, \begin{pmatrix} 1 \\ 0 \end{pmatrix}, \begin{pmatrix} 0 & 0 \\ 0 & 0 \end{pmatrix}) \cdot ((6, \begin{pmatrix} 0 \\ 1 \end{pmatrix}, \begin{pmatrix} 0 & 0 \\ 0 & 0 \end{pmatrix})) \\ &= (6, \begin{pmatrix} 6 \\ 1 \end{pmatrix}, \begin{pmatrix} 0 & 1 \\ 1 & 0 \end{pmatrix}). \end{aligned}$$

The result is $f(x) = 6$, $\nabla f(x) = \begin{pmatrix} 6 \\ 1 \end{pmatrix}$, and $\nabla^2 f(x) = \begin{pmatrix} 0 & 1 \\ 1 & 0 \end{pmatrix}$ for $x = \begin{pmatrix} 1 \\ 2 \end{pmatrix}$.

For an elementary function $s : I\!R \to I\!R$ and $U = (u_f, u_g, u_h)$, we define

$$W = s(U) \Rightarrow \begin{cases} w_f = s(u_f), \\ w_g = s'(u_f) \cdot u_g, \\ w_h = s''(u_f) \cdot u_g \cdot u_g^{\mathrm{T}} + s'(u_f) \cdot u_h, \end{cases} \tag{12.5}$$

Here $s' : I\!R \to I\!R$ and $s'' : I\!R \to I\!R$ are the first and second derivatives of $s$, assuming they exist.

Rules (12.1)–(12.5) have assumed exact arithmetic. To get enclosures for the true values of the function, its gradient, and its Hessian, we use a differentiation arithmetic based on interval arithmetic. That is, the components $u_f$, $u_g$, and $u_h$ are replaced by the corresponding interval values, and all arithmetic operations and function evaluations are executed in an interval arithmetic. The evaluation of a function $f : I\!R^n \to I\!R$ for an argument $[x] \in I\,I\!R^n$ using interval differentiation arithmetic delivers

$$f(X) = ([f_f], [f_g], [f_h])$$

satisfying

$$f([x]) \subseteq [f_f], \quad \nabla f([x]) \subseteq [f_g], \quad \text{and} \quad \nabla^2 f([x]) \subseteq [f_h].$$

Alternately, let $f : \mathbb{R}^n \to \mathbb{R}^n$ be a vector-valued and differentiable function, and let us compute the Jacobian matrix

$$J_f(x) = \begin{pmatrix} \frac{\partial f_1}{\partial x_1}(x) & \frac{\partial f_1}{\partial x_2}(x) & \dots & \frac{\partial f_1}{\partial x_n}(x) \\ \frac{\partial f_2}{\partial x_1}(x) & \frac{\partial f_2}{\partial x_2}(x) & \dots & \frac{\partial f_2}{\partial x_n}(x) \\ \vdots & \vdots & \vdots & \vdots \\ \frac{\partial f_n}{\partial x_1}(x) & \frac{\partial f_n}{\partial x_2}(x) & \dots & \frac{\partial f_n}{\partial x_n}(x) \end{pmatrix}.$$

We can do this using our gradient arithmetic by computing the gradient for each component $f_i$ with $i = 1, 2, \dots, n$. In this case, we need not compute the Hessian components in our Rules (12.1)–(12.5).

## 12.2 Algorithmic Description

We now describe the algorithms for the elementary operators $+$, $-$, $\cdot$, and $/$, and for an elementary function $s \in \{$sqr, sqrt, power, exp, ln, sin, cos, tan, cot, asin, acos, atan, acot, sinh, cosh, tanh, coth, asinh, acosh, atanh, acoth$\}$ of a gradient and Hessian arithmetic for a twice continuously differentiable function $f : \mathbb{R}^n \to \mathbb{R}$. For the operands and arguments $U$, we use the triple representation $U := ([u_f], [u_g], [u_h])$ with $[u_f] \in I\mathbb{R}$, $[u_g] \in I\mathbb{R}^n$, and $[u_h] \in I\mathbb{R}^{n\times n}$. According to the symmetry of the Hessian matrix we really have to compute only the components $[u_h]_{ij}$ with $i = 1, \dots, n$ and $j = 1, \dots, i$.

**Algorithm 12.1:** Operator + $([U], [V])$

1. $[w_f] := [u_f] + [v_f]$; {Function value}
2. **for** $i := 1$ **to** $n$ **do**
   (a) $[w_g]_i := [u_g]_i + [v_g]_i$; {Gradient components}
   (b) **for** $j := 1$ **to** $i$ **do** {Hessian components}
   $[w_h]_{ij} := [u_h]_{ij} + [v_h]_{ij}$;
3. **return** $[W]$;

**Algorithm 12.2:** Operator − $([U], [V])$

1. $[w_f] := [u_f] - [v_f]$; {Function value}
2. **for** $i := 1$ **to** $n$ **do**
   (a) $[w_g]_i := [u_g]_i - [v_g]_i$; {Gradient components}
   (b) **for** $j := 1$ **to** $i$ **do** {Hessian components}
   $[w_h]_{ij} := [u_h]_{ij} - [v_h]_{ij}$;
3. **return** $[W]$;

**Algorithm 12.3: Operator $\cdot$ ([U], [V])**

1. $[w_f] := [u_f] \cdot [v_f]$; {Function value}
2. **for** $i := 1$ **to** $n$ **do**
   (a) $[w_g]_i := [v_f] \cdot [u_g]_i + [u_f] \cdot [v_g]_i$; {Gradient components}
   (b) **for** $j := 1$ **to** $i$ **do** {Hessian components}
   $[w_h]_{ij} := [v_f] \cdot [u_h]_{ij} + [u_g]_i \cdot [v_g]_j + [v_g]_i \cdot [u_g]_j + [u_f] \cdot [v_h]_{ij}$;
3. **return** $[W]$;

In Algorithm 12.4, we do not take care of the case $0 \in [v_f]$, because it does not make sense to go any further in computations when this case occurs. In an implementation, the standard error handling (runtime error) should be invoked if a division by zero occurs while computing the function value. We chose a special form of the rule for the differentiation of a quotient to be close to our implementation, where this form can save some computation time.

**Algorithm 12.4: Operator / ([U], [V])**

1. $[w_f] := [u_f]/[v_f]$; {Function value}
2. **for** $i := 1$ **to** $n$ **do**
   (a) $[w_g]_i := ([u_g]_i - [w_f] \cdot [v_g]_i)/[v_f]$; {Gradient components}
   (b) **for** $j := 1$ **to** $i$ **do** {Hessian components}
   $[w_h]_{ij} := ([u_h]_{ij} - [w_g]_i \cdot [v_g]_j - [v_g]_i \cdot [w_g]_j - [w_f] \cdot [v_h]_{ij})/[v_f]$;
3. **return** $[W]$;

In Algorithm 12.5, the rules from calculus for derivatives of the elementary functions are applied to compute the temporary values. These functions are listed in Table 5.1 in Chapter 5. As in the case of division by zero, we do not have to treat specially the case when the interval argument $[u_f]$ does not lie in the domain specified for the interval functions $s \in E$ (see page 74). In an implementation of Algorithm 12.5, the standard error handling of interval arithmetic should be invoked for domain violations in Steps 1 and 2. See also the discussion of Table 5.1 in Chapter 5.

**Algorithm 12.5: s([U])**

1. $[w_f] := s([u_f])$; {Function value}
2. $[h_1] := s'([u_f])$; $[h_2] := s''([u_f])$; {Temporary values}
3. **for** $i := 1$ **to** $n$ **do**
   (a) $[w_g]_i := [h_1] \cdot [u_g]_i$; {Gradient components}
   (b) **for** $j := 1$ **to** $i$ **do** {Hessian components}
   $[w_h]_{ij} := [h_2] \cdot [u_g]_i \cdot [u_g]_j + [h_1] \cdot [u_h]_{ij}$;
4. **return** s $:= [W]$;

# 12.3 Implementation and Examples

## 12.3.1 C++ Program Code

In the following sections, we present two modules for multi-dimensional automatic differentiation. We give a module supplying differentiation arithmetic for gradients *and* Hessians for scalar-valued functions in Section 12.3.1.1. The module can also be used to compute Jacobians of vector-valued functions. We give a special gradient arithmetic *grad_ari* for computing only gradients for scalar-valued functions and Jacobians for vector-valued functions without the storage overhead for the Hessian components in Section 12.3.1.2.

### 12.3.1.1 Module hess_ari

The header file of module *hess_ari* supplies the definitions of the three classes *LowTriMatrix*, *HessType*, and *HTvector* including operators and elementary functions for an interval differentiation arithmetic for gradients and Hessians. The data type *LowTriMatrix* represents a lower triangular matrix. It is used to store the Hessian matrix of an object of type *HessType*. The type names *HTscalar_FctPtr* and *HTvector_FctPtr* can be used to declare pointers to scalar-valued and vector-valued functions with an argument of type *HTvector*, respectively.

```
//----------------------------------------------------------------------------
// File: hess_ari (header)
// Purpose: Definition of a multi-dimensional interval differentiation
//    arithmetic which allows function evaluation with automatic differen-
//    tiation up to second order (i.e. Hessian matrix).
// Types:
//    HTscalar_FctPtr          : pointer for a scalar valued function
//    HTvector_FctPtr          : pointer for a vector valued function
// Class LowTriMatrix:
//    LowTriMatrix()           : constructors
//    ~LowTriMatrix()          : destructor
//    Resize()                 : for resizing to a fixed dimension
//    operator =               : assignment operators for triangular matrices
//                               and real values
//    operator []              : component access
// Class HessType:
//    HessType()               : constructors
//    Resize()                 : for resizing to a fixed dimension
//    operator =               : assignment operators for arguments of types
//                               HessType, interval, and real
//    Dim()                    : returns the dimension of the Hessian matrix
//    HessVar()                : to define HessType variables
//    fValue(), gradValue(),
//    hessValue()              : to get function and derivative values
//    operators +, -, *, /     : operators of diff. arithmetic
//    sqr(), sqrt(), power(),
//    exp(), sin(), cos(), ...: elementary functions of diff. arithmetic
//    fEvalH()                 : to compute function value only
//    fgEvalH()                : to compute function and first derivative
//                               value (gradient)
//    fghEvalH()               : to compute function value, gradient, and
//                               Hessian matrix value
// Class HTvector:
```

```
//    HTvector()                : constructors
//    ~HTvector()               : destructor
//    operator =                : assignment operator
//    operator []               : component access
//    Dim()                     : to get the actual dimension
//    fValue(), JacValue()      : to get function and derivative values
//    fEvalJ()                  : to compute function value only
//    fJEvalJ()                 : to compute function and first derivative
//                                value (Jacobian)
//----------------------------------------------------------------------------
#ifndef __HESS_ARI_HPP
#define __HESS_ARI_HPP

#include <interval.hpp>     // Interval arithmetic
#include <imatrix.hpp>      // Interval matrix/vector arithmetic
#include <mvi_util.hpp>     // Interval matrix/vector utility functions

class LowTriMatrix {
    int      dim;
    ivector *rows;

    void init ( int );
    friend void Resize ( LowTriMatrix&, index );

  public:
    LowTriMatrix ( );
    LowTriMatrix ( index );
    LowTriMatrix ( LowTriMatrix& );
    ~LowTriMatrix ( );

    LowTriMatrix& operator= ( LowTriMatrix& );
    LowTriMatrix& operator= ( const real& );
    ivector& operator[] ( int );
};

class HessType;
class HTvector;

typedef HessType (*HTscalar_FctPtr)(HTvector&);
typedef HTvector (*HTvector_FctPtr)(HTvector&);

class HessType {
    int              nmax;
    interval         f;
    ivector          g;
    LowTriMatrix h;

    friend void Resize ( HessType&, index );
    friend void TestSize ( HessType&, HessType&, const char* );

  public:
    HessType ( );
    HessType ( index );
    HessType ( HessType& );

    HessType& operator= ( HessType& );
    HessType& operator= ( interval& );
    HessType& operator= ( const real& );

    int  Dim ( ) { return nmax; };

    friend HTvector HessVar ( ivector& );
    friend HTvector HessVar ( rvector& );
```

```
    friend interval fValue    ( HessType& );
    friend ivector  gradValue ( HessType& );
    friend imatrix  hessValue ( HessType& );

    friend HessType operator+ ( HessType&);
    friend HessType operator- ( HessType&);
    friend HessType operator+ ( HessType&, HessType&);
    friend HessType operator- ( HessType&, HessType&);
    friend HessType operator* ( HessType&, HessType&);
    friend HessType operator/ ( HessType&, HessType&);
    friend HessType operator+ ( HessType&, interval&);
    friend HessType operator- ( HessType&, interval&);
    friend HessType operator* ( HessType&, interval&);
    friend HessType operator/ ( HessType&, interval&);
    friend HessType operator+ ( interval&, HessType&);
    friend HessType operator- ( interval&, HessType&);
    friend HessType operator* ( interval&, HessType&);
    friend HessType operator/ ( interval&, HessType&);
    friend HessType operator+ ( HessType&, const real&);
    friend HessType operator- ( HessType&, const real&);
    friend HessType operator* ( HessType&, const real&);
    friend HessType operator/ ( HessType&, const real&);
    friend HessType operator+ ( const real&, HessType&);
    friend HessType operator- ( const real&, HessType&);
    friend HessType operator* ( const real&, HessType&);
    friend HessType operator/ ( const real&, HessType&);

    friend HessType sqr   ( HessType& );
    friend HessType power ( HessType&, const int );
    friend HessType sqrt  ( HessType& );
    friend HessType exp   ( HessType& );
    friend HessType ln    ( HessType& );
    friend HessType sin   ( HessType& );
    friend HessType cos   ( HessType& );
    friend HessType tan   ( HessType& );
    friend HessType cot   ( HessType& );
    friend HessType asin  ( HessType& );
    friend HessType acos  ( HessType& );
    friend HessType atan  ( HessType& );
    friend HessType acot  ( HessType& );
    friend HessType sinh  ( HessType& );
    friend HessType cosh  ( HessType& );
    friend HessType tanh  ( HessType& );
    friend HessType coth  ( HessType& );
    friend HessType asinh ( HessType& );
    friend HessType acosh ( HessType& );
    friend HessType atanh ( HessType& );
    friend HessType acoth ( HessType& );

    friend void fEvalH   ( HTscalar_FctPtr, ivector, interval& );
    friend void fgEvalH  ( HTscalar_FctPtr, ivector, interval&, ivector& );
    friend void fghEvalH ( HTscalar_FctPtr, ivector, interval&, ivector&,
                                                                imatrix& );
};

class HTvector {
    int       nmax;
    HessType *ht;

    friend void TestSize ( HTvector&, HTvector&, const char* );

  public:
    HTvector  ( index );
    HTvector  ( HTvector& );
```

```
    ~HTvector ( );

    HTvector& operator=  ( HTvector& );
    HessType& operator[] ( index );

    int Dim ( ) { return nmax; };

    friend ivector  fValue   ( HTvector& );
    friend imatrix  JacValue ( HTvector& );

    friend void fEvalJ  ( HTvector_FctPtr, ivector, ivector& );
    friend void fJEvalJ ( HTvector_FctPtr, ivector, ivector&, imatrix& );
};
#endif
```

The rows of a lower triangular matrix are stored in the data member *rows* of the class *LowTriMatrix*. The $i$th row of such a matrix has $i$ components and can be accessed by *rows[i]*.

The static variable *HessOrder* is used to select the highest order of derivative desired. This enables the user to save computation time in computing only the function value or the gradient and no Hessian, although storage for Hessian elements is still allocated. The default value of *HessOrder* is 2, so normally the gradient and the Hessian are computed.

The functions *fEvalH()*, *fgEvalH()*, *fghEvalH()*, *fEvalJ()*, and *fJEvalJ()* simplify the mechanism of function evaluating and automate the setting and resetting of the *HessOrder* variable. The ... *EvalH()* functions can be applied to scalar-valued functions (result type *HessType*), whereas the ... *EvalJ()* functions can be applied to vector-valued functions (result type *HTvector*).

For a scalar-valued function of type *HessType*, *fEvalH()* computes and returns only the function value by setting *HessOrder* to 0 before the evaluation is done. If the gradient also is desired, *fgEvalH()* can be used to set *HessOrder* to 1 before the evaluation is done. The function *fghEvalH()* uses the default value of *HessOrder*, computes, and returns the values of $f(x)$, $\nabla f(x)$, and $\nabla^2 f(x)$. For a vector-valued function of type *HTvector*, *fEvalJ()* computes and returns only the function value by setting *HessOrder* to 0 before the evaluation is done. If the Jacobian matrix also is desired, *fJEvalJ()* can be used to set *HessOrder* to 1 before the evaluation is done.

```
//----------------------------------------------------------------------------
// File: hess_ari (implementation)
// Purpose: Definition of a multi-dimensional interval differentiation
//    arithmetic which allows function evaluation with automatic differen-
//    tiation up to second order (i.e. Hessian matrix).
// Method: Overloading of operators and elementary functions for operations
//    of data types 'HessType' and 'HTvector'. To represent a Hessian matrix
//    by a lower triangular matrix, the type 'LowTriMatrix' is supplied.
// Types:
//    HTscalar_FctPtr        : pointer for a scalar valued function
//    HTvector_FctPtr        : pointer for a vector valued function
// Class LowTriMatrix:
//    LowTriMatrix()         : constructors
//    ~LowTriMatrix()        : destructor
//    Resize()               : for resizing to a fixed dimension
//    operator =             : assignment operators for triangular matrices
//                             and real values
//    operator []            : component access
```

```
// Class HessType:
//    HessType()                 : constructors
//    Resize()                   : for resizing to a fixed dimension
//    operator =                 : assignment operators for arguments of types
//                                 HessType, interval, and real
//    Dim()                      : returns the dimension of the Hessian matrix
//    HessVar()                  : to define HessType variables
//    fValue(), gradValue(),
//    hessValue()                : to get function and derivative values
//    operators +, -, *, /       : operators of diff. arithmetic
//    sqr(), sqrt(), power(),
//    exp(), sin(), cos(), ...: elementary functions of diff. arithmetic
//    fEvalH()                   : to compute function value only
//    fgEvalH()                  : to compute function and first derivative
//                                 value (gradient)
//    fghEvalH()                 : to compute function value, gradient, and
//                                 Hessian matrix value
// Class HTvector:
//    HTvector()                 : constructors
//    ~HTvector()                : destructor
//    operator =                 : assignment operator
//    operator []                : component access
//    Dim()                      : to get the actual dimension
//    fValue(), JacValue()       : to get function and derivative values
//    fEvalJ()                   : to compute function value only
//    fJEvalJ()                  : to compute function and first derivative
//                                 value (Jacobian)
//----------------------------------------------------------------------------
#include <stdlib.h>         // For function 'exit()'
#include <iostream.h>       // I/O Handling
#include <imath.hpp>        // Interval mathematical functions
#include <i_util.hpp>       // Interval utility functions
#include <hess_ari.hpp>

// The local variable 'HessOrder' is used to select the highest order of
// derivative which is computed. Its default value is 2, and normally
// the gradient and the Hessian matrix are computed.
//----------------------------------------------------------------------------
static int HessOrder = 2;

//----------------------------------------------------------------------------
// Constructors, destructor, and resize function for the class 'LowTriMatrix'.
//----------------------------------------------------------------------------
void LowTriMatrix::init ( int n )         // Initializes all components of the
{                                         // lower triangular matrix with 0.0
  if (n <= 0)                             //---------------------------------
    { dim = 0; rows = NULL; return; }
  dim = n;
  rows = new ivector[n];
  for (int i = 0; i < n; i++)                  // i-th row of the matrix takes
    { Resize(rows[i],i+1); rows[i] = 0.0; }    // i components, i = 1,2,...,n
}                                              //----------------------------

LowTriMatrix::LowTriMatrix ( )                      // Default: no matrix rows
  { dim = 0; rows = NULL; }                         //------------------------


LowTriMatrix::LowTriMatrix ( index n )             // Constructor: dimension n
  { init(n._int()); }                              //-------------------------

LowTriMatrix::~LowTriMatrix ( )                                   // Destructor
  { delete [] rows; }                                             //-----------

void Resize ( LowTriMatrix& a, index n )           // Resizes matrix to n rows
```

```
  { delete [] a.rows; a.init(n._int()); }          //-------------------------

//---------------------------------------------------------------------------
// Assignment operators and copy constructor of the class 'LowTriMatrix'.
//---------------------------------------------------------------------------
LowTriMatrix& LowTriMatrix::operator= ( LowTriMatrix& a )   // matrix = matrix
{                                                           //----------------
  if (this == &a) return *this;
  delete [] rows;
  dim = a.dim;
  if (dim == 0) { rows = NULL; return *this; }
  rows = new ivector[dim];
  for (int i = 0; i < dim; i++) rows[i] = a.rows[i];
  return *this;
}

LowTriMatrix& LowTriMatrix::operator= ( const real& r )       // matrix = real
{                                                             //--------------
  for (int i = 0; i < dim; i++) rows[i] = r;
  return *this;
}

LowTriMatrix::LowTriMatrix ( LowTriMatrix& a )             // Copy constructor
{                                                          //-----------------
  rows = NULL;
  *this = a;   // use previously defined assignment operator
}

//---------------------------------------------------------------------------
// Component access for the type 'LowTriMatrix'
//---------------------------------------------------------------------------
ivector& LowTriMatrix::operator[] ( int i )
{
  if ( (i < 1) || (i > dim) ) {
    cout << "Index out of range in "
         << "'ivector& LowTriMatrix::operator[] ( int )'!" << endl;
    exit(-1);
  }
  return rows[i-1];       // Index range starts at 1
}

//---------------------------------------------------------------------------
// Constructors, destructor, and resize function for the classes 'HessType'
// and 'HTvector'.
//---------------------------------------------------------------------------
HessType::HessType ( )                               // Default: no components
  { nmax = 0; }                                      //-----------------------

HessType::HessType ( index n )                 // Constructor: resizes internal
{                                              // arrays to dimension 'nmax'
  nmax = n._int();                             //-----------------------------
  if (nmax <= 0) { nmax = 0; return; }
  Resize(g,nmax);
  Resize(h,nmax);
}

void Resize ( HessType& u, index n )
{
  int dim = n._int();
  u.nmax = dim;
  Resize(u.g,dim);
  Resize(u.h,dim);
}
```

```
HTvector::HTvector ( index n )                       // Constructor: allocates 'n'
{                                                    // 'HessType' components of
  nmax = n._int();                                   // dimension 'n'
  if (nmax <= 0)                                     //---------------------------
    { nmax = 0; ht = NULL; return; }
  ht = new HessType[nmax];
  for (int i = 0; i < nmax; i++) Resize(ht[i],nmax);
}

HTvector::~HTvector ( )                                              // Destructor
  { delete [] ht; }                                                  //-----------

//----------------------------------------------------------------------------
// Assignment operators and copy constructors of the classes 'HessType' and
// 'HTvector'.
//----------------------------------------------------------------------------
HessType& HessType::operator= ( HessType& u )               // HessType = HessType
{                                                           //--------------------
  if (this == &u) return *this;
  nmax = u.nmax;
  if (nmax > 0) { f = u.f;  g = u.g;  h = u.h; }
  return *this;
}

HessType::HessType ( HessType& u )                             // Copy constructor
{                                                              //-----------------
  nmax = u.nmax;
  *this = u;   // use previously defined assignment operator
}

HessType& HessType::operator= ( interval& x )              // HessType =  interval
{                                                          //---------------------
  f = x;  g = 0.0;  h = 0.0;
  return *this;
}

HessType& HessType::operator= ( const real& r )                // HessType =  real
{                                                              //-----------------
  *this = _interval(r);
  return *this;
}

HTvector& HTvector::operator= ( HTvector& v )               // HTvector = HTvector
{                                                           //--------------------
  if (this == &v) return *this;
  delete [] ht;
  nmax = v.nmax;
  if (nmax == 0) { ht = NULL; return *this; }
  ht = new HessType[nmax];
  for (int i = 0; i < nmax; i++) ht[i] = v.ht[i];
  return *this;
}

HTvector::HTvector ( HTvector& v )                             // Copy constructor
{                                                              //-----------------
  ht = NULL;
  *this = v;   // use previously defined assignment operator
}

//----------------------------------------------------------------------------
// Component access for the type 'HTvector'.
//----------------------------------------------------------------------------
HessType& HTvector::operator[] ( index ind )
{
```

```
  int n = ind._int();
  if ( (n < 1) || (n > nmax) ) {
    cout << "Index out of range in "
         << "'HessType& HTvector::operator[] ( index )'!" << endl;
    exit(-1);
  }
  return ht[n-1];           // Index range starts at 1
}

//----------------------------------------------------------------------------
// Functions for range check of 'HessType' and 'HTvector' objects.
//----------------------------------------------------------------------------
void TestSize ( HessType& a, HessType& b, const char *fname )
{
  if (a.nmax != b.nmax) {
    cout << "Parameters must be of same size in '" << fname
         << "'!" << endl;
    exit(-1);
  }
}

void TestSize ( HTvector& v, HTvector& w, const char *fname)
{
  if (v.nmax != w.nmax) {
    cout << "Parameters must be of same size in '" << fname
         << "'!" << endl;
    exit(-1);
  }
}

//----------------------------------------------------------------------------
// Transfer functions for variables
//----------------------------------------------------------------------------
HTvector HessVar ( ivector& x )                          // Generate variable
{                                                        //------------------
  int       d = -Lb(x)+1;
  int       ubd = Ub(x)+d;
  HTvector  hht(ubd);

  for (int i = 1; i <= ubd; i++) {
    hht[i].f = x[i-d];
    for (int k = 1; k <= ubd; k++)
      hht[i].g[k] = (i == k) ? 1.0 : 0.0;
    hht[i].h = 0.0;
  }
  return hht;
}

HTvector HessVar ( rvector& v )                          // Generate variable
{                                                        //------------------
  int     i, n = Lb(v), m = Ub(v);
  ivector u(n,m);

  for (i = n; i <= m; i++) u[i] = v[i];
  return HessVar(u);
}

//----------------------------------------------------------------------------
// Access functions for the function value, the gradient, or the Hessian
//----------------------------------------------------------------------------
interval fValue ( HessType& u )                         // Get function value
  { return u.f; }                                       //-------------------

ivector gradValue ( HessType& u )                       // Get gradient value
```

```
  { return u.g; }                                           //------------------

imatrix hessValue ( HessType& u )                            // Get Hessian value
{                                                            //------------------
  imatrix hm(u.nmax,u.nmax);

  for (int i = 1; i <= u.nmax; i++) {
    hm[i][i] = u.h[i][i];
    for (int j=1; j <= i-1; j++) {
      hm[i][j] = u.h[i][j];
      hm[j][i] = u.h[i][j];
    }
  }
  return hm;
}

//----------------------------------------------------------------------------
// Access functions for vector-valued functions (function value or Jacobian)
//----------------------------------------------------------------------------
ivector fValue ( HTvector& u )                          // Get function value of
{                                                       // n-dimensional vector-
  ivector hv(u.nmax);                                   // valued function
                                                        //----------------------
  for (int i = 1; i <= u.nmax; i++) hv[i] = fValue(u[i]);
  return hv;
}

imatrix JacValue ( HTvector& u )                        // Get Jacobian value of
{                                                       // n-dimensional scalar-
  imatrix hm(u.nmax,u.nmax);                            // valued function
                                                        //----------------------
  for (int i = 1; i <= u.nmax; i++) hm[i] = gradValue(u[i]);
  return hm;
}

//----------------------------------------------------------------------------
// Unary operators + and - for HessType operands
//----------------------------------------------------------------------------
HessType operator+ ( HessType& u )
  { return u; }

HessType operator- ( HessType& u )
{
  HessType umin(u.nmax);

  umin.f = -u.f;
  if (HessOrder > 0)
    for (int i = 1; i <= u.nmax; i++) {
      umin.g[i] = -u.g[i];
      if (HessOrder > 1)
        for (int j = 1; j <= i; j++) umin.h[i][j] = -u.h[i][j];
    }
  return umin;
}

//----------------------------------------------------------------------------
// Operators +, -, *, and / for two HessType operands
//----------------------------------------------------------------------------
HessType operator+ ( HessType& u, HessType& v )
{
  HessType add(u.nmax);

  TestSize(u,v,"operator+ ( HessType&, HessType& )");
  add.f = u.f + v.f;
```

```
  if (HessOrder > 0)
    for (int i = 1; i <= u.nmax; i++) {
      add.g[i] = u.g[i] + v.g[i];
      if (HessOrder > 1)
        for (int j = 1; j <= i; j++) add.h[i][j] = u.h[i][j] + v.h[i][j];
    }
  return add;
}

HessType operator- ( HessType& u, HessType& v )
{
  HessType sub(u.nmax);

  TestSize(u,v,"operator- ( HessType&, HessType& )");
  sub.f = u.f - v.f;
  if (HessOrder > 0)
    for (int i=1; i <= u.nmax; i++) {
      sub.g[i] = u.g[i] - v.g[i];
      if (HessOrder > 1)
        for (int j = 1; j <= i; j++) sub.h[i][j] = u.h[i][j] - v.h[i][j];
    }
  return sub;
}

HessType operator* ( HessType& u, HessType& v )
{
  HessType mul(u.nmax);

  TestSize(u,v,"operator* ( HessType&, HessType& )");
  mul.f = u.f * v.f;
  if (HessOrder > 0)
    for (int i = 1; i <= u.nmax; i++) {
      mul.g[i] = v.f*u.g[i] + u.f*v.g[i];
      if (HessOrder > 1)
        for (int j = 1; j <= i; j++)
          mul.h[i][j] = v.f*u.h[i][j] + u.g[i]*v.g[j] + v.g[i]*u.g[j]
                                                      + u.f*v.h[i][j];
    }
  return mul;
}

HessType operator/ ( HessType& u,HessType& v )
{
  HessType div(u.nmax);

  TestSize(u,v,"operator/ ( HessType&, HessType& )");
  div.f = u.f / v.f;              // Can propagate 'division by zero' error
  if (HessOrder > 0)              //---------------------------------------
    for (int i = 1; i <= u.nmax; i++) {
      div.g[i] = (u.g[i] - div.f*v.g[i]) / v.f;
      if (HessOrder > 1)
        for (int j = 1; j <= i; j++)
          div.h[i][j] = (u.h[i][j] - div.g[i]*v.g[j] - v.g[i]*div.g[j]
                                                     - div.f*v.h[i][j]) / v.f;
    }
  return div;
}

//----------------------------------------------------------------------------
// Operators +, -, *, and / for one interval and one HessType operand
//----------------------------------------------------------------------------
HessType operator+ ( HessType& u, interval& b )
{
  HessType add(u.nmax);
```

```
  add = u; add.f += b;
  return add;
}

HessType operator- ( HessType& u, interval& b )
{
  HessType sub(u.nmax);

  sub = u; sub.f -= b;
  return sub;
}

HessType operator* ( HessType& u, interval& b )
{
  HessType mul(u.nmax);

  mul.f = u.f * b;
  if (HessOrder > 0)
    for (int i = 1; i <= u.nmax; i++) {
      mul.g[i] = b * u.g[i];
      if (HessOrder > 1)
        for (int j = 1; j <= i; j++) mul.h[i][j] = b * u.h[i][j];
    }
  return mul;
}

HessType operator/ ( HessType& u, interval& b )
{
  HessType div(u.nmax);

  div.f = u.f / b;                 // Can propagate 'division by zero' error
  if (HessOrder > 0)               //---------------------------------------
    for (int i = 1; i <= u.nmax; i++) {
      div.g[i] = u.g[i] / b;
      if (HessOrder > 1)
        for (int j = 1; j <= i; j++) div.h[i][j] = u.h[i][j] / b;
    }
  return div;
}

HessType operator+ ( interval& a, HessType& v )
{
  HessType add(v.nmax);

  add = v; add.f = a + v.f;
  return add;
}

HessType operator- ( interval& a, HessType& v )
{
  HessType sub(v.nmax);

  sub.f = a - v.f;
  if (HessOrder > 0)
    for (int i = 1; i <= v.nmax; i++) {
      sub.g[i] = -v.g[i];
      if (HessOrder > 1)
        for (int j = 1; j <= i; j++) sub.h[i][j] = -v.h[i][j];
    }
  return sub;
}

HessType operator* ( interval& a, HessType& v )
```

```
{
  HessType mul(v.nmax);

  mul.f = a * v.f;
  if (HessOrder > 0)
    for (int i = 1; i <= v.nmax; i++) {
      mul.g[i] = a * v.g[i];
      if (HessOrder > 1)
        for (int j = 1; j <= i; j++) mul.h[i][j] = a * v.h[i][j];
    }
  return mul;
}

HessType operator/ ( interval& a, HessType& v )
{
  HessType div(v.nmax);
  interval p, q;

  div.f = a / v.f;                // Can propagate 'division by zero' error
  if (HessOrder > 0) {            //---------------------------------------
    p = -div.f / v.f;
    q = (-2.0*p) / v.f;
    for (int i = 1; i <= v.nmax; i++) {
      div.g[i] = p * v.g[i];
      if (HessOrder > 1)
        for (int j = 1; j <= i; j++)
          div.h[i][j] = p*v.h[i][j] + q*v.g[i]*v.g[j];
    }
  }
  return div;
}

//----------------------------------------------------------------------------
// Operators +, -, * and / for one real and one HessType operand
//----------------------------------------------------------------------------
HessType operator+ ( HessType& u, const real& b )
  { return u + _interval(b); }

HessType operator- ( HessType& u, const real& b )
  { return u - _interval(b); }

HessType operator* ( HessType& u, const real& b )
  { return u * _interval(b); }

HessType operator/ ( HessType& u, const real& b )
  { return u / _interval(b); }   // Can propagate 'division by zero' error
                                 //---------------------------------------
HessType operator+ ( const real& a, HessType& v )
  { return _interval(a) + v; }

HessType operator- ( const real& a, HessType& v )
  { return _interval(a) - v; }

HessType operator* ( const real& a, HessType& v )
  { return _interval(a) * v; }

HessType operator/ ( const real& a, HessType& v )
  { return _interval(a) / v; }   // Can propagate 'division by zero' error
                                 //---------------------------------------

//----------------------------------------------------------------------------
// Elementary function for HessType arguments
//----------------------------------------------------------------------------
HessType sqr ( HessType& u )
```

```
{
  HessType res(u.nmax);
  interval h1;

  res.f = sqr(u.f);                  // Can propagate domain error
  if (HessOrder > 0) {               //---------------------------
    h1 = 2.0 * u.f;
    for (int i = 1; i <= u.nmax; i++) {
      res.g[i] = h1 * u.g[i];
      if (HessOrder > 1)
        for (int j = 1; j <= i; j++)
          res.h[i][j] = h1*u.h[i][j] + 2.0*u.g[i]*u.g[j];
    }
  }
  return res;
}

HessType power ( HessType& u, const int k )
{
  HessType res(u.nmax);
  interval h1;

  if (k == 0)
    { res = 1.0; return res; }
  if (k == 1)
    return u;
  if (k == 2)
    return sqr(u);

  res.f = Power(u.f,k);
  if (HessOrder > 0) {
    h1 = double(k) * Power(u.f,k-1);
    for (int i = 1; i <= u.nmax; i++) {
      res.g[i] = h1 * u.g[i];
      if (HessOrder > 1)
        for (int j = 1; j <= i; j++)
          res.h[i][j] = h1*u.h[i][j] +
                        double(k*(k-1))*Power(u.f,k-2)*u.g[i]*u.g[j];
    }
  }
  return res;
}

HessType sqrt ( HessType& u )
{
  HessType res(u.nmax);
  interval h0, h1, h2;

  h0 = sqrt(u.f);
  res.f = h0;
  if (HessOrder > 0) {
    h1 = 0.5 / h0;
    h2 = -0.5*h1 / u.f;
    for (int i = 1; i <= u.nmax; i++) {
      res.g[i] = h1 * u.g[i];
      if (HessOrder > 1)
        for (int j = 1; j <= i; j++)
          res.h[i][j] = h1*u.h[i][j] + h2*u.g[i]*u.g[j];
    }
  }
  return res;
}

HessType exp ( HessType& u )
```

```
{
  HessType res(u.nmax);
  interval h0;

  h0 = exp(u.f);
  res.f = h0;
  if (HessOrder > 0)
    for (int i = 1; i <= u.nmax; i++) {
      res.g[i] = h0 * u.g[i];
      if (HessOrder > 1)
        for (int j = 1; j <= i; j++)
          res.h[i][j] = h0 * (u.h[i][j] + u.g[i]*u.g[j]);
    }
  return res;
}

HessType ln ( HessType& u )
{
  HessType res(u.nmax);
  interval h1,h2;

  res.f = ln(u.f);                    // Can propagate domain error
  if (HessOrder > 0) {                //-----------------------------
    h1 = 1.0/u.f;
    h2 = -sqr(h1);
    for (int i = 1; i <= u.nmax; i++) {
      res.g[i] = h1 * u.g[i];
      if (HessOrder > 1)
        for (int j = 1; j <= i; j++)
          res.h[i][j] = h1*u.h[i][j] + h2*u.g[i]*u.g[j];
    }
  }
  return res;
}

HessType sin ( HessType& u )
{
  HessType res(u.nmax);
  interval h0, h1, h2;

  h0 = sin(u.f);
  res.f = h0;
  if (HessOrder > 0) {
    h1 = cos(u.f);
    h2 = -h0;
    for (int i = 1; i <= u.nmax; i++) {
      res.g[i] = h1 * u.g[i];
      if (HessOrder > 1)
        for (int j = 1; j <= i; j++)
          res.h[i][j] = h1*u.h[i][j] + h2*u.g[i]*u.g[j];
    }
  }
  return res;
}

HessType cos ( HessType& u )
{
  HessType res(u.nmax);
  interval h0, h1, h2;

  h0 = cos(u.f);
  res.f = h0;
  if (HessOrder > 0) {
    h1 = -sin(u.f);
```

```
    h2 = -h0;
    for (int i = 1; i <= u.nmax; i++) {
      res.g[i] = h1 * u.g[i];
      if (HessOrder > 1)
        for (int j = 1; j <= i; j++)
          res.h[i][j] = h1*u.h[i][j] + h2*u.g[i]*u.g[j];
    }
  }
  return res;
}

HessType tan ( HessType& u )
{
  HessType res(u.nmax);
  interval h0, h1, h2;

  h0 = tan(u.f);                // Can propagate domain error
  res.f = h0;                   //---------------------------
  if (HessOrder > 0) {
    h1 = sqr(h0) + 1.0;                    // The subdistributive law implies
    h2 = 2.0*h0*h1;                        //    h0 * (h0↑2 + 1) <= h0↑3 + h0.
    for (int i = 1; i <= u.nmax; i++) {    // So, the first form is used.
      res.g[i] = h1 * u.g[i];              //---------------------------------
      if (HessOrder > 1)
        for (int j = 1; j <= i; j++)
          res.h[i][j] = h1*u.h[i][j] + h2*u.g[i]*u.g[j];
    }
  }
  return res;
}

HessType cot ( HessType& u )
{
  HessType res(u.nmax);
  interval h0, h1, h2;

  h0 = cot(u.f);                // Can propagate domain error
  res.f = h0;                   //---------------------------
  if (HessOrder > 0) {
    h1 = -(sqr(h0) + 1.0);                 // The subdistributive law implies
    h2 = -2.0*h0*h1;                       //    h0 * (h0↑2 + 1) <= h0↑3 + h0.
    for (int i = 1; i <= u.nmax; i++) {    // So, the first form is used.
      res.g[i] = h1 * u.g[i];              //---------------------------------
      if (HessOrder > 1)
        for (int j = 1; j <= i; j++)
          res.h[i][j] = h1*u.h[i][j] + h2*u.g[i]*u.g[j];
    }
  }
  return res;
}

HessType asin ( HessType& u )
{
  HessType res(u.nmax);
  interval h0, h1, h2;

  res.f = asin(u.f);                // Can propagate domain error
  if (HessOrder > 0) {              //---------------------------
    h0 = 1.0-sqr(u.f);
    h1 = 1.0/sqrt(h0);
    h2 = u.f*h1/h0;
    for (int i = 1; i <= u.nmax; i++) {
      res.g[i] = h1 * u.g[i];
      if (HessOrder > 1)
```

```
        for (int j = 1; j <= i; j++)
          res.h[i][j] = h1*u.h[i][j] + h2*u.g[i]*u.g[j];
    }
  }
  return res;
}

HessType acos ( HessType& u )
{
  HessType res(u.nmax);
  interval h0, h1, h2;

  res.f = acos(u.f);                // Can propagate domain error
  if (HessOrder > 0) {              //---------------------------
    h0 = 1.0-sqr(u.f);
    h1 = -1.0/sqrt(h0);
    h2 = u.f*h1/h0;
    for (int i = 1; i <= u.nmax; i++) {
      res.g[i] = h1 * u.g[i];
      if (HessOrder > 1)
        for (int j = 1; j <= i; j++)
          res.h[i][j] = h1*u.h[i][j] + h2*u.g[i]*u.g[j];
    }
  }
  return res;
}

HessType atan ( HessType& u )
{
  HessType res(u.nmax);
  interval h1, h2;

  res.f = atan(u.f);
  if (HessOrder > 0) {
    h1 = 1.0/(1.0+sqr(u.f));
    h2 = -2.0*u.f*sqr(h1);
    for (int i = 1; i <= u.nmax; i++) {
      res.g[i] = h1 * u.g[i];
      if (HessOrder > 1)
        for (int j = 1; j <= i; j++)
          res.h[i][j] = h1*u.h[i][j] + h2*u.g[i]*u.g[j];
    }
  }
  return res;
}

HessType acot ( HessType& u )
{
  HessType res(u.nmax);
  interval h1, h2;

  res.f = acot(u.f);                // Can propagate domain error
  if (HessOrder > 0) {              //---------------------------
    h1 = -1.0/(1.0+sqr(u.f));
    h2 = 2.0*u.f*sqr(h1);
    for (int i = 1; i <= u.nmax; i++) {
      res.g[i] = h1 * u.g[i];
      if (HessOrder > 1)
        for (int j = 1; j <= i; j++)
          res.h[i][j] = h1*u.h[i][j] + h2*u.g[i]*u.g[j];
    }
  }
  return res;
}
```

```
HessType sinh ( HessType& u )
{
  HessType res(u.nmax);
  interval h0, h1, h2;

  h0 = sinh(u.f);
  res.f = h0;
  if (HessOrder > 0) {
    h1 = cosh(u.f);
    h2 = h0;
    for (int i = 1; i <= u.nmax; i++) {
      res.g[i] = h1 * u.g[i];
      if (HessOrder > 1)
        for (int j = 1; j <= i; j++)
          res.h[i][j] = h1*u.h[i][j] + h2*u.g[i]*u.g[j];
    }
  }
  return res;
}

HessType cosh ( HessType& u )
{
  HessType res(u.nmax);
  interval h0, h1, h2;

  h0 = cosh(u.f);
  res.f = h0;
  if (HessOrder > 0) {
    h1 = sinh(u.f);
    h2 = h0;
    for (int i = 1; i <= u.nmax; i++) {
      res.g[i]  = h1 * u.g[i];
      if (HessOrder > 1)
        for (int j = 1; j <= i; j++)
          res.h[i][j] = h1*u.h[i][j] + h2*u.g[i]*u.g[j];
    }
  }
  return res;
}

HessType tanh ( HessType& u )
{
  HessType res(u.nmax);
  interval h0, h1, h2;

  h0 = tanh(u.f);
  res.f = h0;
  if (HessOrder > 0) {
    h1 = 1.0-sqr(h0);                      // The subdistributive law implies
    h2 = -2.0*h0*h1;                       //   h0 * (h0↑2 - 1) <= h0↑3 - h0.
    for (int i = 1; i <= u.nmax; i++) {    // So, the first form is used.
      res.g[i]  = h1 * u.g[i];             //---------------------------------
      if (HessOrder > 1)
        for (int j = 1; j <= i; j++)
          res.h[i][j] = h1*u.h[i][j] + h2*u.g[i]*u.g[j];
    }
  }
  return res;
}

HessType coth ( HessType& u )
{
  HessType res(u.nmax);
```

```
  interval h0, h1, h2;

  h0 = coth(u.f);                 // Can propagate domain error
  res.f = h0;                     //---------------------------
  if (HessOrder > 0) {
    h1 = 1.0-sqr(h0);                        // The subdistributive law implies
    h2 = -2.0*h0*h1;                         //   h0 * (h0↑2 - 1) <= h0↑3 - h0.
    for (int i = 1; i <= u.nmax; i++) {      // So, the first form is used.
      res.g[i] = h1 * u.g[i];                //--------------------------------
      if (HessOrder > 1)
        for (int j = 1; j <= i; j++)
          res.h[i][j] = h1*u.h[i][j] + h2*u.g[i]*u.g[j];
    }
  }
  return res;
}

HessType asinh ( HessType& u )
{
  HessType res(u.nmax);
  interval h0, h1, h2;

  res.f = asinh(u.f);             // Can propagate domain error
  if (HessOrder > 0) {            //---------------------------
    h0 = 1.0+sqr(u.f);
    h1 = 1.0/sqrt(h0);
    h2 = -u.f*h1/h0;
    for (int i = 1; i <= u.nmax; i++) {
      res.g[i] = h1 * u.g[i];
      if (HessOrder > 1)
        for (int j = 1; j <= i; j++)
          res.h[i][j] = h1*u.h[i][j] + h2*u.g[i]*u.g[j];
    }
  }
  return res;
}

HessType acosh ( HessType& u )
{
  HessType res(u.nmax);
  interval h0, h1, h2;

  res.f = acosh(u.f);             // Can propagate domain error
  if (HessOrder > 0) {            //---------------------------
    h0 = sqr(u.f)-1.0;
    h1 = 1.0/sqrt(h0);
    h2 = -u.f*h1/h0;
    for (int i = 1; i <= u.nmax; i++) {
      res.g[i] = h1 * u.g[i];
      if (HessOrder > 1)
        for (int j = 1; j <= i; j++)
          res.h[i][j] = h1*u.h[i][j] + h2*u.g[i]*u.g[j];
    }
  }
  return res;
}

HessType atanh ( HessType& u )
{
  HessType res(u.nmax);
  interval h1, h2;

  res.f = atanh(u.f);             // Can propagate domain error
  if (HessOrder > 0) {            //---------------------------
```

```
    h1 = 1.0/(1.0-sqr(u.f));
    h2 = 2.0*u.f*sqr(h1);
    for (int i = 1; i <= u.nmax; i++) {
      res.g[i] = h1 * u.g[i];
      if (HessOrder > 1)
        for (int j = 1; j <= i; j++)
          res.h[i][j] = h1*u.h[i][j] + h2*u.g[i]*u.g[j];
    }
  }
  return res;
}

HessType acoth ( HessType& u )
{
  HessType res(u.nmax);
  interval h1,h2;

  res.f = acoth(u.f);                // Can propagate domain error
  if (HessOrder > 0) {               //---------------------------
    h1 = 1.0/(1.0-sqr(u.f));
    h2 = 2.0*u.f*sqr(h1);
    for (int i = 1; i <= u.nmax; i++) {
      res.g[i] = h1 * u.g[i];
      if (HessOrder > 1)
        for (int j = 1; j <= i; j++)
          res.h[i][j] = h1*u.h[i][j] + h2*u.g[i]*u.g[j];
    }
  }
  return res;
}

//----------------------------------------------------------------------------
// Predefined routines for evaluation of HessType-functions
//----------------------------------------------------------------------------

//----------------------------------------------------------------------------
// Purpose: Evaluation of function 'f' for argument 'x' in differentiation
//    arithmetic computing only the function value.
// Parameters:
//    In : 'f' : function of 'HessType'
//         'x' : argument for evaluation of 'f'
//    Out: 'fx': returns the function value 'f(x)'
// Description: This function sets 'HessOrder' to 0, evaluates 'f(x)' in
//    differentiation arithmetic, and returns the function value only.
//----------------------------------------------------------------------------
void fEvalH ( HTscalar_FctPtr f, ivector x, interval& fx )
{
  HessOrder = 0;
  fx        = fValue(f(HessVar(x)));
  HessOrder = 2;
}

//----------------------------------------------------------------------------
// Purpose: Evaluation of function 'f' for argument 'x' in differentiation
//    arithmetic computing the function value and the gradient value.
// Parameters:
//    In : 'f' : function of 'HessType'
//         'x' : argument for evaluation of 'f'
//    Out: 'fx': returns the function value 'f(x)'
//         'gx': returns the gradient value 'grad f(x)'
// Description: This function sets 'HessOrder' to 1, evaluates 'f(x)' in
//    differentiation arithmetic, and returns the function value and the
//    value of the gradient.
//----------------------------------------------------------------------------
```

```
void fgEvalH ( HTscalar_FctPtr f, ivector x, interval& fx, ivector& gx )
{
  // HessType fxH(Ub(x));      // not implemented in HP C++ V2.0!
  int       n = Ub(x);
  HessType fxH(n);

  HessOrder = 1;
  fxH       = f(HessVar(x));
  fx        = fValue(fxH);
  gx        = gradValue(fxH);
  HessOrder = 2;
}

//----------------------------------------------------------------------------
// Purpose: Evaluation of function 'f' for argument 'x' in differentiation
//    arithmetic computing the function value, the gradient value, and the
//    Hessian matrix value.
// Parameters:
//    In : 'f' : function of 'HessType'
//         'x' : argument for evaluation of 'f'
//    Out: 'fx': returns the function value 'f(x)'
//         'gx': returns the gradient value 'grad f(x)'
//         'hx': returns the Hessian matrix value 'hess f(x)'
// Description: This function keeps 'HessOrder' = 2, evaluates 'f(x)' in
//    differentiation arithmetic, and returns the function value, the
//    value of the gradient, and the value of the Hessian matrix.
//----------------------------------------------------------------------------
void fghEvalH ( HTscalar_FctPtr f, ivector x, interval& fx, ivector& gx,
                                                           imatrix& hx )
{
  // HessType fxH(Ub(x));      // not implemented in HP C++ V2.0!
  int       n = Ub(x);
  HessType fxH(n);

  fxH = f(HessVar(x));
  fx  = fValue(fxH);
  gx  = gradValue(fxH);
  hx  = hessValue(fxH);
}

//----------------------------------------------------------------------------
// Predefined routines for evaluation of HTvector-functions (with Jacobians)
//----------------------------------------------------------------------------

//----------------------------------------------------------------------------
// Purpose: Evaluation of function 'f' for argument 'x' in differentiation
//    arithmetic computing only the function value.
// Parameters:
//    In : 'f' : function of type 'HTvector'
//         'x' : argument for evaluation of 'f'
//    Out: 'fx': returns the vector function value 'f(x)'
// Description: This function sets 'HessOrder' to 0, evaluates 'f(x)' in
//    differentiation arithmetic, and returns the function value only.
//----------------------------------------------------------------------------
void fEvalJ ( HTvector_FctPtr f, ivector x, ivector& fx )
{
  HessOrder = 0;
  fx        = fValue(f(HessVar(x)));
  HessOrder = 2;
}

//----------------------------------------------------------------------------
// Purpose: Evaluation of function 'f' for argument 'x' in differentiation
//    arithmetic computing the function value and the Jacobian matrix value.
```

```
// Parameters:
//    In : 'f' : function of type 'HTvector'
//         'x' : argument for evaluation of 'f'
//    Out: 'fx': returns the function value 'f(x)'
//         'Jx': returns the Jacobian value 'Jf(x)'
// Description: This function sets 'HessOrder' to 1, evaluates 'f(x)' in
//    differentiation arithmetic, and returns the function value and the
//    value of the Jacobian matrix.
//----------------------------------------------------------------------------
void fJEvalJ ( HTvector_FctPtr f, ivector x, ivector& fx, imatrix& Jx )
{
  // HTvector fxGTv(Ub(x));        // not implemented in HP C++ V2.0!
  int      n = Ub(x);
  HTvector fxGTv(n);

  HessOrder = 1;
  fxGTv     = f(HessVar(x));
  fx        = fValue(fxGTv);
  Jx        = JacValue(fxGTv);
  HessOrder = 2;
}
```

### 12.3.1.2 Module grad_ari

If we only want to compute gradients or Jacobians, the module presented in Section 12.3.1.1 has the disadvantage of allocating storage that is not used for the Hessian matrix. We now present the module *grad_ari* whose header file supplies the definitions of the classes *GradType* and *GTvector*. These classes include operators and elementary functions for an interval differentiation arithmetic for gradients. The module can also be used to compute Jacobians of vector-valued functions. The type names *GTscalar_FctPtr* and *GTvector_FctPtr* can be used to declare pointers to scalar-valued and vector-valued functions with an argument of type *GTvector*, respectively.

```
//----------------------------------------------------------------------------
// File: grad_ari (header)
// Purpose: Definition of a multi-dimensional interval differentiation
//    arithmetic which allows function evaluation with automatic differen-
//    tiation up to first order (i.e. gradient or Jacobian matrix).
// Types:
//    GTscalar_FctPtr          : pointer for a scalar valued function
//    GTvector_FctPtr          : pointer for a vector valued function
// Class GradType:
//    GradType()               : constructors
//    Resize()                 : for resizing to a fixed dimension
//    operator =               : assignment operators for arguments of types
//                               GradType, interval, and real
//    GradVar()                : to define GradType variables
//    fValue(), gradValue()    : to get function and derivative values
//    operators +, -, *, /     : operators of diff. arithmetic
//    sqr(), sqrt(), power(),
//    exp(), sin(), cos(), ...: elementary functions of diff. arithmetic
//    fEvalG()                 : to compute function value only
//    fgEvalG()                : to compute function and first derivative
//                               value (gradient)
// Class GTvector:
//    GTvector()               : constructors
//    ~GTvector()              : destructor
//    Dim()                    : to get the actual dimension
```

```
//    operator =                  : assignment operator
//    operator []                 : component access
//    fValue(), JacValue()        : to get function and derivative values
//    fEvalJ()                    : to compute function value only
//    fJEvalJ()                   : to compute function and first derivative
//                                  value (Jacobian matrix)
//-----------------------------------------------------------------------------
#ifndef __GRAD_ARI_HPP
#define __GRAD_ARI_HPP

#include <interval.hpp>      // Interval arithmetic
#include <imatrix.hpp>       // Interval matrix/vector arithmetic
#include <mvi_util.hpp>      // Interval matrix/vector utility functions

class GradType;
class GTvector;

typedef GradType (*GTscalar_FctPtr)(GTvector&);
typedef GTvector (*GTvector_FctPtr)(GTvector&);

class GradType {
    int      nmax;
    ivector g;

    friend void TestSize ( GradType&, GradType&, const char* );

  public:
    GradType ( );
    GradType ( index );
    GradType ( GradType& );

    friend void Resize ( GradType&, index );
    int  Dim ( ) { return nmax; };

    interval& operator[] ( int );

    GradType& operator= ( GradType& );
    GradType& operator= ( interval& );
    GradType& operator= ( const real& );


    friend GTvector GradVar ( ivector& );
    friend GTvector GradVar ( rvector& );

    friend interval fValue    ( GradType& );
    friend ivector  gradValue ( GradType& );

    friend GradType operator+ ( GradType&);
    friend GradType operator- ( GradType&);
    friend GradType operator+ ( GradType&, GradType&);
    friend GradType operator- ( GradType&, GradType&);
    friend GradType operator* ( GradType&, GradType&);
    friend GradType operator/ ( GradType&, GradType&);
    friend GradType operator+ ( GradType&, interval&);
    friend GradType operator- ( GradType&, interval&);
    friend GradType operator* ( GradType&, interval&);
    friend GradType operator/ ( GradType&, interval&);
    friend GradType operator+ ( interval&, GradType&);
    friend GradType operator- ( interval&, GradType&);
    friend GradType operator* ( interval&, GradType&);
    friend GradType operator/ ( interval&, GradType&);
    friend GradType operator+ ( GradType&, const real&);
    friend GradType operator- ( GradType&, const real&);
    friend GradType operator* ( GradType&, const real&);
```

```
    friend GradType operator/ ( GradType&, const real&);
    friend GradType operator+ ( const real&, GradType&);
    friend GradType operator- ( const real&, GradType&);
    friend GradType operator* ( const real&, GradType&);
    friend GradType operator/ ( const real&, GradType&);
    friend GradType sqr   ( GradType& );
    friend GradType power ( GradType&, const int );
    friend GradType sqrt  ( GradType& );
    friend GradType exp   ( GradType& );
    friend GradType ln    ( GradType& );
    friend GradType sin   ( GradType& );
    friend GradType cos   ( GradType& );
    friend GradType tan   ( GradType& );
    friend GradType cot   ( GradType& );
    friend GradType asin  ( GradType& );
    friend GradType acos  ( GradType& );
    friend GradType atan  ( GradType& );
    friend GradType acot  ( GradType& );
    friend GradType sinh  ( GradType& );
    friend GradType cosh  ( GradType& );
    friend GradType tanh  ( GradType& );
    friend GradType coth  ( GradType& );
    friend GradType asinh ( GradType& );
    friend GradType acosh ( GradType& );
    friend GradType atanh ( GradType& );
    friend GradType acoth ( GradType& );

    friend void fEvalG   ( GTscalar_FctPtr, ivector, interval& );
    friend void fgEvalG  ( GTscalar_FctPtr, ivector, interval&, ivector& );
};

class GTvector {
    int      nmax;
    GradType *gt;

    friend void TestSize ( GTvector&, GTvector&, const char* );

  public:
    GTvector  ( index );
    GTvector  ( GTvector& );
    ~GTvector ( );

    int Dim ( ) { return nmax; };

    GTvector& operator=  ( GTvector& );

    GradType& operator[] ( index );

    friend ivector  fValue   ( GTvector& );
    friend imatrix  JacValue ( GTvector& );

    friend void fEvalJ  ( GTvector_FctPtr, ivector, ivector& );
    friend void fJEvalJ ( GTvector_FctPtr, ivector, ivector&, imatrix& );
};
#endif
```

The static variable *GradOrder* is used to select the highest order of derivative desired. This enables the user to save computation time in computing only the function value and no gradient. The default value of *GradOrder* is 1, so normally the gradient is computed.

The functions *fEvalG()*, *fgEvalG()*, *fEvalJ()*, and *fJEvalJ()* simplify the mechanism of function evaluating and automate the setting and resetting of the *GradOrder* variable. The ... *EvalG()* functions can be applied to scalar-valued functions (result type *GradType*), whereas the ... *EvalJ()* functions can be applied to vector-valued functions (result type *GTvector*).

For a scalar-valued function of type *GradType*, *fEvalG()* computes and returns only the function value by setting *GradOrder* to 0 before the evaluation is done. The function *fgEvalG()* uses the default value of *GradOrder*, computes, and returns the values of $f(x)$ and $\nabla f(x)$. For a vector-valued function of type *GTvector*, *fEvalJ()* computes and returns only the function value by setting *GradOrder* to 0 before the evaluation is done. If the Jacobian matrix also is desired, *fJEvalJ()* can be used.

```
//----------------------------------------------------------------------------
// File: grad_ari (implementation)
// Purpose: Definition of a multi-dimensional interval differentiation
//    arithmetic which allows function evaluation with automatic differen-
//    tiation up to first order (i.e. gradient or Jacobian matrix).
// Method: Overloading of operators and elementary functions for operations
//    of data types 'GradType' and 'GTvector'.
// Class GradType:
//    GradType()                  : constructors
//    Resize()                    : for resizing to a fixed dimension
//    operator =                  : assignment operators for arguments of types
//                                  GradType, interval, and real
//    GradVar()                   : to define GradType variables
//    fValue(), gradValue()       : to get function and gradient values
//    operators +, -, *, /        : operators of diff. arithmetic
//    sqr(), sqrt(), power(),
//    exp(), sin(), cos(), ...    : elementary functions of diff. arithmetic
//    fEvalG()                    : to compute function value only
//    fgEvalG()                   : to compute function and first derivative
//                                  value (gradient)
// Class GTvector:
//    GTvector()                  : constructors
//    ~GTvector()                 : destructor
//    Dim()                       : to get the actual dimension
//    operator =                  : assignment operator
//    operator []                 : component access
//    fValue(), JacValue()        : to get function and derivative values
//    fEvalJ()                    : to compute function value only
//    fJEvalJ()                   : to compute function and first derivative
//                                  value (Jacobian matrix)
//----------------------------------------------------------------------------
#include <stdlib.h>          // For function 'exit()'
#include <iostream.h>        // I/O Handling
#include <imath.hpp>         // Interval mathematical functions
#include <i_util.hpp>        // Interval utility functions
#include <grad_ari.hpp>

// The local variable 'GradOrder' is used to select the highest order of
// derivative which is computed. Its default value is 1, and normally
// the gradient or the Jacobian matrix are computed.
//----------------------------------------------------------------------------
static int GradOrder = 1;

//----------------------------------------------------------------------------
// Constructors and destructors
//----------------------------------------------------------------------------
GradType::GradType ( )                                          // Constructor
  { nmax = -1; }                                                //------------
```

```
GradType::GradType ( index n )                             // Constructor
{                                                          //------------
  nmax = n._int();
  Resize(g,0,n._int());
}

GradType::GradType ( GradType& A )                    // Copy constructor
{                                                     //-----------------
  nmax = A.nmax;
  g    = A.g;
}

GTvector::GTvector ( index n )                             // Constructor
{                                                          //------------
  nmax = n._int();
  gt = new GradType[nmax];
  for (int i = 0; i < nmax; i++) Resize(gt[i],nmax);
}

GTvector::GTvector ( GTvector& v )                    // Copy constructor
{                                                     //-----------------
  nmax = v.nmax;
  gt = new GradType[nmax];
  for (int i = 0; i < nmax; i++)
    { Resize(gt[i],nmax); gt[i] = v.gt[i];}
}

GTvector::~GTvector ( )                                     // Destructor
{                                                           //-----------
  nmax = 0;
  delete [] gt;
}

//----------------------------------------------------------------------------
// Functions for resizing and range check
//----------------------------------------------------------------------------
void Resize ( GradType& A, index n )
{
  A.nmax = n._int();
  Resize(A.g,0,n._int());
}

void TestSize ( GradType& A, GradType& b, const char *fname )
{
  if (A.nmax != b.nmax) {
    cout << "Parameters must be of same size in '" << fname
         << "'!" << endl;
    exit(-1);
  }
}

void TestSize ( GTvector& v, GTvector& w, const char *fname)
{
  if (v.nmax != w.nmax) {
    cout << "Parameters must be of same size in '" << fname
         << "'!" << endl;
    exit(-1);
  }
}

//----------------------------------------------------------------------------
// Operators for component access
//----------------------------------------------------------------------------
```

```
interval& GradType::operator[] ( int i )              // Private [] operator!
{                                                     //---------------------
  if ( (i < 0) || (i > nmax) ) {
    cout << "Index out of range in 'interval& GradType::operator[] ( int )'!"
         << endl;
    exit(-1);
  }
  return g[i];
}

GradType& GTvector::operator[] ( index ind )           // Public [] operator!
{                                                     //--------------------
  int n = ind._int();
  if ( (n < 1) || (n > nmax) ) {
    cout << "Index out of range in "
         << "'GradType& GTvector::operator[] ( index )'!" << endl;
    exit(-1);
  }
  return gt[n-1];
}

//----------------------------------------------------------------------------
// Assignment operators for 'GradType' and 'GTvector' variables with the
// right-hand side of type ...
//----------------------------------------------------------------------------
GradType& GradType::operator= ( GradType& A )                   // ... GradType
{                                                               //-------------
  TestSize(*this,A,"operator= ( GradType&, GradType& )");
  g = A.g;
  return *this;
}

GradType& GradType::operator= ( interval& x )                   // ... interval
{                                                               //-------------
  g = 0.0;
  g[0] = x;
  return *this;
}

GradType& GradType::operator= ( const real& r )                     // ... real
{                                                                   //---------
  *this = _interval(r);
  return *this;
}

GTvector& GTvector::operator= ( GTvector& v )                   // ... GTvector
{                                                               //-------------
  // Check if left-hand side is already allocated!
  //-----------------------------------------------
  TestSize(*this,v,"operator= ( GTvector&, GTvector& )");
  for (int i = 0; i < nmax; i++) gt[i] = v.gt[i];
  return *this;
}

//----------------------------------------------------------------------------
// Transfer functions for variables
//----------------------------------------------------------------------------
GTvector GradVar ( ivector& x )                            // Generate variable
{                                                          //------------------
  int      d = -Lb(x)+1;
  int      ubd = Ub(x)+d;
  GTvector hht(ubd);

  for (int i = 1; i <= ubd; i++) {
```

```
    hht[i][0] = x[i-d];     // GradVar[][0] = x;
    for (int k = 1; k <= ubd; k++)
      if (i == k) hht[i][k] = 1.0; else hht[i][k] = 0.0;
  }
  return hht;
}

GTvector GradVar ( rvector& v )                              // Generate variable
{                                                            //------------------
  int     i, n = Lb(v), m = Ub(v);
  ivector u(n,m);

  for (i = n; i <= m; i++) u[i] = v[i];
  return GradVar(u);
}

//----------------------------------------------------------------------------
// Access functions for the function value and the gradient
//----------------------------------------------------------------------------
interval fValue ( GradType& f )                             // Get function value
  { return f[0]; }                                          //-------------------

ivector gradValue ( GradType& f )                           // Get gradient value
{                                                           //-------------------
  ivector hv(f.nmax);
  for (int i = 1; i <= f.nmax; i++) hv[i] = f[i];
  return hv;
}

//----------------------------------------------------------------------------
// Access functions for vector-valued functions (function value or Jacobian)
//----------------------------------------------------------------------------
ivector fValue ( GTvector& f )                           // Get function value of
{                                                        // n-dimensional vector-
  ivector hv(f.nmax);                                    // valued function
                                                         //----------------------
  for (int i = 1; i <= f.nmax; i++)
    hv[i] = f[i][0];
  return hv;
}

imatrix JacValue ( GTvector& f )                         // Get Jacobian value of
{                                                        // n-dimensional scalar-
  imatrix hm(f.nmax,f.nmax);                             // valued function
                                                         //----------------------
  for (int i = 1; i <= f.nmax; i++)
    for (int j = 1; j <= f.nmax; j++)
      hm[i][j] = f[i][j];
  return hm;
}

//----------------------------------------------------------------------------
// Unary operators + and - for GradType operands
//----------------------------------------------------------------------------
GradType operator+ ( GradType& u )
  { return u; }

GradType operator- ( GradType& u )
{
  GradType umin(u.nmax);

  umin[0] = -u[0];
  if (GradOrder > 0)
    for (int i = 1; i <= u.nmax; i++)
```

```
      umin[i] = -u[i];
  return umin;
}

//----------------------------------------------------------------------------
// Operators +, -, *, and / for two GradType operands
//----------------------------------------------------------------------------
GradType operator+ ( GradType& u, GradType& v)
{
  GradType add(u.nmax);

  TestSize(u,v,"operator+ ( GradType&, GradType& )");
  add[0] = u[0] + v[0];
  if (GradOrder > 0)
    for (int i = 1; i <= u.nmax; i++)
      add[i] = u[i] + v[i];
  return add;
}

GradType operator- ( GradType& u, GradType& v )
{
  GradType sub(u.nmax);

  TestSize(u,v,"operator- ( GradType&, GradType& )");
  sub[0] = u[0] - v[0];
  if (GradOrder > 0)
    for (int i=1; i <= u.nmax; i++)
      sub[i] = u[i] - v[i];
  return sub;
}

GradType operator* ( GradType& u, GradType& v )
{
  GradType mul(u.nmax);

  TestSize(u,v,"operator* ( GradType&, GradType& )");
  mul[0] = u[0] * v[0];
  if (GradOrder > 0)
    for (int i = 1; i <= u.nmax; i++)
      mul[i] = v[0]*u[i] + u[0]*v[i];
  return mul;
}

GradType operator/ ( GradType& u, GradType& v )
{
  GradType div(u.nmax);

  TestSize(u,v,"operator/ ( GradType&, GradType& )");
  div[0] = u[0]/v[0];    // Can propagate 'division by zero' error
  if (GradOrder > 0)     //---------------------------------------
    for (int i = 1; i <= u.nmax; i++)
      div[i] = (u[i] - div[0]*v[i]) / v[0];
  return div;
}

//----------------------------------------------------------------------------
// Operators +, -, *, and / for one interval and one GradType operand
//----------------------------------------------------------------------------
GradType operator+ ( GradType& u, interval& b )
{
  GradType add(u.nmax);

  add = u; add[0] += b;
  return add;
```

```
}

GradType operator- ( GradType& u, interval& b )
{
  GradType sub(u.nmax);

  sub = u; sub[0] -= b;
  return sub;
}

GradType operator* ( GradType& u, interval& b )
{
  GradType mul(u.nmax);

  mul[0] = u[0]*b;
  if (GradOrder > 0)
    for (int i = 1; i <= u.nmax; i++)
      mul[i] = b*u[i];
  return mul;
}

GradType operator/ ( GradType& u, interval& b )
{
  GradType div(u.nmax);

  div[0] = u[0]/b;          // Can propagate 'division by zero' error
  if (GradOrder > 0)        //--------------------------------------
    for (int i = 1; i <= u.nmax; i++)
      div[i] = u[i]/b;
  return div;
}

GradType operator+ ( interval& a, GradType& v )
{
  GradType add(v.nmax);

  add = v; add[0] = a + v[0];
  return add;
}

GradType operator- ( interval& a, GradType& v )
{
  GradType sub(v.nmax);

  sub[0] = a - v[0];
  if (GradOrder > 0)
    for (int i = 1; i <= v.nmax; i++)
      sub[i] = -v[i];
  return sub;
}

GradType operator* ( interval& a, GradType& v )
{
  GradType mul(v.nmax);

  mul[0] = a*v[0];
  if (GradOrder > 0)
    for (int i = 1; i <= v.nmax; i++)
      mul[i] = a*v[i];
  return mul;
}

GradType operator/ ( interval& a, GradType& v )
{
```

```
  GradType div(v.nmax);
  interval p;

  div[0] = a/v[0];           // Can propagate 'division by zero' error
  if (GradOrder > 0) {       //---------------------------------------
    p = -div[0]/v[0];
    for (int i = 1; i <= v.nmax; i++)
      div[i] = p*v[i];
  }
  return div;
}

//----------------------------------------------------------------------------
// Operators +, -, *, and / for one real and one GradType operand
//----------------------------------------------------------------------------
GradType operator+ ( GradType& u, const real& b )
  { return u + _interval(b); }

GradType operator- ( GradType& u, const real& b )
  { return u - _interval(b); }

GradType operator* ( GradType& u, const real& b )
  { return u * _interval(b); }

GradType operator/ ( GradType& u, const real& b )
  { return u / _interval(b); }    // Can propagate 'division by zero' error
                                  //---------------------------------------
GradType operator+ ( const real& a, GradType& v )
  { return _interval(a) + v; }

GradType operator- ( const real& a, GradType& v )
  { return _interval(a) - v; }

GradType operator* ( const real& a, GradType& v )
  { return _interval(a) * v; }

GradType operator/ ( const real& a, GradType& v )
  { return _interval(a) / v; }    // Can propagate 'division by zero' error
                                  //---------------------------------------

//----------------------------------------------------------------------------
// Elementary function for GradType arguments
//----------------------------------------------------------------------------
GradType sqr ( GradType& u )
{
  GradType res(u.nmax);
  interval h1;

  res[0] = sqr(u[0]);        // Can propagate domain error
  if (GradOrder > 0) {       //---------------------------
    h1 = 2.0*u[0];
    for (int i = 1; i <= u.nmax; i++)
      res[i] = h1*u[i];
  }
  return res;
}

GradType power ( GradType& u, const int k )
{
  GradType res(u.nmax);
  interval h1;

  if (k == 0)
    { res = 1.0; return res; }
```

```
  if (k == 1)
    return u;
  if (k == 2)
    return sqr(u);

  res[0] = Power(u[0],k);
  if (GradOrder > 0) {
    h1 = (double)k * Power(u[0],k-1);
    for (int i = 1; i <= u.nmax; i++)
      res[i] = h1 * u[i];
  }
  return res;
}

GradType sqrt ( GradType& u )
{
  GradType res(u.nmax);
  interval h0, h1;

  h0 = sqrt(u[0]);
  res[0] = h0;
  if (GradOrder > 0) {
    h1 = 0.5/h0;
    for (int i = 1; i <= u.nmax; i++)
      res[i] = h1*u[i];
  }
  return res;
}

GradType exp ( GradType& u )
{
  GradType res(u.nmax);
  interval h0;

  h0 = exp(u[0]);
  res[0] = h0;
  if (GradOrder > 0)
    for (int i = 1; i <= u.nmax; i++)
      res[i] = h0*u[i];
  return res;
}

GradType ln ( GradType& u )
{
  GradType res(u.nmax);
  interval h1;

  res[0] = ln(u[0]);          // Can propagate domain error
  if (GradOrder > 0) {        //---------------------------
    h1 = 1.0/u[0];
    for (int i = 1; i <= u.nmax; i++)
      res[i] = h1*u[i];
  }
  return res;
}

GradType sin ( GradType& u )
{
  GradType res(u.nmax);
  interval h1;

  res[0] = sin(u[0]);
  if (GradOrder > 0) {
    h1 = cos(u[0]);
```

```
    for (int i = 1; i <= u.nmax; i++)
      res[i] = h1*u[i];
  }
  return res;
}

GradType cos ( GradType& u )
{
  GradType res(u.nmax);
  interval h0, h1;

  res[0] = cos(u[0]);
  if (GradOrder > 0) {
    h1 = -sin(u[0]);
    for (int i = 1; i <= u.nmax; i++)
      res[i] = h1*u[i];
  }
  return res;
}

GradType tan ( GradType& u )
{
  GradType res(u.nmax);
  interval h0, h1;

  h0 = tan(u[0]);                  // Can propagate domain error
  res[0] = h0;                     //----------------------------
  if (GradOrder > 0) {
    h1 = sqr(h0)+1.0;                          // The subdistributive law implies
                                               //   h0 * (h0↑2 + 1) <= h0↑3 + h0.
    for (int i = 1; i <= u.nmax; i++)          // So, the first form is used.
      res[i] = h1*u[i];                        //--------------------------------
  }
  return res;
}

GradType cot ( GradType& u )
{
  GradType res(u.nmax);
  interval h0, h1;

  h0 = cot(u[0]);                  // Can propagate domain error
  res[0] = h0;                     //----------------------------
  if (GradOrder > 0) {
    h1 = -(sqr(h0)+1.0);                       // The subdistributive law implies
                                               //   h0 * (h0↑2 + 1) <= h0↑3 + h0.
    for (int i = 1; i <= u.nmax; i++)          // So, the first form is used.
      res[i] = h1*u[i];                        //--------------------------------
  }
  return res;
}

GradType asin ( GradType& u )
{
  GradType res(u.nmax);
  interval h, h1;

  res[0] = asin(u[0]);          // Can propagate domain error
  if (GradOrder > 0) {          //----------------------------
    h  = 1.0-sqr(u[0]);
    h1 = 1.0/sqrt(h);
    for (int i = 1; i <= u.nmax; i++)
      res[i] = h1*u[i];
  }
```

```
  return res;
}

GradType acos ( GradType& u )
{
  GradType res(u.nmax);
  interval h, h1;

  res[0] = acos(u[0]);        // Can propagate domain error
  if (GradOrder > 0) {        //---------------------------
    h  = 1.0-sqr(u[0]);
    h1 = -1.0/sqrt(h);
    for (int i = 1; i <= u.nmax; i++)
      res[i] = h1*u[i];
  }
  return res;
}

GradType atan ( GradType& u )
{
  GradType res(u.nmax);
  interval h1;

  res[0] = atan(u[0]);
  if (GradOrder > 0) {
    h1 = 1.0/(1.0+sqr(u[0]));
    for (int i = 1; i <= u.nmax; i++)
      res[i] = h1*u[i];
  }
  return res;
}

GradType acot ( GradType& u )
{
  GradType res(u.nmax);
  interval h1;

  res[0] = acot(u[0]);        // Can propagate domain error
  if (GradOrder > 0) {        //---------------------------
    h1 = -1.0/(1.0+sqr(u[0]));
    for (int i = 1; i <= u.nmax; i++)
      res[i] = h1*u[i];
  }
  return res;
}

GradType sinh ( GradType& u )
{
  GradType res(u.nmax);
  interval h0, h1;

  h0 = sinh(u[0]);
  res[0] = h0;
  if (GradOrder > 0) {
    h1 = cosh(u[0]);
    for (int i = 1; i <= u.nmax; i++)
      res[i] = h1*u[i];
  }
  return res;
}

GradType cosh ( GradType& u )
{
  GradType res(u.nmax);
```

```
  interval h0, h1;

  res[0] = cosh(u[0]);
  if (GradOrder > 0) {
    h1 = sinh(u[0]);
    for (int i = 1; i <= u.nmax; i++)
      res[i]  = h1*u[i];
  }
  return res;
}

GradType tanh ( GradType& u )
{
  GradType res(u.nmax);
  interval h0, h1;

  h0 = tanh(u[0]);
  res[0] = h0;
  if (GradOrder > 0) {
    h1 = 1.0-sqr(h0);                        // The subdistributive law implies
                                             //   h0 * (h0↑2 - 1) <= h0↑3 - h0.
    for (int i = 1; i <= u.nmax; i++)        // So, the first form is used.
      res[i]  = h1*u[i];                     //--------------------------------
  }
  return res;
}

GradType coth ( GradType& u )
{
  GradType res(u.nmax);
  interval h0, h1;

  h0 = coth(u[0]);              // Can propagate domain error
  res[0] = h0;                  //---------------------------
  if (GradOrder > 0) {
    h1 = 1.0-sqr(h0);                        // The subdistributive law implies
                                             //   h0 * (h0↑2 - 1) <= h0↑3 - h0.
    for (int i = 1; i <= u.nmax; i++)        // So, the first form is used.
      res[i] = h1*u[i];                      //--------------------------------
  }
  return res;
}

GradType asinh ( GradType& u )
{
  GradType res(u.nmax);
  interval h, h1;

  res[0] = asinh(u[0]);       // Can propagate domain error
  if (GradOrder > 0) {        //---------------------------
    h  = 1.0+sqr(u[0]);
    h1 = 1.0/sqrt(h);
    for (int i = 1; i <= u.nmax; i++)
      res[i] = h1*u[i];
  }
  return res;
}

GradType acosh ( GradType& u )
{
  GradType res(u.nmax);
  interval h, h1;

  res[0] = acosh(u[0]);       // Can propagate domain error
```

```
  if (GradOrder > 0) {      //--------------------------
    h  = sqr(u[0])-1.0;
    h1 = 1.0/sqrt(h);
    for (int i = 1; i <= u.nmax; i++)
      res[i] = h1*u[i];
  }
  return res;
}

GradType atanh ( GradType& u )
{
  GradType res(u.nmax);
  interval h1;

  res[0] = atanh(u[0]);     // Can propagate domain error
  if (GradOrder > 0) {      //--------------------------
    h1 = 1.0/(1.0-sqr(u[0]));
    for (int i = 1; i <= u.nmax; i++)
      res[i] = h1*u[i];
  }
  return res;
}

GradType acoth ( GradType& u )
{
  GradType res(u.nmax);
  interval h1;

  res[0] = acoth(u[0]);     // Can propagate domain error
  if (GradOrder > 0) {      //--------------------------
    h1 = 1.0/(1.0-sqr(u[0]));
    for (int i = 1; i <= u.nmax; i++)
      res[i] = h1*u[i];
  }
  return res;
}

//----------------------------------------------------------------------------
// Predefined routines for evaluation of GradType-functions
//----------------------------------------------------------------------------

//----------------------------------------------------------------------------
// Purpose: Evaluation of function 'f' for argument 'x' in differentiation
//    arithmetic computing only the function value.
// Parameters:
//    In : 'f' : function of 'GradType'
//         'x' : argument for evaluation of 'f'
//    Out: 'fx': returns the function value 'f(x)'
// Description: This function sets 'GradOrder' to 0, evaluates 'f(x)' in
//    differentiation arithmetic, and returns the function value only.
//----------------------------------------------------------------------------
void fEvalG ( GTscalar_FctPtr f, ivector x, interval& fx )
{
  GradOrder = 0;
  fx        = fValue(f(GradVar(x)));
  GradOrder = 1;
}

//----------------------------------------------------------------------------
// Purpose: Evaluation of function 'f' for argument 'x' in differentiation
//    arithmetic computing the function value and the gradient value.
// Parameters:
//    In : 'f' : function of 'GradType'
//         'x' : argument for evaluation of 'f'
```

```
//    Out: 'fx': returns the function value 'f(x)'
//         'gx': returns the gradient value 'grad f(x)'
// Description: This function keeps 'GradOrder' = 1, evaluates 'f(x)' in
//    differentiation arithmetic, and returns the function value and the
//    value of the gradient.
//----------------------------------------------------------------------------
void fgEvalG ( GTscalar_FctPtr f, ivector x, interval& fx, ivector& gx )
{
  // GradType fxG(Ub(x));      // not implemented in HP C++ V2.0!
  int      n = Ub(x);
  GradType fxG(n);

  fxG      = f(GradVar(x));
  fx       = fValue(fxG);
  gx       = gradValue(fxG);
}

//----------------------------------------------------------------------------
// Predefined routines for evaluation of GTvector-functions (with Jacobians)
//----------------------------------------------------------------------------

//----------------------------------------------------------------------------
// Purpose: Evaluation of function 'f' for argument 'x' in differentiation
//    arithmetic computing only the function value.
// Parameters:
//    In : 'f' : function of type 'GTvector'
//         'x' : argument for evaluation of 'f'
//    Out: 'fx': returns the vector function value 'f(x)'
// Description: This function sets 'GradOrder' to 0, evaluates 'f(x)' in
//    differentiation arithmetic, and returns the function value only.
//----------------------------------------------------------------------------
void fEvalJ ( GTvector_FctPtr f, ivector x, ivector& fx )
{
  GradOrder = 0;
  fx        = fValue(f(GradVar(x)));
  GradOrder = 1;
}

//----------------------------------------------------------------------------
// Purpose: Evaluation of function 'f' for argument 'x' in differentiation
//    arithmetic computing the function value and the Jacobian matrix value.
// Parameters:
//    In : 'f' : function of type 'GTvector'
//         'x' : argument for evaluation of 'f'
//    Out: 'fx': returns the function value 'f(x)'
//         'Jx': returns the Jacobian value 'Jf(x)'
// Description: This function keeps 'GradOrder' = 1, evaluates 'f(x)' in
//    differentiation arithmetic, and returns the function value and the
//    value of the Jacobian matrix.
//----------------------------------------------------------------------------
void fJEvalJ ( GTvector_FctPtr f, ivector x, ivector& fx, imatrix& Jx )
{
  // GTvector fxGTv(Ub(x));      // not implemented in HP C++ V2.0!
  int      n = Ub(x);
  GTvector fxGTv(n);

  fxGTv    = f(GradVar(x));
  fx       = fValue(fxGTv);
  Jx       = JacValue(fxGTv);
}
```

### 12.3.2 Examples

In this section, we illustrate applications of automatic differentiation. The well-known approximate Newton method applied to optimization and root-finding problems makes effective use of automatic differentiation, but we do not treat sophisticated methods in that area in this section. A detailed discussion of methods with automatic result verification using Newton's method is given in Chapters 13 and 14.

We first illustrate the use of our differentiation arithmetic in a program to solve Example 12.1 on page 246.

**Example 12.2** We must define a C++ function for the function $f(x) = x_1 \cdot (4+x_2)$:

```
HessType f ( HTvector& x )
{
  return x[1] * (4.0 + x[2]);
}
```

With the declarations

```
HTvector  x(2);
HessType  fx(2);
ivector   y(2), grady(2);
interval  fy;
imatrix   hessy(2,2);
```

we can compute function values and derivative values directly by the sequence

```
 y[1] = 123.0;  y[2] = 456.0;
    x = HessVar(y);
   fx = f(x);
   fy = fValue(fx);         // Function value f(y)
grady = gradValue(fx);      // Gradient g(y)
hessy = hessValue(fx);      // Hessian H(y)
```

Alternatively, we can use the evaluating function *fghEvalH()*:

```
y[1] = 123.0;  y[2] = 456.0;
fghEvalH(f,y,fy,grady,hessy);
```

**Example 12.3** The well known method of Newton for computing an approximation of a zero of a system of nonlinear equations (see [89] for example) can also be used for computing a stationary point of a twice continuously differentiable multi-dimensional function. Newton's method is applied to the gradient of the function, i.e. to the equation $\nabla f(x) = 0$, starting from an approximation $x^{(0)} \in I\!R^n$. The method can be stated as

$$\left.\begin{array}{rcl} R^{(k)} & := & (\nabla^2 f(x^{(k)}))^{-1} \\ x^{(k+1)} & := & x^{(k)} - R^{(k)} \cdot \nabla f(x^{(k)}) \end{array}\right\} k = 0, 1, 2, \ldots$$

In our sample program for this method, all real operations are replaced by interval operations, because we make use of the interval differentiation arithmetic. That is, we iterate

$$\left.\begin{array}{rcl} R^{(k)} & := & (m(\nabla^2 f([x]^{(k)})))^{-1} \\ [x]^{(k+1)} & := & [x]^{(k)} - R^{(k)} \cdot \nabla f([x]^{(k)}) \end{array}\right\} k = 0, 1, 2, \ldots$$

The influence of rounding errors is directly demonstrated by the increasing interval diameters. The approximate inverse $R^{(k)}$ is computed by applying the function *MatInv()* from Chapter 10 to the midpoint matrix of the interval Hessian evaluation.

We terminate the iteration if an interval evaluation of $\nabla f$ over the current interval iterate $[x]^{(k)}$ contains zero, or if the number of iterations is greater than or equal to 100 to prevent endless iterations. Although interval operations are applied, the result is an enclosure of the sequence of values $x^{(k+1)}$ that would be computed in exact real arithmetic. The result is *not* a validated enclosure of the zero of the gradient because this approach does *not* enclose the truncation errors in the Newton iteration. For a more sophisticated interval method for finding *all* global minimizers of a nonlinear function with guarantee and high accuracy, refer to Chapter 14.

We apply our iteration to the problem of finding a stationary point of Rosenbrock's function [81] $f(x) = 100 \cdot (x_2 - x_1^2)^2 + (x_1 - 1)^2$, which has become accepted as a difficult test problem for optimization methods.

```
//----------------------------------------------------------------------------
// Example: Automatic differentiation for Hessians
// This is an implementation of Newton's Method for computing a stationary
// point of a twice continuously differentiable multi-dimensional function.
//
//    given:      the function f(x) and the starting value x0
//    iteration:  x[n+1] := x[n] - InvHf(x[n])*Gf(x[n]) , n = 0,1,...
//
// where InvHf(x) denotes the inverse of the Hessian matrix of f(x).
// All real operations are replaced by interval operations, all function and
// derivative evaluations are calculated by differentiation arithmetic.
//----------------------------------------------------------------------------
#include <matinv.hpp>        // Matrix inversion
#include <hess_ari.hpp>      // Hessian differentiation arithmetic

const int
  fDim = 2,
  nmax = 100;

interval  fx;
ivector   x(fDim), Gfx(fDim);
imatrix   Hfx(fDim,fDim);
rmatrix   InvHfx(fDim,fDim);
int       n, Error;

HessType f ( HTvector& x )
{
  return 100.0*sqr(x[2] - sqr(x[1])) + sqr(x[1] - 1.0);
}

int main ( )
{
  cout << SetPrecision(23,15) << Scientific;   // Output format

  cout << "Newton-Method for finding a stationary point of Rosenbrock's"
       << endl
       << "function:  f(x) = 100*sqr(x[2] - sqr(x[1])) + sqr(x[1] - 1)"
       << endl << endl;
  cout << "Starting vector x = "; cin >> x;
  cout << endl << "Iteration:" << endl;
  fghEvalH(f,x,fx,Gfx,Hfx);
  n = 0;
```

```
  do {
    n++;
    cout << "x: " << endl << x << "Gf(x): " << endl << Gfx << endl;
    MatInv(mid(Hfx),InvHfx,Error);
    if (!Error) {
      x = x - InvHfx * Gfx;
      fghEvalH(f,x,fx,Gfx,Hfx);
    }
  } while( !( (in(0.0,Gfx[1]) && in(0.0,Gfx[2])) || (n >= nmax) || Error ) );

  if (!Error) {
    cout << endl
         << "Stationary point:  " << endl << x
         << "Gradient value:    " << endl << Gfx << endl
         << "Expected solution: " << endl
         << " 1.0 " << endl << " 1.0 " << endl;
  }
  else
    cout << MatInvErrMsg(Error) << endl;

  return 0;
}
```

If we run our sample program for Newton's method we get the following runtime output.

```
Newton-Method for finding a stationary point of Rosenbrock's
function:  f(x) = 100*sqr(x[2] - sqr(x[1])) + sqr(x[1] - 1)

Starting vector x = [3,3] [3,3]

Iteration:
x:
[ 3.000000000000000E+000, 3.000000000000000E+000]
[ 3.000000000000000E+000, 3.000000000000000E+000]
Gf(x):
[ 7.204000000000000E+003, 7.204000000000000E+003]
[-1.200000000000000E+003,-1.200000000000000E+003]

x:
[ 2.998334721065778E+000, 2.998334721065779E+000]
[ 8.990008326394672E+000, 8.990008326394675E+000]
Gf(x):
[ 3.999995379607563E+000, 3.999995379616088E+000]
[-5.546307857429156E-004,-5.546307843218301E-004]

x:
[ 1.001107723574369E+000, 1.001107723582889E+000]
[-2.986699005308295E+000,-2.986699005257204E+000]
Gf(x):
[ 1.597335933603855E+003, 1.597335933644754E+003]
[-7.977831359051213E+002,-7.977831358914919E+002]

x:
[ 1.001106336789176E+000, 1.001106336840377E+000]
[ 1.002213897506430E+000, 1.002213897711126E+000]
Gf(x):
[ 2.212612838605913E-003, 2.212735960794468E-003]
[-3.110560697905385E-008, 3.033631124083059E-008]

x:
[ 9.999999384393309E-001, 1.000000061561520E+000]
[ 9.999986525566056E-001, 9.999988994828021E-001]
```

```
Gf(x):
[ 3.908352000696407E-004, 5.883497343004194E-004]
[-2.941132875244890E-004,-1.954791727376559E-004]


Statinary point:
[ 9.999012668993396E-001, 1.000098733094591E+000]
[ 9.998022868724866E-001, 1.000197713115381E+000]
Gradient value:
[-1.582809001175031E-001, 1.582886994519247E-001]
[-7.903781298350677E-002, 7.903391369521451E-002]

Expected solution:
 1.0
 1.0
```

This approximate method using interval operations converges, but the intervals grow wider. This shows the importance of using true *interval* algorithms, as opposed to *point* algorithms in interval arithmetic. The algorithm presented in Chapter 14 uses a method to generate a sequence of intervals that is guaranteed to enclose the global minimizer, and can achieve high accuracy.

**Example 12.4** In this example, we demonstrate the use of *grad_ari* to compute Jacobians. We again use the method of Newton, now applied to the vector-valued function itself, i.e. to the problem $f(x) = 0$, starting from an approximation $x^{(0)} \in \mathbb{R}^n$. Newton's method is

$$\left.\begin{array}{rcl} R^{(k)} & := & (J_f(x^{(k)}))^{-1} \\ x^{(k+1)} & := & x^{(k)} - R^{(k)} \cdot f(x^{(k)}) \end{array}\right\} \quad k = 0, 1, 2, \ldots$$

In the following sample program for this method, all real operations are replaced by interval operations, because we make use of the interval differentiation arithmetic. That is, we start with $[x]_0$ and iterate

$$\left.\begin{array}{rcl} R^{(k)} & := & (m(J_f([x]^{(k)})))^{-1} \\ [x]^{(k+1)} & := & [x]^{(k)} - R^{(k)} \cdot f([x]^{(k)}) \end{array}\right\} \quad k = 0, 1, 2, \ldots$$

The influence of rounding errors is directly demonstrated by the increasing interval diameters. The approximate inverse $R^{(k)}$ is computed by applying the function *MatInv()* from Chapter 10 to the midpoint matrix of the interval Jacobian evaluation.

We terminate the iteration if an interval evaluation of $f$ over the current interval iterate $[x]^{(k)}$ contains zero, or if the number of iterations is greater than or equal to 100. Although, this approximate method can *not* guarantee to find and enclose a zero of $f$ in the last iterate. For a more sophisticated interval method for finding *all* zeros of a nonlinear system of equations with guarantee and high accuracy, refer to Chapter 13.

We apply our iteration to the problem of finding a zero of Hansen's function [32]

$$f(x) = \begin{pmatrix} 6x_1^5 - 25.2x_1^3 + 24x_1 - 6x_2 \\ 12x_2 - 6x_1 \end{pmatrix}.$$

```
//----------------------------------------------------------------------------
// Example: Automatic differentiation for gradients
// This is an implementation of Newton's Method for computing a zero of
// a continuously differentiable multi-dimensional function.
//
//    given:      the function f(x) and the starting value x[0]
//    iteration:  x[n+1] := x[n] - InvJf(x[n])*f(x[n]) , n = 0,1,...
//
// where InvJf(x) denotes the inverse of the Jacobian matrix of f(x).
// All real operations are replaced by interval operations, all function and
// derivative evaluations are calculated by differentiation arithmetic.
//----------------------------------------------------------------------------
#include <matinv.hpp>       // Matrix inversion
#include <grad_ari.hpp>     // Gradient differentiation arithmetic

const int
  fDim = 2,
  nmax = 100;

ivector  fx(fDim), x(fDim);
imatrix  Jfx(fDim,fDim);
rmatrix  InvJfx(fDim,fDim);
int      n, Error;

GTvector f ( GTvector& x )
{
  GradType x1sqr(fDim);
  GTvector result(fDim);

  x1sqr = sqr(x[1]);
  result[1] = ((6.0*x1sqr - _interval(252.0)/10.0)*x1sqr + 24.0)*x[1]
              - 6.0*x[2];
  result[2] = 12.0*x[2] - 6.0*x[1];
  return result;
}

int main ( )
{
  cout << SetPrecision(23,15) << Scientific;   // Output format

  cout << "Newton's method for finding a zero of Hansen's function:" << endl;
  cout << "f1(x) = 6 x1↑5 - 25.2 x1↑3 + 24 x1 - 6 x2" << endl;
  cout << "f2(x) = 12 x2 - 6 x1" << endl << endl;
  cout << "Starting vector x: " << endl;
  cin  >> x;
  cout << "Iteration" << endl;

  fJEvalJ(f, x, fx, Jfx);

  n = 0;
  do {
    n++;
    cout << "x:" << endl << x << "f(x):" << endl << fx << endl;
    MatInv(mid(Jfx), InvJfx, Error);
    if (!Error) {
      x = x - InvJfx * fx;
      fJEvalJ(f, x, fx, Jfx);
    }
  } while ( !( (in(0.0,fx[1]) && in(0.0,fx[2])) || (n >= nmax) || Error ) );
  cout << endl;

  if (!Error) {
    cout << "Zero:" << endl << x;
    cout << "Function value:" << endl << fx << endl;
```

```
    cout << "Expected zeros:" << endl;
    cout << "-1.7475... or -1.0705... or 0.0 or 1.0705... or 1.7475..."
         << endl;
    cout << "-0.8737...    -0.5352...    0.0    0.5352... or 0.8737..."
         << endl;
  }
  else
    cout << MatInvErrMsg(Error) << endl;

  return 0;
}
```

If we run our sample program for Newton's method, we get the report

```
Newton's method for finding a zero of Hansen's function:
f1(x) = 6 x1^5 - 25.2 x1^3 + 24 x1 - 6 x2
f2(x) = 12 x2 - 6 x1

Starting vextor x: [0.5,0.5] [0.25,0.25]

Iteration
x:
[ 5.000000000000000E-001, 5.000000000000000E-001]
[ 2.500000000000000E-001, 2.500000000000000E-001]
f(x):
[ 7.537499999999999E+000, 7.537500000000002E+000]
[-0.000000000000000E+000, 0.000000000000000E+000]

x:
[-1.396226415094340E+000,-1.396226415094339E+000]
[-6.981132075471700E-001,-6.981132075471696E-001]
f(x):
[ 7.433516327255063E+000, 7.433516327255163E+000]
[-5.329070518200752E-015, 5.329070518200752E-015]

x:
[-7.951902474521197E-001,-7.951902474521103E-001]
[-3.975951237260601E-001,-3.975951237260549E-001]
f(x):
[-5.935598109066682E+000,-5.935598109066246E+000]
[-5.950795411990840E-014, 5.950795411990840E-014]

x:
[-1.196005361090201E+000,-1.196005361090157E+000]
[-5.980026805451034E-001,-5.980026805450755E-001]
f(x):
[ 3.312990631505093E+000, 3.312990631508398E+000]
[-2.957634137601418E-013, 2.957634137601418E-013]

x:
[-1.067378601459598E+000,-1.067378601459414E+000]
[-5.336893007298207E-001,-5.336893007296851E-001]
f(x):
[-8.293597505247741E-002,-8.293597504056116E-002]
[-1.363353874239693E-012, 1.363353874239693E-012]

x:
[-1.070545193619334E+000,-1.070545193618642E+000]
[-5.352725968097762E-001,-5.352725968092117E-001]
f(x):
[ 7.613889939950269E-005, 7.613894476232730E-005]
[-5.458744567476970E-012, 5.456968210637570E-012]
```

```
Zero:
[-1.070542291827323E+000,-1.070542291824694E+000]
[-5.352711459141216E-001,-5.352711459118869E-001]
Function value:
[-2.492317463520522E-011, 1.481588185470173E-010]
[-2.129141307705141E-011, 2.129052489863171E-011]

Expected zeros:
-1.7475... or -1.0705... or 0.0 or 1.0705... or 1.7475...
-0.8737...    -0.5352...    0.0    0.5352... or 0.8737...
```

This runtime output shows a very important fact in connection with the local convergence property of the classical Newton's method. Using the starting point $x^{(0)} = (0.5, 0.25)^T$, we might expect the method to converge to the zero $(0,0)^T$ or to the zero $(1.0705..., 0.5352...)^T$. However, the iteration stops at $(-1.0705..., -0.5352...)^T$. An approximation method cannot guarantee that we reach the "nearest" zero of $x^{(0)}$. We again refer the reader to Chapter 13, where we introduce a method for finding *all* zeros, which generates a sequence of intervals that is guaranteed to enclose the roots, and can achieve high accuracy.

### 12.3.3 Restrictions and Hints

The implementations in modules *hess_ari* and *grad_ari* use the standard error handling of the interval library employed, if the interval argument of an elementary function does not lie in the domain specified for this interval function. The same holds for an interval containing zero as the second operand of the division operator. These cases can also occur during a function evaluation using differentiation arithmetic because of the known overestimation effects of interval arithmetic (see Section 3.1).

Note that the rules for getting true enclosures in connection with conversion errors (see Section 3.7) also apply to interval differentiation arithmetic. To compute enclosures for the values of the gradients or Hessians of function $f$ in Example 12.2 at point $x$ with $x_1 = 0.1$, for example, you must assure that the machine interval argument $[x]_1$ used as argument for the interval function evaluation satisfies $x_1 \in [x]_1$. The same rule applies to real coefficients in the definition of $f$. That is why we expressed 25.2 as `_interval(252.0)/10.0` in Example 12.4.

## 12.4 Exercises

**Exercise 12.1** Implement a *real* version of module *hess_ari* by replacing all *interval* data types by the corresponding *real* data types. Then use the new module and apply it to Example 12.3.

**Exercise 12.2** Implement a *real* version of module *grad_ari* by replacing all *interval* data types by the corresponding *real* data types. Then use the new module and apply it to Example 12.4.

## 12.5 References and Further Reading

The implementations in this chapter have followed those of Rall [73] in PASCAL–SC. We use the *forward mode* of automatic differentiation in which the values of the derivative objects are propagated through the code list in the same order as intermediate values during the evaluation of $f$ in ordinary arithmetic (see [72] or [74] for example).

It is possible to optimize the time complexity of computing gradients or Hessians by using the *backward or reverse mode* (also called fast automatic differentiation, see [18] for example). The reverse mode is faster than the forward mode for computing most large gradients, but it requires more storage space and is trickier to program.

Automatic differentiation methods can also be used to compute interval slopes (see [70]). Further applications and differentiation arithmetics can be found in [22], [23], and [75]. A large bibliography on automatic differentiation is given in [12].

# Chapter 13

# Nonlinear Systems of Equations

In Chapter 6, we considered the problem of finding zeros (or roots) of nonlinear functions of a single variable. Now, we consider its generalization, the problem of finding the solution vectors of a system of nonlinear equations. We give a method for finding *all* solutions of a nonlinear system of equations $f(x) = 0$ for a continuously differentiable function $f : I\!R^n \to I\!R^n$ in a given interval vector (box). Our method computes close bounds on the solution vectors, and it delivers information about existence and uniqueness of the computed solutions. The method we present is a variant of the *interval Gauss-Seidel method* based on the method of Hansen and Sengupta [3], [32], and a modification of Ratz [79]. Our method makes use of the extended interval operations defined in Section 3.3.

In classical numerical analysis, methods for solving nonlinear systems start from some approximate trial solution and iterate to (hopefully) improve the approximation until some convergence or termination criterion is satisfied. These non-interval methods have their difficulties finding *all* solutions of the given problem. It is a commonly held myth that it is impossible for a numerical method to find all solutions to a nonlinear system in a specified region. The myth is true if an algorithm only uses information from point evaluations of $f$. However, the algorithm of this chapter uses values of $f$ and its derivatives evaluated at interval arguments to show that the myth is *false.*

## 13.1 Theoretical Background

Let $f : I\!R^n \to I\!R^n$ be a continuously differentiable function, and let $x \in [x]$ with $[x] \in II\!R^n$. We address the problem of finding all solutions of the equation

$$f(x) = 0. \tag{13.1}$$

Let $J_f$ denote the Jacobian matrix of $f$. We will compute $J_f$ using automatic differentiation as described in Chapter 12. Interval Newton methods are used in general to sharpen bounds on the solutions of (13.1). They can be derived from the mean value form

$$f(m([x])) - f(x^*) = J_f(\xi)(m([x]) - x^*),$$

where $x^* \in [x]$, $\xi = (\xi_1, \ldots, \xi_n)$, and $\xi_i \in [x]$ for $i = 1, \ldots, n$. If we assume $x^*$ to be a zero of $f$, we get

$$f(m([x])) = J_f(\xi) \cdot (m([x]) - x^*).$$

If we assume $J_f(\xi) \in I\!R^{n\times n}$ and all real matrices in $J_f([x]) \in I\,I\!R^{n\times n}$ to be nonsingular, we have

$$\begin{aligned} x^* &= m([x]) - (J_f(\xi))^{-1} \cdot f(m([x])) \\ &\in \underbrace{m([x]) - (J_f([x]))^{-1} \cdot f(m([x]))}_{=:\,N([x])}. \end{aligned}$$

$N([x])$ is the multi-dimensional interval Newton operator. Every zero of $f$ in $[x]$ also lies in $N([x])$ and therefore in $N([x]) \cap [x]$. However, this method is only applicable if the interval matrix $J_f([x])$ is regular, i.e. if every real matrix $B \in J_f([x])$ is regular.

Thus, we are lead to another form of the interval Newton method that relaxes the requirement that $J_f([x])$ be nonsingular. We compute an enclosure of the set of solutions $x^*$ of

$$f(m([x])) = J_f([x]) \cdot (m([x]) - x^*). \tag{13.2}$$

This method works better if we first *precondition* (13.2), i.e. if we multiply by a real matrix $R \in I\!R^{n\times n}$ to obtain

$$R \cdot f(m([x])) = R \cdot J_f([x]) \cdot (m([x]) - x^*).$$

Frequently, the inverse of the midpoint matrix $m(J_f([x]))$ is used as a preconditioner $R$. We will do so, too. Define

$$R := (m(J_f([x])))^{-1}, \quad b := R \cdot f(m([x])), \quad \text{and} \quad [A] := R \cdot J_f([x]), \tag{13.3}$$

and consider the interval linear equation

$$b = [A] \cdot (m([x]) - x^*). \tag{13.4}$$

The method we describe for solving Equation (13.4) is based on the interval *Gauss-Seidel iteration*, which can also be applied to singular systems (cf. [70]).

### 13.1.1 Gauss-Seidel Iteration

We are interested in the solution set

$$S := \{x \in [x] \mid A \cdot (c - x) = b, \quad \text{for } A \in [A]\}$$

of the interval linear equation

$$[A] \cdot (c - x) = b$$

with fixed $c \approx m([x])$. In fact, we can use an arbitrary $c \in [x]$. Gauss-Seidel iteration is based on writing the linear system $A \cdot (c - x) = b$ explicitly in components as

$$\sum_{j=1}^{n} A_{ij} \cdot (c_j - x_j) = b_i, \quad i = 1, \ldots, n,$$

and solving the $i$th equation for the $i$th variable, assuming that $A_{ii} \neq 0$. Then we have

$$\begin{aligned} x_i &= c_i - \Big( b_i + \sum_{\substack{j=1 \\ j \neq i}}^{n} A_{ij} \cdot (x_j - c_j) \Big) \Big/ A_{ii} \\ &\in \underbrace{c_i - \Big( b_i + \sum_{\substack{j=1 \\ j \neq i}}^{n} [A]_{ij} \cdot ([x]_j - c_j) \Big) \Big/ [A]_{ii}}_{=: [z]_i} \end{aligned} \tag{13.5}$$

for $i = 1, \dots, n$ if $0 \notin [A]_{ii}$ for all $i$. We obtain a new enclosure $[z]$ for $x$ by computing the interval vector components $[z]_i$ according to (13.5) yielding

$$S \subseteq [z] \cap [x].$$

Moreover, it is possible to improve the enclosure $[z]$ using the fact that in the $i$th step improved enclosures $[z]_1, \dots, [z]_{i-1}$ are already available. Thus, if we compute

$$\left.\begin{aligned} [y] &:= [x] \\ [y]_i &:= \Big( c_i - \Big( b_i + \sum_{\substack{j=1 \\ j \neq i}}^{n} [A]_{ij} \cdot ([y]_j - c_j) \Big) \Big/ [A]_{ii} \Big) \cap [y]_i \\ & \quad i = 1, \dots, n \\ N_{\mathrm{GS}}([x]) &:= [y] \end{aligned}\right\} \tag{13.6}$$

we have

$$S \subseteq N_{\mathrm{GS}}([x]) \subseteq [z] \cap [x],$$

and every zero of $f$ which lies in $[x]$ also lies in $N_{\mathrm{GS}}([x])$. In (13.6), it is not necessary to compute the $[y]_i$ in fixed order $i = 1, \dots, n$.

In summary, one interval Newton Gauss-Seidel step for the interval vector $[x]$ yields $N_{\mathrm{GS}}([x])$ by first computing $b$ and $[A]$ according to (13.3) and then computing $[y]$ according to (13.6). The interval Newton Gauss-Seidel iteration starts with an interval vector $[x]^{(0)}$ and iterates according to

$$[x]^{(k+1)} := N_{\mathrm{GS}}([x]^{(k)}), \quad k = 0, 1, 2, \dots \tag{13.7}$$

The intersections performed in (13.6) prevent the method from diverging. If an empty intersection occurs, we set $N_{\mathrm{GS}}([x]^{(k)}) := [y]^{(k)} := \emptyset$, and we know that $[x]^{(k)}$ contains no zero of $f$.

If $0 \in [A]_{ii}$ for some $i$, the method can be used if extended interval arithmetic (see Chapters 3 and 6) is applied. In this case, a gap can be produced in the corresponding components $[y]_i^{(k)}$ of $[y]^{(k)}$, i.e. $[y]_i^{(k)}$ is given by one or two intervals resulting from the extended interval division and the succeeding intersection with the old value $[x]_i^{(k)}$. Therefore, $N_{\mathrm{GS}}([x]^{(k)})$ is given by one or more interval vectors, and the next step of the interval Newton Gauss-Seidel iteration must be applied to each of these boxes which possibly contain solutions of (13.1). Thus, it is possible to compute *all* zeros of $f$ in the starting interval vector $[x]^{(0)}$.

In the following theorem, we summarize the most important properties of the interval Newton Gauss-Seidel method.

**Theorem 13.1** *Let $f : D \subseteq \mathbb{R}^n \to \mathbb{R}^n$ be a continuously differentiable function, and let $[x] \in I\mathbb{R}^n$ be an interval vector with $[x] \subseteq D$. Then $N_{\mathrm{GS}}([x])$ defined by (13.6) has the following properties:*

1. *Every zero $x^* \in [x]$ of $f$ satisfies $x^* \in N_{\mathrm{GS}}([x])$.*
2. *If $N_{\mathrm{GS}}([x]) = \emptyset$, then there exists no zero of $f$ in $[x]$.*
3. *If $N_{\mathrm{GS}}([x]) \overset{\circ}{\subset} [x]$, then there exists a unique zero of $f$ in $[x]$ and hence in $N_{\mathrm{GS}}([x])$.*

For proofs see [31] or [70].

**Remark:** The conditions of Theorem 13.1 can be checked on a computer. For example, if $N_{\Diamond\mathrm{GS}}([x])$ denotes the machine interval computation of $N_{\mathrm{GS}}([x])$ and if condition 3 is satisfied for $N_{\Diamond\mathrm{GS}}$, then we have

$$N_{\mathrm{GS}}([x]) \subseteq N_{\Diamond\mathrm{GS}}([x]) \overset{\circ}{\subset} [x].$$

Thus, the condition is fulfilled for $N_{\mathrm{GS}}([x])$, too. On the other hand, if we cannot fulfill conditions 2 and 3 of the Theorem, then $[x]$ *may* contain one or more zeros. This is especially the case if the problem has infinitely many solutions. Our interval method can find and bound an infinite set of solutions consisting of a continuum of points, but of course, none will be unique.

## 13.2 Algorithmic Description

The main algorithm AllNLSS (Algorithm 13.4) consists of two parts. The first part is the extended interval Newton iteration itself, including the interval Newton Gauss-Seidel step with intermediate checks for the uniqueness of a zero in a computed enclosure. The second part is an additional verification step which tries to verify the local uniqueness for enclosures that have not already been marked as enclosing a unique zero.

Algorithm 13.1 describes the interval Newton Gauss-Seidel step, where (13.6) is applied to the box $[y] \in I\mathbb{R}^n$ resulting in at most $n+1$ boxes $[V]_i$. After some initializations, we use Algorithm 10.1 (MatInv) of Chapter 10 to compute an approximate inverse of the midpoint matrix of the interval Jacobian $J_f([y])$. In Step 4, we compute $[A]$, $[b]$, and $[y_c]$ followed by the initialization for the loop. Interval arithmetic must be used to compute $f(c)$ to bound all rounding errors (denoted by $f_\Diamond(c)$ in Step 4).

We first perform the single component steps of the Gauss-Seidel step for all $i$ with $0 \notin [A]_{ii}$ (Step 6) and then for the remaining indices with $0 \in [A]_{ii}$ (Step 7). Using this strategy, it is possible that the intervals $[y]_i$ become smaller by the intersections with the old values $[y]_i$ in Step 6(b) before the first splitting is produced in Step 7(b). If a gap is produced in this case, we store one part of the actual box $[y]$ in the $p$th row of the interval matrix $[V]$, and we continue the iteration for the other part of $[y]$. An empty intersection in Step 6(c) or Step 7(c) signals that no solution exists in $[y]$ . Thus, we set $p := 0$ and terminate.

**Algorithm 13.1:** NewtonStep $(f, [y], [V], p)$

1. $c := m([y])$; {Compute midpoint of $[y]$}
2. MatInv $(m(J_f([y])), R, InvErr)$; {Invert the midpoint matrix}
3. **if** ($InvErr \neq$ "No Error") **then** $R := I$;
4. $[A] := R \cdot J_f([y])$; $[b] := R \cdot f_{\diamond}(c)$; $[y_c] := [y] - c$;
5. $p := 0$; {Initialization for loop}
6. **for** $i := 1$ **to** $n$ **do** {Interval Gauss-Seidel step for $0 \notin [A]_{ii}$}
   (a) **if** $(0 \in [A]_{ii})$ **then next**$_i$;
   (b) $[y]_i := \left( c_i - \left( [b]_i + \sum_{\substack{j=1 \\ j\neq i}}^{n} [A]_{ij} \cdot [y_c]_j \right) / [A]_{ii} \right) \cap [y]_i$;
   (c) **if** $[y]_i = \emptyset$ **then return** $p := 0$; {No solution possible}
   (d) $[y_c]_i := [y]_i - c_i$;
7. **for** $i := 1$ **to** $n$ **do** {Extended interval Gauss-Seidel step for $0 \in [A]_{ii}$}
   (a) **if** $(0 \notin [A]_{ii})$ **then next**$_i$;
   (b) $[z] := \left( c_i - \left( [b]_i + \sum_{\substack{j=1 \\ j\neq i}}^{n} [A]_{ij} \cdot [y_c]_j \right) / [A]_{ii} \right) \cap [y]_i$; $\{[z] = [z]_1 \cup [z]_2\}$
   (c) **if** $([z] = \emptyset)$ **then return** $p := 0$; {No solution possible}
   (d) $[y]_i := [z]_1$; $[y_c]_i := [y]_i - c_i$;
   (e) **if** $([z]_2 \neq \emptyset)$ **then**
   $p := p + 1$; $[V]_p := [y]$; $[V]_{pi} := [z]_2$; {Store part of $[y]$ in $[V]_p$}
8. $p := p + 1$; $[V]_p := [y]$;
9. **return** $[V], p$;

Algorithm 13.2 is a recursive formulation of the extended interval Newton Gauss-Seidel iteration for the function $f$. The input parameter $[y]$ specifies the actual box. The input parameter $\varepsilon$ corresponds to the desired relative accuracy or tolerance for the resulting interval enclosures of the zeros. Algorithm AllNLSS guarantees that $\varepsilon$ is not chosen less than the relative machine accuracy (1 ulp). The input parameter $yUnique$ signals whether we have already verified that the incoming interval vector $[y]$ contains a locally unique zero.

If $0 \notin f([y])$, then we know that no zero can lie within the interval vector $[y]$. Hence, a single interval function evaluation of $f$ can guarantee that a complete range of real values cannot contain a zero of $f$. If $f([y])$ contains zero, then the extended interval Newton step given by (13.7) is applied to $[y]$ resulting in $p$ non-empty boxes $[V]_i$ with $i = 1, \ldots, p$ and $p \leq n+1$ (Step 2).

If the Newton step results in only one interval vector and if the local uniqueness of a zero in $[y]$ is not already proven, then the algorithm checks Condition 3 of Theorem 13.1 for the resulting box $[V]_1$ and sets the flags *InnerIncl* and *yUnique* in Step 3. In Step 4 the actual box $[V]_i$ gets stored in the rectangular interval matrix $[Sol]$ if the tolerance criterion is satisfied. Otherwise (else-branch of Step 4), the box $[V]_i$ is bisected with respect to the component with maximum diameter. Then Algorithm XINewton is recursively called for both halves in Steps 4(e) and 4(g). The corresponding information on the uniqueness of the zero is stored in the flag vector *Info*. $N$ represents the number of enclosures of zeros stored in $[Sol]$.

Algorithm XINewton terminates when no more recursive calls are necessary, that is if $0 \notin f([y])$ or if $(d_{\text{rel},\infty}([V]_i) \leq \varepsilon)$ for $i = 1, \ldots, p$. The bisection step just mentioned guarantees that this second condition is fulfilled at some stage of the recursion.

**Algorithm 13.2:** XINewton $(f, [y], \varepsilon, yUnique, [Sol], Info, N)$

1. **if** $0 \notin f([y])$ **then return** ;
2. NewtonStep $(f, [y], [V], p)$;
3. **if** $p = 1$ **then**

   *InnerIncl* := $([V]_1 \mathring{\subset} [y])$;

   *yUnique* := *yUnique* **or** *InnerIncl*; {Inner inclusion ⇒ uniqueness}

   **else**

   *InnerIncl* := *false*; *yUnique* := *false*;
4. **for** $i := 1$ **to** $p$ **do**

   **if** $(d_{\text{rel},\infty}([V]_i) \leq \varepsilon)$ **then** {Store enclosure of zero and uniqueness info}

   **if** $(0 \in f([V]_i))$ **then**

   $N := N + 1$; $[Sol]_N := [V]_i$; $Info_N :=$ *yUnique*;

   **else** {Bisect the box and apply XINewton recursively}

   **if not** *InnerIncl* **then**

   (a) *yUnique* := *false*;

   (b) Find $l$ with $d([y]_l) = d_\infty([y])$; {Component with max. diameter}

   (c) $c_l := m([V]_{il})$; $s_l := \overline{V}_{il}$; {Center and supremum of $l$th component}

   (d) $[V]_{il} := [\underline{V}_{il}, c_l]$;

   (e) XINewton $(f, [V]_i, \varepsilon, yUnique, [Sol], Info, N)$; {1st half of the box}

   (f) $[V]_{il} := [c_l, s_l]$;

(g) XINewton $(f, [V]_i, \varepsilon, yUnique, [Sol], Info, N)$; {2nd half of the box}

5. **return** $[Sol]$, $Info$, $N$;

Algorithm 13.3 describes an additional verification step which checks the local uniqueness of the solution of the nonlinear system of equations enclosed in the interval vector $[y]$. The function $f$ may have many zeros in the original box $[x]$. We attempt to find subboxes $[y]$, each of which contains a single zero of $f$. That is what we refer to as local uniqueness. Algorithm 13.3 can be used for boxes which have not yet been guaranteed to enclose a locally unique zero. This is done according to condition 3 of Theorem 13.1 by applying interval Newton steps including an epsilon inflation of the iterates $[y]$, which can help to verify zeros lying on the edge of $[y]$. We use $k_{\max} = 10$ as the maximum number of iterations, $\varepsilon = 0.25$ as the starting value for the epsilon inflation, and a factor of 8 to increase $\varepsilon$ within the iterations. It turned out that these are good values for minimizing the effort if no verification is possible (see also [52]).

**Algorithm 13.3:** VerificationStep $(f, [y], yUnique)$

1. $k_{\max} := 10$; $k := 0$; $[y_{\text{in}}] := [y]$; $\varepsilon := 0.25$; $yUnique := false$; {Initializations}
2. **while** (**not** $yUnique$) **and** $(k < k_{\max})$ **do** {Do $k_{\max}$ loops to achieve inclusion}
   (a) $k := k + 1$; $[y_{\text{old}}] := [y] \bowtie \varepsilon$; {Epsilon inflation of $[y]$}
   (b) NewtonStep $(f, [y_{\text{old}}], [V], p)$;
   (c) **if** $p \neq 1$ **then exit**$_{\text{while-loop}}$; {No verification possible}
   (d) $[y] := [V]_1$;
   (e) $yUnique := ([V]_1 \overset{\circ}{\subset} [y_{\text{old}}])$; {Inner inclusion $\Rightarrow$ uniqueness}
   (f) **if** $[y] = [y_{\text{old}}]$ **then** $\varepsilon := \varepsilon \cdot 8$; {Increase $\varepsilon$}
3. **if not** $yUnique$ **then** $[y] := [y_{\text{in}}]$; {Reset $[y]$ to starting value}
4. **return** $[y]$, $yUnique$;

Algorithm 13.4 now combines Algorithm 13.2 and Algorithm 13.3 to compute enclosures for all solutions of equation $f(x) = 0$ within the input interval vector $[x]$ and tries to prove the local uniqueness of a solution in each enclosure computed. The desired accuracy (relative diameter) of the interval enclosures is specified by the input parameter $\varepsilon$. 1 ulp accuracy is chosen if the specified value of $\varepsilon$ is too small (for example 0). The enclosures for the solutions are returned row by row in the interval matrix $[Sol]$. The corresponding information on the local uniqueness of the solution is returned in the Boolean vector *Info*. The number of enclosures computed is returned in the integer variable $N$.

We use a function called CheckParameters as an abbreviation for the error checks for the parameters of AllNLSS which are necessary in an implementation. If no error occurs, AllNLSS delivers the $N$ enclosures $[Sol]_i$, $i = 1, 2, \ldots, N$, satisfying

if $Info_i$ = *true*, then $[Sol]_i$ encloses a locally unique zero of $f$,
if $Info_i$ = *false*, then $[Sol]_i$ may enclose a zero of $f$.

If $N = 0$, then it is guaranteed that there is no zero of $f$ in the starting box $[x]$.

**Algorithm 13.4:** AllNLSS $(f, [x], \varepsilon, [Sol], Info, N, Err)$

1. $Err$ := CheckParameters;
2. **if** $Err \neq$ "No Error" **then return** $Err$;
3. $N := 0$;
4. Set $\varepsilon$ to "1 ulp accuracy" if $\varepsilon$ is too small;
5. XINewton $(f, [x], \varepsilon, false, [Sol], Info, N)$;
6. **for** $i := 1$ **to** $N$ **do**
   **if** $Info_i \neq true$ **then** VerificationStep $(f, [Sol]_i, Info_i)$;
7. **return** $[Sol]$, $Info$, $N$, $Err$;

## Applicability of the Algorithm

To keep our algorithm and implementation as simple as possible, Algorithm 13.2 uses only the very simple stopping criteria

$$d_{\mathrm{rel},\infty}([V]_i) \leq \varepsilon.$$

Therefore, if the interval Newton steps do not improve the actual interval vector iterate, the method corresponds to a bisection method. The same applies to problems where the solution set is given by a continuum of points. For more sophisticated stopping criteria, see [31]. On the other hand, if the actual interval vector iterate $[y]$ causes no more extended interval divisions in NewtonStep, then the asymptotic rate of convergence to a zero of $f$ in $[y]$ is quadratic.

The algorithm cannot verify the existence and the uniqueness of a multiple zero $x^*$ of $f$ in the enclosing interval vector. Nevertheless, the zero is bounded to the desired accuracy specified by $\varepsilon$. In this case, the corresponding component of the *Info*-vector is *false*. As a consequence of the splitting of the intervals, it may happen that a zero lying exactly on the splitting point is enclosed in several boxes. The method can be extended by a function to determine a single interval vector representing a set of abutting boxes (see [79] for example).

# 13.3 Implementation and Examples

## 13.3.1 C++ Program Code

Our implementation of Algorithm 13.4 uses the extended interval arithmetic module *xi_ari* (Section 3.3 and Chapter 6) and the automatic differentiation module *grad_ari* (Chapter 12).

#### 13.3.1.1 Module nlinsys

The header file of module *nlinsys* supplies the interface of the function *AllNLSS()* for computing all solutions of a system of nonlinear equations and the interface of the function *AllNLSSErrMsg()* which can be used to get an error message for the error code returned by *AllNLSS()*.

```
//----------------------------------------------------------------------------
// File: nlinsys (header)
// Purpose: Computing enclosures for all solutions of systems of nonlinear
//    equations given by continuously differentiable functions.
// Global functions:
//    AllNLSS()      : computes enclosures for all solutions
//    AllNLSSErrMsg(): delivers an error message text
//----------------------------------------------------------------------------
#ifndef __NLINSYS_HPP
#define __NLINSYS_HPP

#include <intvecto.hpp>      // Integer vector type
#include <imatrix.hpp>       // Interval matrix/vector arithmetic
#include <mv_util.hpp>       // Real matrix/vector utility functions
#include <mvi_util.hpp>      // Interval matrix/vector utility functions
#include <grad_ari.hpp>      // Gradient differentiation arithmetic

const int MaxCount = 10000; // Maximum count of result components

extern char* AllNLSSErrMsg ( int );
extern void  AllNLSS ( GTvector_FctPtr, ivector, real, imatrix&, intvector&,
                       int&, int&, int = MaxCount );
#endif
```

The function *AllNLSS()* is an implementation of Algorithm 13.4. The static functions *NewtonStep()*, *XINewton()*, and *VerificationStep()* are implementations of the Algorithms 13.1 to 13.3. All derivatives are evaluated in the differentiation arithmetic provided by module *grad_ari* (see Chapter 12).

The function *AllNLSS()* uses the *GTvector* function *f()* and the starting interval vector *Start* as input parameters and stores all computed enclosures in the interval matrix *SoluVector* (row by row) which is also passed to and from the function *XINewton()*. The information about the uniqueness is stored in the integer vector *InfoVector*. The number of rows of the matrix *SoluVector* and the length of the vector *InfoVector* are automatically adapted to the number of solutions located. The function *AllNLSS()* has an optional last parameter *MaxNumberOfSolus* by which you can specify the maximum number of solutions to be computed. Its default value is given by the local constant *MaxCount* = 10000 which should be big enough to compute *all* solutions in most cases. *AllNLSS()* reports a warning if there are more solutions than specified by parameter *MaxNumberOfSolus*. The value of the desired accuracy *Epsilon* is checked to be greater than or equal to *MinEpsilon* (the relative machine accuracy). If this is not true, then *Epsilon* is reset to 1 ulp accuracy.

```
//----------------------------------------------------------------------------
// File: nlinsys (implementation)
// Purpose: Computing enclosures for all solutions of systems of nonlinear
//    equations given by continuously differentiable functions.
// Method: Extended interval Newton Gauss-Seidel method.
// Global functions:
```

```
//    AllNLSS()      : computes enclosures for all solutions
//    AllNLSSErrMsg(): delivers an error message text
//----------------------------------------------------------------------------
#include <string.h>          // String handling
#include <xi_ari.hpp>        // Extended interval arithmetic
#include <matinv.hpp>        // Matrix inversion
#include <nlinsys.hpp>

static int MaxSoluNo;        // Maximum number of solutions to be computed

//----------------------------------------------------------------------------
// Error codes used in this module.
//----------------------------------------------------------------------------
const int
  NoError      = 0,  // No error occurred
  IllMaxSoluNo = 1,  // Illegal value for the maximum number of solutions
  NotAllSolus  = 2;  // Not all solutions computed

//----------------------------------------------------------------------------
// Error messages depending on the error code.
//----------------------------------------------------------------------------
char* AllNLSSErrMsg ( int Err )
{
  static char Msg[80] = "";

  switch (Err) {
    case NoError:
      break;  // Blank message
    case IllMaxSoluNo:
      sprintf(Msg,"Error: Parameter for maximum number of solutions "
                  "must lie in 1,...,%1d!",MaxCount); break;
    case NotAllSolus:
      sprintf(Msg,"Warning: Not all solutions found due to the user "
                  "limit of %1d solution(s)!",MaxSoluNo); break;
    default:
      strcpy(Msg,"Error: Code not defined!");
  }
  return(Msg);
}

//----------------------------------------------------------------------------
// Purpose: Execution of one single interval Newton Gauss-Seidel step for the
//    interval vector 'Y' and the function 'f'.
// Parameters:
//    In    : 'f'         : must be declared for the type 'GTvector' to
//                          enable the internal use of the differentiation
//                          arithmetic 'grad_ari'
//            'Y'         : starting interval
//            'JfY'       : Jacobian matrix of 'f(Y)', already computed
//                          outside of 'NewtonStep'
//            'Epsilon'   : desired relative accuracy
//    In/Out: 'V'         : enclosures 'V[i]' for the splittings
//                          generated by the Newton step
//            'p'         : number of non-empty interval vectors 'V[i]'
// Description:
//    'NewtonStep' executes the extended interval Newton Gauss-Seidel step
//    for 'Y' with result interval vector(s) 'V[i]' which can be empty.
//    'p' gives the number of non-empty interval vectors stored as vectors
//    'V[1]',...,'V[p]'.
//----------------------------------------------------------------------------
static void NewtonStep ( GTvector_FctPtr f,
                         ivector         Y,
                         imatrix&        JfY,
                         real            Epsilon,
```

```
                              imatrix&        V,
                              int&            p )
{
  const int      n = Ub(Y);
  rvector        c(n);
  ivector        fC(n), b(n), Y_minus_c(n);
  rmatrix        R(n,n);
  imatrix        A(n,n);
  int            i, i0, j, InvErr;
  idotprecision  Accu;
  interval       h;
  ivector        z(2);

  c = mid(Y);  fEvalJ(f,_ivector(c),fC);        // Midpoint evaluation of 'f'
                                                //---------------------------
  MatInv(mid(JfY),R,InvErr);                    // Invert the midpoint matrix
  if (InvErr) R = Id(R);                        //---------------------------

  A = R * JfY;  b = R * fC;             // Compute data for Gauss-Seidel step
  Y_minus_c = Y - c;                    //-----------------------------------

  p = 0;  i0 = 0;                       // Initializations, A[i0,i0] contains zero
                                        //----------------------------------------
  for (i = 1; i <= n; i++ ) {                    // Interval-Gauss-Seidel-step for
                                                 // non-zero A[i,i] elements
                                                 //------------------------------
    if ( in(0.0,A[i][i]) || (RelDiam(Y[i]) <= Epsilon) ) {
      i0 = i;        // Largest i with 0 in A[i,i]
      continue;      // Next i
    }

    Accu = b[i];
    for (j = 1; j < i; j++) accumulate(Accu,A[i][j],Y_minus_c[j]);
    for (j = i+1; j <= n; j++) accumulate(Accu,A[i][j],Y_minus_c[j]);
    rnd(Accu,h);
    h = c[i] - h / A[i][i];

    if ( Disjoint(Y[i],h) ) { p = 0; return; }         // No solution possible
                                                       //---------------------
    Y[i] = Y[i] & h;  Y_minus_c[i] = Y[i] - c[i];
  } // for (i...

  for (i = 1; i <= i0; i++ ) {                   // Interval-Gauss-Seidel-step for
                                                 // zero A[i,i] elements
                                                 //------------------------------
    if ( !in(0.0,A[i][i]) || (RelDiam(Y[i]) <= Epsilon) ) continue;  // Next i
                                                                     //-------
    Accu = b[i];
    for (j = 1; j < i; j++) accumulate(Accu,A[i][j],Y_minus_c[j]);
    for (j = i+1; j <= n; j++) accumulate(Accu,A[i][j],Y_minus_c[j]);
    rnd(Accu,h);
    z = Y[i] & (c[i] - h % A[i][i]);              // Extended interval division
                                                  //---------------------------
    if (z[1] == EmptyIntval())                  // z[1] == z[2] == EmptyIntval()
      { p = 0; return; }                        // i.e. no solution possible
                                                //-----------------------------
    Y[i] = z[1];  Y_minus_c[i] = Y[i] - c[i];                  // Compute new 'Y'
    if (z[2] != EmptyIntval())                                 //---------------
      { p++; V[p] = Y;  V[p][i] = z[2]; }          // Store further splittings
  } // for (i...                                   //-------------------------

  p++;  V[p] = Y;
} // NewtonStep
```

```
//----------------------------------------------------------------------
// Purpose: Recursive function for the execution of the extended interval
//    Newton Gauss_Seidel method for the function 'f'.
// Parameters:
//    In    : 'f'          : must be declared for the type 'GTvector' to
//                           enable the internal use of the differentiation
//                           arithmetic 'grad_ari'
//            'y'          : starting interval
//            'Epsilon'    : desired relative accuracy
//                           (interval diameter) of the result intervals
//            'yUnique'    : signals whether it is already verified that the
//                           actual interval 'y' contains a unique zero
//    Out   : 'SoluVector': enclosures for the zeros of 'f'
//            'InfoVector': corresponding information on the uniqueness of
//                           the zero in each of these enclosures
//    In/Out: 'SoluNo'     : represents the number of the zero computed last
//                           (in) and the total number of enclosures
//                           computed at the end of the recursion (out)
// Description:
//    The function 'XINewton()' is recursively called whenever the extended
//    interval Newton step results in a splitting of the actual interval 'y'
//    in two or more intervals, and the tolerance condition is not fulfilled
//    yet. Otherwise, the enclosures for the zeros of 'f' are stored in the
//    interval matrix 'SoluVector' row by row, the corresponding information
//    on the uniqueness of the zero in each of these enclosures is stored in
//    the integer vector 'InfoVector'. The number of enclosures computed is
//    returned in the variable 'SoluNo'. The maximum number of enclosures to
//    be computed is given by the value of the global variable 'MaxSoluNo'.
//----------------------------------------------------------------------
static void XINewton ( GTvector_FctPtr f,
                       ivector&        y,
                       real&           Epsilon,
                       int             yUnique,
                       imatrix&        SoluVector,
                       intvector&      InfoVector,
                       int&            SoluNo )
{
  if (SoluNo > MaxSoluNo) return;         // Stop due to user specified limit
                                          //---------------------------------
  const int n = Ub(y);
  ivector   fy(n);
  rvector   dY(n);
  imatrix   Jfy(n,n);
  imatrix   V(n+1,n);
  int       i, p, k, l;
  real      my, supy;
  int       InnerIncl;

  fJEvalJ(f, y, fy, Jfy);                 // Compute f(y) and Jf(y)
  if ( !in(0,fy) ) return;                // Start if 0 in f(y), else do nothing
                                          //-----------------------------------
  NewtonStep(f,y,Jfy,Epsilon,V,p);        // Extended interval Newton step with
                                          // results V[i], i = 1,...,p
  if (p == 1) {                           //-----------------------------------
    InnerIncl = in(V[1],y);               // Inner inclusion ===> uniqueness
    yUnique = yUnique || InnerIncl;       //--------------------------------
  }
  else
    InnerIncl = yUnique = 0;

  for (i = 1; i <= p; i++) {
    if (MaxRelDiam(V[i]) <= Epsilon) {      // No more Newton steps

      fEvalJ(f, V[i], fy);                  // Compute f(V[i])
```

```
      if ( in(0,fy) ) {                    // Store enclosure and info
        SoluNo++;                          //-------------------------

        if (SoluNo > MaxSoluNo) return;    // Stop due to user specified limit
                                           //---------------------------------
        if (SoluNo > Ub(SoluVector,1))
          { DoubleSize(SoluVector); DoubleSize(InfoVector); }
        SoluVector[SoluNo] = V[i];         // Store enclosure of the zero
        InfoVector[SoluNo] = yUnique;      // Store uniqueness info
      }                                    //----------------------------
    }
    else { // Recursive call of 'XINewton()' for interval 'V[i]'
      if ( !InnerIncl ) {                                  // Bisect the box
        yUnique = 0;                                       //---------------
        dY = diam(V[i]);
        l = 1;
        for (k = 2; k <= n; k++)
          if (dY[k] > dY[l]) l = k;
        my = mid(V[i][l]);                // Compute midpoint of l-th component
        supy = Sup(V[i][l]);              // Save supremum of l-th component
        SetSup(V[i][l],my);               // Set supremum to midpoint
                                          //-----------------------------------
        XINewton(f,V[i],Epsilon,yUnique,                // Call 'XINewton()' for
                 SoluVector,InfoVector,SoluNo);         // first half of 'V[i]'
                                                        //---------------------
        SetSup(V[i][l],supy);                  // Reset supremum to old value
        SetInf(V[i][l],my);                    // Set infimum to midpoint
      }                                        //----------------------------
      XINewton(f,V[i],Epsilon,yUnique,                 // Call 'XINewton()' for
               SoluVector,InfoVector,SoluNo);          // (second half of) 'V[i]'
    } // else                                          //------------------------
  } // for
} // XINewton

//----------------------------------------------------------------------------
// Purpose: Execution of a verification step including the use of an epsilon
//    inflation.
// Parameters:
//    In    : 'f'       : function of type 'GTvector'
//    Out   : 'yUnique' : returns 'true' if the verification is successful
//    In/Out: 'y'       : interval enclosure to be verified
// Description: This function checks the uniqueness of the zero enclosed in
//    the variable 'y' by an additional verification step including the use
//    of an epsilon inflation of the iterates.
//----------------------------------------------------------------------------
static void VerificationStep ( GTvector_FctPtr f, ivector& y, int& yUnique )
{
  const int kmax = 10, n = Ub(y);
  ivector   yIn(n), fY(n), yOld(n);
  imatrix   JfY(n,n);
  imatrix   V(n+1,n);
  rvector   m(n);
  int       k, p;
  real      eps;

  yIn = y;  k = 0;  eps = 0.25;                             // Initializations
  yUnique = 0;                                              //----------------
  while ( !yUnique && (k < kmax) ) {     // Do kmax loops to achieve inclusion
    yOld = Blow(y,eps);                  // Epsilon inflation of 'y'
                                         //-----------------------------------
    k++;  fJEvalJ(f, yOld, fY, JfY);     // Perform interval Newton step
    NewtonStep(f, yOld, JfY, 0.0, V, p); //-----------------------------------

    if (p != 1) break;                              // No verification possible
```

```
                                                        //-------------------------
    if (V[1] == yOld)                                   // Increase 'eps'
      eps = eps * 8.0;                                  //-------------------------
    else
      { y = V[1]; yUnique = in(y,yOld); }   // Inner inclusion ===> uniqueness
  }                                         //-------------------------------

  if (!yUnique) y = yIn;
} // VerificationStep

//----------------------------------------------------------------------------
// Purpose: Computation of enclosures for all zeros of a continuously
//    differentiable multi-dimensional, vector-valued function.
// Parameters:
//    In : 'f'                : objective function, must be declared for the
//                              'GTvector' to enable the internal use of
//                              the differentiation arithmetic 'grad_ari'.
//         'Start'            : starting interval.
//         'Epsilon'          : desired relative accuracy
//                              (interval diameter) of the result intervals.
//    Out: 'SoluVector'       : stores and returns the enclosures for the
//                              zeros of 'f'.
//         'InfoVector'       : corresponding information on the uniqueness
//                              of the zeros in each of these enclosures.
//         'NumberOfSolus'    : number of enclosures computed.
//         'Err'              : error code.
//         'MaxNumberOfSolus' : maximum number of zeros to be computed
//                              (default value: 'MaxCount' = compute all)
// Description:
//    The enclosures for the zeros of 'f' are computed by calling function
//    'XINewton'. Then an additional verification step is applied to those
//    enclosures which have not been verified. If an error occurs the value
//    of 'Err' is different from 0. The output parameters 'SoluVector' and
//    'InfoVector' are automatically resized to standard index range with
//    lower index bounds 1. The result arrays 'SoluVector' and 'InfoVector'
//    are automatically resized to standard index range with lower index
//    bounds 1. The maximum number of zeros which can be computed is given by
//    'MaxNumberOfSolus'. It is bounded by 'MaxCount'. If your system can
//    handle more than 'MaxCount' zeros, you can increase its value.
//----------------------------------------------------------------------------
void AllNLSS ( GTvector_FctPtr f,
               ivector         Start,
               real            Epsilon,
               imatrix&        SoluVector,
               intvector&      InfoVector,
               int&            NumberOfSolus,
               int&            Err,
               int             MaxNumberOfSolus )
{
  int     MinSize, i, n = VecLen(Start);
  real    MinEpsilon;

  if (1 <= MaxNumberOfSolus && MaxNumberOfSolus <= MaxCount) {
    MaxSoluNo = MaxNumberOfSolus;   // Initialize global static variable

    // Set index range of the start vector to standard index range with
    // lower bound 1 and resize the result arrays to 'MinSize' components.
    //----------------------------------------------------------------------
    SetLb(Start,1);

    MinSize = (MaxNumberOfSolus == MaxCount) ? 10 : MaxNumberOfSolus;

    Resize(SoluVector,MinSize,n);
    Resize(InfoVector,MinSize);
```

```
    // Start extended interval Newton method
    //-------------------------------------
    Err = NoError;  NumberOfSolus = 0;

    MinEpsilon = succ(1.0) - 1.0;       // Relative machine accuracy (1 ulp)
    SetLb(Start,1);                     // Change to standard bounds 1..n
                                        //---------------------------------
    if (Epsilon < MinEpsilon)
      Epsilon = MinEpsilon;             // Set 'Epsilon' to 1 ulp accuracy

    XINewton(f,Start,Epsilon,0,SoluVector,InfoVector,NumberOfSolus);

    if (NumberOfSolus > MaxNumberOfSolus)
      { Err = NotAllSolus;  NumberOfSolus = MaxNumberOfSolus; }
  }
  else {
    Err = IllMaxSoluNo;
    NumberOfSolus = 0;
  }

  // Resize the result matrix and the info vector to appropriate
  // number of rows if no error has occurred. Otherwise, resize
  // them to minimum dimensions with undefined value.
  //-----------------------------------------------------------
  if (NumberOfSolus > 0) {
    Resize(SoluVector,NumberOfSolus,n);
    Resize(InfoVector,NumberOfSolus);
  }
  else {
    Resize(SoluVector,1,1);      // Values undefined
    Resize(InfoVector,1);        //-----------------
  }

  for (i = 1; i <= NumberOfSolus; i++)                       // Verification step
    if (InfoVector[i] == 0)                                  // for the enclosures
      VerificationStep(f,SoluVector[i],InfoVector[i]);       //-------------------
} // AllNLSS
```

### 13.3.2 Example

Our sample program uses the module *nlinsys* to compute all zeros of the function $f : I\!R^n \to I\!R^n$ with

$$f_i(x) = 0.6 \cdot x_i - 2 + 0.49 \cdot x_i \cdot \sum_{j=1}^{n} x_j^2, \quad i = 1, \dots, n$$

in a specified starting interval vector. The function is taken from [32], where it is chosen because it is easily programmable for arbitrary dimension. Its zero $x = (x_i)$ is given by

$$x_i := \frac{4\sqrt{5}}{7\sqrt{n}} \cdot \sinh\left(\frac{1}{3} \cdot \operatorname{asinh}\left(\frac{7\sqrt{5n}}{2}\right)\right), \quad i = 1, \dots, n. \tag{13.8}$$

```
//----------------------------------------------------------------------------
// This program uses module 'nlinsys' to compute the solution of a system of
// nonlinear equations given by Hansen. A starting interval vector and a
// tolerance must be entered.
//----------------------------------------------------------------------------
#include <nlinsys.hpp>     // Nonlinear system solver
#include <stacksz.hpp>     // To increase stack size for some
                           // special C++ compilers

const char* name = "Hansen\'s Function";

//----------------------------------------------------------------------------
// Definition of Hansen's function in 'fDim' variables.
//----------------------------------------------------------------------------
GTvector f ( GTvector& x )
{
  int      i, n = x.Dim();
  GradType SqrSum(n);
  GTvector Result(n);

  SqrSum = 0.0;
  for (i = 1; i <= n; ++i)
    SqrSum = SqrSum + sqr(x[i]);
  for (i = 1; i <= n; ++i)
    Result[i] = 6.0*x[i]/10.0 - 2.0 + 49.0*x[i]*SqrSum/100.0;
  return Result;
}


int main ( )
{
  int  fDim;

  // Get the actual value of 'fDim'
  //------------------------------
  cout << "Desired dimension    : ";  cin >> fDim;

  ivector   SearchInterval(fDim);
  real      Tolerance;
  imatrix   Solu;
  intvector Unique;
  int       NumberOfSolus, i, Error;

  cout << endl
       << "Computing all solutions for " << name << " in " << fDim
       << " variables" << endl << endl;
  cout << "Search interval      : ";  cin >> SearchInterval;
  cout << "Tolerance (relative) : ";  cin >> Tolerance;
  cout << endl;

  AllNLSS(f, SearchInterval, Tolerance, Solu, Unique, NumberOfSolus, Error);

  cout << SetPrecision(23,15) << Scientific;   // Output format
  for (i = 1; i <= NumberOfSolus; i++) {
    cout << Solu[i];
    if (Unique[i])
      cout << "encloses a locally unique zero!";
    else
      cout << "may contain a zero (not verified unique)!";
    cout << endl << endl;
  }

  cout << NumberOfSolus << " interval enclosure(s)" << endl;
```

```
  if (Error)
    cout << endl << AllNLSSErrMsg(Error) << endl;
  else if ( (NumberOfSolus == 1) && Unique[1] )
    cout << endl << "We have validated that there is a globally unique zero!"
         << endl;

  return 0;
}
```

If we execute this program for $n = 2$ and $n = 6$, we get the following runtime outputs:

```
Desired dimension    : 2

Computing all solutions for Hansen's Function in 2 variables

Search interval      : [-1,1] [-1,1]
Tolerance (relative) : 1e-10

0 interval enclosure(s)



Desired dimension    : 6

Computing all solutions for Hansen's Function in 6 variables

Search interval      : [-1,1] [-1,1] [-1,1] [-1,1] [-1,1] [-1,1]
Tolerance (relative) : 1e-10

[ 8.023509103292852E-001, 8.023509103360037E-001]
[ 8.023509103313217E-001, 8.023509103339674E-001]
[ 8.023509103321313E-001, 8.023509103331578E-001]
[ 8.023509103324553E-001, 8.023509103328336E-001]
[ 8.023509103325909E-001, 8.023509103326981E-001]
[ 8.023509103326380E-001, 8.023509103326511E-001]
encloses a locally unique zero!

1 interval enclosure(s)

We have validated that there is a globally unique zero!
```

For $n = 2$, we know from (13.8) that the solution $x$ satisfies $x_i > 1$ for $i = 1, 2$. Thus, our routine delivers no solution in the search interval and therefore confirms the fact that the solution given in [32] does not exist. For $n = 6$, our algorithm guarantees that there is only one zero of our sample function within the specified starting box.

### 13.3.3 Restrictions, Hints, and Improvements

The function $f$ can only contain operators and elementary functions supported by module *grad_ari*.

If *AllNLSS()* reports a warning that not all solutions were found due to a user limit for the number of solutions to be computed, then you may call *AllNLSS()* again to find additional solutions.

The method is not very fast if a very small value of $\varepsilon$ (*Epsilon*) is used and the interval Newton step does not improve the actual iterates because of rounding and overestimation effects of the machine interval arithmetic. In this case, the method is equivalent with a bisection method. The method can be improved by incorporating a local iteration procedure, i.e. a non-interval Newton method which tries to find an approximation of a zero starting from the midpoint of the current interval vector iterate. The computed approximation can be used as the point of expansion in the interval Gauss-Seidel step (see [31] or [78] for details on this improvement).

In *XINewton()*, the evaluation of the function with differentiation arithmetic can cause a runtime error if the interval argument of an elementary function does not lie in the domain specified for this interval function or if a division by an interval containing zero occurs. This is also due to the known overestimation effects of interval arithmetic (see Section 3.1). To get rid of these errors, the user may try to split the starting interval vector in several parts and call *AllNLSS()* for each of these parts.

The rules for getting true enclosures in the presence of conversion errors (see Section 3.7) also apply here. That is why we expressed $0.49x_i$ as `49.0*x[i]/100.0` in our example in Section 13.3.2.

## 13.4 Exercises

**Exercise 13.1** Use the function *AllNLSS()* to compute the zeros of the function $f : I\!R^2 \to I\!R^2$ with

$$\begin{aligned} f_1(x) &= x_1^2 - 20x_1 + x_2^2 - 2x_2 + 100, \\ f_2(x) &= x_1^2 - 22x_1 + x_2^2 - 2x_2 + 121. \end{aligned}$$

These zeros correspond to the intersection points of the circle with midpoint $(10, 1)^T$ and radius 1 and the circle with midpoint $(11, 1)^T$ and radius 1. They lie within the box $([10, 12], [0, 3])^T$ which can be used as starting box.

**Exercise 13.2** Use the function *AllNLSS()* to verify that $y \approx (1.6641, 1.6641)^T$ is a zero of the function $f : I\!R^2 \to I\!R^2$ with

$$\begin{aligned} f_1(x) &= \left(\frac{x_1}{2}\right)^2 + \left(\frac{x_2}{3}\right)^2 - 1, \\ f_2(x) &= \left(\frac{x_1}{3}\right)^2 + \left(\frac{x_2}{2}\right)^2 - 1, \end{aligned}$$

by simply taking a small interval vector $[y]$ with $y \in [y]$ and $d_{\mathrm{rel},\infty}([y]) \approx 10^{-4}$ as input data for the function.

## 13.5 References and Further Reading

The method we discussed in this chapter is an *a priori* method because the iteration starts with a (possibly large) interval vector enclosing all the solutions which have to be found. Here, the iterates of the method are subintervals of the previous iterates. There are also methods for finding (and bounding) a single solution of a system of nonlinear equations called *a posteriori* methods. These methods start with an approximation of a zero and apply a test procedure for a neighborhood interval of the approximation to verify a zero within that interval. Our method presented in this chapter can also be applied to verify such an approximation, if we start the process with a small interval containing the approximation.

A large number of authors (Alefeld, Hansen, Moore, Neumaier, Kearfott, Rump, and many others) have considered the problem of computing enclosures for the solutions of a system of nonlinear equations. For further references in the field of *a priori* methods and for improvements for our method, see [3], [8], [31], [68], [69], [70], [77], or [78]. For *a posteriori* methods, see [11], [42], [43], [56], [59], [61], [66], [70], [83], [84], or [85]. In the field of preconditioning of the interval Gauss-Seidel step, interesting work and improvements can be found in [45] or [46].

# Chapter 14

# Global Optimization

In Chapter 7, we considered the problem of finding the global minimizers of one-dimensional nonlinear functions. Now, we consider its generalization, the problem of finding the global minimizers of multi-dimensional nonlinear functions. Our algorithm is based on the method of Hansen [29]. The algorithm computes enclosures for all global minimizers and for the global minimum value of a twice continuously differentiable function in the interior of a given interval vector.

Classical numerical global optimization methods for the multi-dimensional case start from some approximate trial points and iterate. Thus, classical optimization methods sample the objective function at only a finite number of points. There is no way to guarantee that the function does not have some unexpectedly small values between these trial points.

Hansen's algorithm uses interval arithmetic to evaluate the objective function and its first- and second-order partial derivatives over a continuum of points, including those points that are not finitely representable on a computer. Interval analysis supplies the prerequisite for solving the global optimization problem with *automatic result verification*, i.e. with the guarantee that the global minimum points and the global minimum values have been found.

## 14.1 Theoretical Background

Let $f : I\!R^n \rightarrow I\!R$ be a twice continuously differentiable function, and let $[x] \in II\!R^n$. We address the problem of finding all points $x^*$ in the interior of $[x]$ such that

$$f(x^*) = \min_{x \in [x]} f(x). \tag{14.1}$$

We are interested in both the global minimizers $x^*$ and the minimum value $f^* = f(x^*)$.

We use the method of Hansen [29, 31] with the modifications of Ratz [79]. Our algorithm does not find non-stationary minima on the boundary of $[x]$. A sophisticated method for handling this case can be found in [79]. Starting from the initial interval vector (box) $[x]$, our algorithm subdivides $[x]$ and stores the subboxes $[y] \subset [x]$ in a list $L$. Subintervals which are guaranteed not to contain a global minimizer of $f$ are discarded from that list, while the remaining subintervals get subdivided again until the desired accuracy (relative diameter) of the intervals in the list is achieved. The power and speed of self-validating methods for global optimization comes not so much from the ability to find the answer as from the ability to discard from consideration regions where the answer is not. The tests we use to discard pending subboxes are multi-dimensional analogues of those we used for one-dimensional global optimization in Chapter 7. We use the four tests

- midpoint test,
- monotonicity test,
- concavity test, and
- interval Newton Gauss-Seidel step.

In the following sections, we consider each of these methods in detail. We use the notation $\underline{f_y}$ as abbreviation for the lower interval bound of the interval function evaluation $[f_y] := f_{[]}([y])$ for $[y] \in I\mathbb{R}^n$.

### 14.1.1 Midpoint Test

If we are able to determine an upper bound $\widetilde{f}$ for the global minimum value $f^*$, then we can delete all subintervals $[y]$ for which

$$\underline{f_y} > \widetilde{f} \geq f^*. \tag{14.2}$$

The midpoint test first determines or improves such an upper bound for $f^*$. Initially, let $\widetilde{f} = +\infty$. We choose a box $[y]$ from the list $L$ which satisfies $\underline{f_y} \leq \underline{f_z}$ for all intervals $[z]$ in the list $L$. That is, $[y]$ has the smallest lower bound for the range of $f$. Hence, it is a likely candidate to contain a minimizer. Let $c = m([y])$ (or any other point in $[y]$), and compute $\widetilde{f} = \min\{f(c), \widetilde{f}\}$. Such an upper bound can also be computed on a computer when rounding errors occur: we compute $f_\diamond(c)$ and use the upper bound of the resulting interval as the possibly new value $\widetilde{f}$.

Now, with a possibly improved (decreased) value of $\widetilde{f}$, we can discard all intervals $[z]$ from the list $L$ for which $\widetilde{f} < \underline{f_z}$. The midpoint test is relatively inexpensive, and it often allows us to discard from consideration large portions of the original interval $[x]$. Figure 7.1 in Chapter 7 illustrates this procedure, which deletes the intervals $[y]_2$, $[y]_3$, $[y]_8$, and $[y]_9$ in this special case. The midpoint test remains valid if an arbitrary $c' \in [y]$ is used instead of $c = m([y])$. Our algorithm could be extended by using a local approximate search to find a $c' \in [y]$ that is likely to give a smaller upper bound for $f$ than $c$ gives.

The value $\widetilde{f}$ is also used for function value checks when entering newly subdivided intervals $[w]$ in our list $L$. If we know that $[w]$ satisfies $\underline{f_w} > \widetilde{f}$, then $[w]$ cannot contain a global minimizer. Thus, we must only enter intervals $[w]$ in the list $L$ if $\underline{f_w} \leq \widetilde{f}$ holds.

### 14.1.2 Monotonicity Test

The monotonicity test determines whether the function $f$ is *strictly monotone* in an entire subbox $[y] \subset [x]$, in which case $[y]$ cannot contain a global minimizer (stationary point). Therefore, if $[g] := \nabla f([y])$ satisfies

$$0 \notin [g]_i \quad \text{for some } i = 1, \dots, n, \tag{14.3}$$

then $f$ is strictly monotone over the subbox $[y]$ with respect to the $i$th coordinate. Thus, $[y]$ can be deleted. Figure 7.2 in Chapter 7 demonstrates the monotonicity test in the one-dimensional case.

We emphasize that if $f$ is monotone *in a single coordinate direction*, that subbox may be discarded. Therefore, a small number (relative to $n$) of interval-valued evaluations of partial derivatives are often enough to discard large portions of the original box $[x]$. We use the differentiation arithmetic of Chapter 12 to compute $\nabla f$.

### 14.1.3 Concavity Test

The concavity test (non-convexity test) detects whether the function $f$ is *not convex* in a subbox $[y] \subset [x]$, in which case $[y]$ cannot contain a global minimizer. The Hessian matrix of a convex function $f$ must be positive semidefinite. We know that a necessary condition for this is that all diagonal elements of the Hessian are non-negative. If $[H] := \nabla^2 f([y])$ satisfies

$$\overline{H}_{ii} < 0 \quad \text{for some } i = 1, \dots, n, \tag{14.4}$$

then $H_{ii} < 0$ for $H = \nabla^2 f(y)$ and all $y \in [y]$. Thus, $f$ cannot be convex over $[y]$, the subbox $[y]$ cannot contain a stationary minimum, and $[y]$ can be deleted. Figure 7.3 in Chapter 7 demonstrates the concavity test in the one-dimensional case. The Hessian for the concavity test is computed by automatic differentiation described in Chapter 12.

### 14.1.4 Interval Newton Step

In our global optimization method, we apply one step of the extended interval Newton Gauss-Seidel method described in Chapter 13 to the nonlinear system

$$\nabla f(y) = 0, \quad y \in [y]. \tag{14.5}$$

The subbox $[y]$ is a candidate for containing a minimizer $x^*$, which we have assumed must satisfy $\nabla f(x^*) = 0$. When we apply the algorithm of Chapter 13, three things may happen. First, we may validate that $[y]$ contains no stationary point, in which case we may discard $[y]$. Second, the Newton step may contract $[y]$ significantly. Subsequently, $f$ can be evaluated on the narrower box $[y]$ with less overestimation, so the midpoint, monotonicity, and concavity tests are likely to be more effective. Third, we may get splittings of the box $[y]$ due to gaps produced by the extended interval divisions applied in the Newton step. We only apply one Newton step because this test is relatively expensive, and the other test (with bisection) may subsequently discard even subboxes containing local minimizers. In addition, a stationary point is not necessarily even a local minimizer.

To apply the extended interval Newton Gauss-Seidel method, we first compute $[A] \in I\mathbb{R}^{n\times n}$ and $b \in \mathbb{R}^n$ by

$$[A] := R \cdot \nabla^2 f([y]) \quad \text{and} \quad b := R \cdot \nabla f(m([y])), \tag{14.6}$$

where $R \approx (m(\nabla^2 f([y])))^{-1}$. Then, we compute $N'_{\mathrm{GS}}([y])$ according to

$$\left.\begin{aligned} [z] &:= [y] \\ [z]_i &:= \left( c_i - \left( b_i + \sum_{\substack{j=1 \\ j\neq i}}^{n} [A]_{ij} \cdot ([z]_j - c_j) \right) \Big/ [A]_{ii} \right) \cap [z]_i \\ & \qquad i = 1, \ldots, n \\ N'_{\mathrm{GS}}([y]) &:= [z] \end{aligned}\right\}. \tag{14.7}$$

If the extended interval division results in a splitting of a component $[z]_i$, we will store one part of the actual box $[z]$, and we use the other part to continue the computing in (14.7). Thus, our interval Newton step results in at most $n+1$ subboxes to place on the pending list $L$.

### 14.1.5 Verification

When all subboxes in the list $L$ have been reduced to have widths less than a specified tolerance, we have a list of small subboxes we could not discard by any of our tests. We attempt to verify the existence and uniqueness of a local minimizer within each subbox by checking two conditions. The first condition is

$$N'_{\mathrm{GS}}([y]) \overset{\circ}{\subset} [y], \tag{14.8}$$

which guarantees the existence and uniqueness of a stationary point of $f$, i.e. a zero of $\nabla f$ in $[y]$ (cf. Theorem 13.1). The second condition (14.9) guarantees that $\nabla^2 f([y])$ is positive definite, i.e. all real symmetric matrices $A \in \nabla^2 f([y])$ are positive definite. If the real matrix $B := I - \|A\|^{-1} \cdot A$ has only eigenvalues with absolute values less than 1, i.e. the spectral radius $\rho(B) < 1$, then the symmetric matrix $A$ is positive definite. Thus, we use the following theorem:

**Theorem 14.1** *Let $[H] \in I\mathbb{R}^{n\times n}$ and $[S]$ be defined by $[S] := I - \frac{1}{\kappa}[H]$ with $\|[H]\|_\infty \leq \kappa \in \mathbb{R}$. If $[S]$ satisfies*

$$[S] \cdot [z] \overset{\circ}{\subset} [z] \tag{14.9}$$

*for an interval vector $[z] \in I\mathbb{R}^n$, then we have $\rho(B) < 1$ for all $B \in [S]$, and all symmetric matrices $A \in [H]$ are positive definite.*

For a proof of this theorem, see [79].

To check the condition of Theorem 14.1, we first compute $[H] = \nabla^2 f([y])$, $\kappa$ with $\|[H]\|_\infty \leq \kappa$, and $[S] := I - \frac{1}{\kappa}[H]$. Starting with an interval vector $[z]^{(0)}$ with every component $[z]_i^{(0)} = [-1, 1]$, we iterate

$$[z]^{(k+1)} := [S] \cdot [z]^{(k)}, \quad k = 0, 1, \ldots$$

until the condition $[z]^{(k+1)} \overset{\circ}{\subset} [z]^{(k)}$ is satisfied. We stop the iteration if a certain number of steps could not satisfy this condition.

We know of no way to verify the uniqueness of a global minimizer in general. The global minimizer can even be one or more continua of points. Hence, we settle for attempting to verify that intervals we compute as candidates for containing a global minimizer contain unique local minimizers. Failure to verify the uniqueness of a local minimizer in a subinterval is not grounds for discarding that subinterval from the list of candidates.

In fact, our method produces a final list containing enclosures for *locally unique candidates for global minimizers*. If we have in the final list exactly one subinterval $[y]$ in which we can validate a local minimizer, then we have validated a unique global minimizer in the starting interval $[x]$. If we have two or more subintervals validated to contain unique local minimizers, then the best we can say is that each contains a candidate for a global minimizer, which need not be unique (but in this case, the global minimizer could not be a continuum of points).

## 14.2 Algorithmic Description

The main algorithm AllGOp (Algorithm 14.7) consists of two parts. The first part is the subdivision method including the tests and the extended interval Newton steps described in Section 14.1. The second part is a verification step which tries to verify the local uniqueness of a minimizer within each of the remaining intervals of the pending list.

The algorithm for the monotonicity test (Algorithm 14.1) uses the interval gradient $[g] = \nabla f([y])$ and returns *true* if no further processing of the actual box $[y]$ is necessary because $f$ is monotonic in $[y]$, or *false*, otherwise.

### Algorithm 14.1: MonotonicityTest ($[g]$)

1. **for** $i := 1$ **to** $n$ **do**

   **if** $0 \notin [g]_i$ **then return** MonotonicityTest := *true*;

2. **return** MonotonicityTest := *false*;

Algorithm 14.2 executes the concavity test using the interval Hessian $[H] = \nabla^2 f([y])$ returning *true* if no further processing of the actual box $[y]$ is necessary because $f$ is not convex in $[y]$, or *false*, otherwise.

**Algorithm 14.2:** ConcavityTest ($[H]$)

1. **for** $i := 1$ **to** $n$ **do**
   **if** $\overline{H}_{ii} < 0$ **then return** ConcavityTest := *true*;
2. **return** ConcavityTest := *false*;

Next, we present the algorithm for the execution of the extended interval Newton Gauss-Seidel step (Algorithm 14.3), where (14.7) is applied to the box $[y] \in I\mathbb{R}^n$ resulting in at most $n+1$ boxes $[V]_i$. After some initializations, we use Algorithm 10.1 (MatInv) of Chapter 10 to compute an approximate inverse of the midpoint matrix of the interval Hessian matrix $[H] = \nabla^2 f([y])$. In Step 4, we compute $[A]$, $[b]$, and $[y_c]$ followed by the initialization for the loop. Interval arithmetic must be used to compute $\nabla f(c)$ for bounding all rounding errors.

We perform the single component steps of the Gauss-Seidel step for all $i$ with $0 \notin [A]_{ii}$ (Step 6) and then for the remaining indices with $0 \in [A]_{ii}$ (Step 7). Using this strategy, it is possible that the intervals $[y]_i$ become smaller by the intersections with the old values $[y]_i$ in Step 6(b), before the first splitting is produced in Step 7(b). If a gap is produced in this case, we store one part of the actual box $[y]$ in the $p$th row of the interval matrix $[V]$, and we continue the iteration for the other part of $[y]$. An empty intersection in Step 6(c) or Step 7(c) signals that no solution exists in $[y]$. Thus, we set $p := 0$ and terminate.

**Algorithm 14.3:** NewtonStep ($f, [y], [H], [V], p$)

1. $c := m([y])$; {Compute midpoint of $[y]$}
2. MatInv ($m([H]), R, InvErr$); {Invert the midpoint matrix}
3. **if** ($InvErr \neq$ "No Error") **then** $R := I$;
4. $[A] := R \cdot [H]$; $[b] := R \cdot \nabla f_\diamond(c)$; $[y_c] := [y] - c$;
5. $p := 0$; {Initialization for loop}
6. **for** $i := 1$ **to** $n$ **do** {Interval Gauss-Seidel step for $0 \notin [A]_{ii}$}
   (a) **if** ($0 \in [A]_{ii}$) **then next**$_i$;
   (b) $[y]_i := \left( c_i - \left( [b]_i + \sum_{\substack{j=1\\j\neq i}}^{n} [A]_{ij} \cdot [y_c]_j \right) / [A]_{ii} \right) \cap [y]_i$;
   (c) **if** $[y]_i = \emptyset$ **then return** $p := 0$; {No solution possible}
   (d) $[y_c]_i := [y]_i - c_i$;
7. **for** $i := 1$ **to** $n$ **do** {Extended interval Gauss-Seidel step for $0 \in [A]_{ii}$}
   (a) **if** ($0 \notin [A]_{ii}$) **then next**$_i$;
   (b) $[z] := \left( c_i - \left( [b]_i + \sum_{\substack{j=1\\j\neq i}}^{n} [A]_{ij} \cdot [y_c]_j \right) / [A]_{ii} \right) \cap [y]_i$; $\{[z] = [z]_1 \cup [z]_2\}$

(c) **if** ($[z] = \emptyset$) **then return** $p := 0$; {No solution possible}

(d) $[y]_i := [z]_1$; $[y_c]_i := [y]_i - c_i$;

(e) **if** ($[z]_2 \neq \emptyset$) **then**

$p := p + 1$; $[V]_p := [y]$; $[V]_{pi} := [z]_2$; {Store part of $[y]$ in $[V]_p$}

8. $p := p + 1$; $[V]_p := [y]$;
9. **return** $[V]$, $p$;

Algorithm 14.4 manages the bisection of subboxes and their insertion in the pending list $L$. The subdivided boxes $[y]$ are stored together with the lower bound of the interval function evaluation $[f_y] := f_{[]}([y])$ as pairs $([y], \underline{f_y})$. Pairs are stored in the list sorted in nondecreasing order of lower bounds $\underline{f_y}$. Therefore, a newly computed pair is stored in the list $L$ according to the ordering rule (cf. [79]):

$$\left.\begin{array}{ll} \bullet \text{ either} & \underline{f_w} \leq \underline{f_y} < \underline{f_z} \text{ holds,} \\ \bullet \text{ or} & \underline{f_y} < \underline{f_z} \text{ holds, and } ([y], \underline{f_y}) \text{ is the first element of the list,} \\ \bullet \text{ or} & \underline{f_w} \leq \underline{f_y} \text{ holds, and } ([y], \underline{f_y}) \text{ is the last element of the list,} \\ \bullet \text{ or} & ([y], \underline{f_y}) \text{ is the only element of the list,} \end{array}\right\} \quad (14.10)$$

where $([w], \underline{f_w})$ is the predecessor and $([z], \underline{f_z})$ is the successor of $([y], \underline{f_y})$ in $L$.

That is, the second components of the list elements may not decrease, and a new pair is entered behind all other pairs with the same second component. Thus, the first element of the list has the smallest second component. We can directly use the corresponding box to compute $f(c)$ for the improvement of $\tilde{f}$ in performing the midpoint test. We can also save some work when deleting elements in the midpoint test, by deleting the rest of the list when we have found the first element to be deleted. At each stage of our algorithm, $\tilde{f}$ is the best known upper bound for the global minimum value $f^*$. The second component $\underline{f_y}$ from the first element of the list is the best known lower bound for $f^*$. Therefore, we might choose to terminate the algorithm on the basis of the interval $[\underline{f_y}, \tilde{f}]$.

Given the list $L$ and a list element $E$, we use the following notation in our algorithms:

| Notation | Meaning |
|---|---|
| $L := \{\}$ | Initialization by an empty list |
| $L := E$ | Initialization by a single element |
| $L := L + E$ | Enter element $E$ in $L$ according to Condition (14.10) |
| $L := L - E$ | Discard element $E$ from $L$ |
| $E :=$ Head $(L)$ | Set $E$ to the first element of $L$ |
| MultiDelete $(L, \tilde{f})$ | Discard all elements from $L$ satisfying Condition (14.2) |
| Length $(L)$ | Delivers the number of elements in $L$ |

In GlobalOptimize, we first compute an upper bound for the global minimum value, and we do some initializations. Step 3 is the main iteration. Here, we first do a bisection of the actual interval $[y]$. Then in Step 3(c), we apply the monotonicity test, a function value check using the centered form, the concavity test, and the interval Newton step to the bisected boxes $[u]_1$ and $[u]_2$. The interval Newton step may result in at most $p$ intervals. We have to handle them all in Step 3(c)viii, where we again apply a monotonicity test and a function value check with centered forms. If the actual box $[V]_j$ has not been discarded, then it is still a candidate for a minimizer, and we store it in $L$.

In Step 3(e), we remove the first element from the list $L$, i.e. the element of $L$ with the smallest lower bound of the interval function evaluation, and we perform the midpoint test. Then, we check the tolerance criterion for the new actual interval. If the desired accuracy is achieved, we store this interval in the result list $L_{\text{res}}$. Otherwise, we go to the bisection step.

When the iteration stops because the pending list $L$ is empty, we compute a final enclosure $[f^*]$ for the global minimum value in Step 4, and we return $L_{\text{res}}$ and $[f^*]$. Algorithm GlobalOptimize terminates because the elements of $L$ move to $L_{\text{res}}$ if $(d_{\text{rel},\infty}([y]) \leq \varepsilon)$ or if $(d_{\text{rel}}([f^*]) \leq \varepsilon)$. The bisection step (Step 3(b)) guarantees that the first condition is fulfilled at some stage of the iteration.

**Algorithm 14.4:** GlobalOptimize $(f, [x], \varepsilon, L_{\text{res}}, [f^*])$

1. $c := m([x])$; $[f_c] := f_\diamond(c)$; $\tilde{f} := \overline{f_c}$; {Compute upper bound for $f^*$}
2. $[y] := [x]$; $L := \{\,\}$; $L_{\text{res}} := \{\,\}$;
3. **repeat** {Start iteration}
   (a) Find $k$ with $d([y]_k) = d_\infty([y])$; {Component with maximum diameter}
   (b) $[u]_1 := [y]$; $[u]_2 := [y]$; $[u]_{1k} := [\underline{y}_k, c_k]$; $[u]_{2k} := [c_k, \overline{y}_k]$; {Bisect $[y]$}
   (c) **for** $i := 1$ **to** $2$ **do**
      i. $[g] := \nabla f([u]_i)$;
      ii. **if** MonotonicityTest $([g])$ **then next**$_i$;
      iii. $[f_u] := (f(c) + [g] \cdot ([u]_i - c)) \cap f([u]_i)$; {Centered form}
      iv. **if** $\tilde{f} < \underline{f_u}$ **then next**$_i$;
      v. $[H] := \nabla^2 f([u]_i)$;
      vi. **if** ConcavityTest $([H])$ **then next**$_i$;
      vii. NewtonStep $(f, [u]_i, [H], [V], p)$; {Interval Newton Gauss-Seidel step}
      viii. **for** $j := 1$ **to** $p$ **do**
         A. $[g] := \nabla f([V]_j)$;
         B. **if** MonotonicityTest $([g])$ **then next**$_j$;
         C. $c_V := m([V]_j)$;
         D. $[f_V] := (f(c_V) + [g] \cdot ([V]_j - c_V)) \cap f([V]_j)$; {Centered form}
         E. **if** $\tilde{f} \geq \underline{f_V}$ **then** $L := L + ([V]_j, \underline{f_V})$; {Store $[V]_j$}
   (d) *Bisect* := *false*;
   (e) **while** $(L \neq \{\,\})$ **and** (**not** *Bisect*) **do**

i. $([y], \underline{f_y}) := \text{Head}(L)$; $L := L - ([y], \underline{f_y})$; $c := m([y])$;
ii. $\tilde{f} := \min\{\tilde{f}, f(c)\}$; MultiDelete $(L, \tilde{f})$; {Midpoint test}
iii. $[f^*] := [\underline{f_y}, \tilde{f}]$;
iv. **if** $(d_{\text{rel}}([f^*]) \leq \varepsilon)$ **or** $(d_{\text{rel},\infty}([y]) \leq \varepsilon)$ **then** $L_{\text{res}} := L_{\text{res}} + ([y], \underline{f_y})$;
**else** $Bisect := true$;

**until** (**not** $Bisect$);

4. $([y], \underline{f_y}) := \text{Head}(L_{\text{res}})$; $[f^*] := [\underline{f_y}, \tilde{f}]$;
5. **return** $L_{\text{res}}$, $[f^*]$;

Algorithm 14.5 applies the test given by Theorem 14.1 to check whether the interval Hessian matrix over the current enclosure $[y]$ is positive definite. We use $k_{\max} = 10$ as the maximum number of iterations, $\varepsilon = 0.25$ as the starting value for the epsilon inflation, and a factor of 8 to increase $\varepsilon$ within the iterations. It turned out that these are good values for minimizing the effort if no verification is possible (see also [52]).

**Algorithm 14.5: PosDef $([H])$**

1. Compute $\kappa \geq \|[H]\|_\infty$; $[S] := \Diamond(I - \frac{1}{\kappa}[H])$;
2. **for** $k := 1$ **to** $n$ **do** $[z]_k := [-1, 1]$;
3. $k_{\max} := 10$; $k := 0$; $\varepsilon := 0.25$;
4. **repeat**

   $[u] := [z] \bowtie \varepsilon$; $[z] := [S] \cdot [u]$; $k := k + 1$; $\varepsilon := 8 \cdot \varepsilon$;

   **until** $([z] \overset{\circ}{\subset} [u])$ **or** $(k = k_{\max})$;
5. **return** PosDef $:= ([z] \overset{\circ}{\subset} [u])$;

Algorithm 14.6 describes the verification step checking the local uniqueness of a minimizer enclosed in the interval $[y]$. The algorithm tries to do a "zero check" for the gradient $\nabla f$ according to condition 3 of Theorem 13.1 by applying interval Newton steps including an epsilon inflation of the iterates $[y]$ (Step 4). We also check the condition of Theorem 14.1 by applying Algorithm PosDef in Step 5. We use the same values for $k_{\max}$ and $\varepsilon$ as in PosDef.

**Algorithm 14.6: VerificationStep $(f, [y], Unique)$**

1. **if not** $Unique$ **then** $Unique := (\underline{y} = \overline{y})$; {Point interval $\Rightarrow$ uniqueness}
2. **if** $Unique$ **then return** $[y]$, $Unique$;
3. $k_{\max} := 10$; $k := 0$; $[y_{\text{in}}] := [y]$; $\varepsilon := 0.25$; {Initializations}
4. **while** (**not** $Unique$) **and** $(k < k_{\max})$ **do** {Do $k_{\max}$ loops to achieve inclusion}

(a) $[y_{\text{old}}] := [y] \bowtie \varepsilon$; {Epsilon inflation of $[y]$}
(b) $[H] := \nabla^2 f([y_{\text{old}}])$;
(c) $k := k + 1$; $c := m([y_{\text{old}}])$;
(d) NewtonStep $(f, [y_{\text{old}}], [H], [V], p)$; {Interval Newton step}
(e) **if** $p \neq 1$ **then exit**$_{\text{while-loop}}$; {No verification possible}
(f) $[y] := [V]_1$;
(g) $Unique := ([y] \overset{\circ}{\subset} [y_{\text{old}}])$; {Inner inclusion $\Rightarrow$ uniqueness}
(h) **if** $[y] = [y_{\text{old}}]$ **then** $\varepsilon := \varepsilon \cdot 8$; {Increase $\varepsilon$}

5. **if** *Unique* **then** *Unique* := PosDef$([H])$;
6. **if not** *Unique* **then** $[y] := [y_{\text{in}}]$;
7. **return** $[y]$, *Unique*;

Algorithm 14.7 now combines the subalgorithms described above to compute enclosures for all global minimizers $x^*$ of the function $f$ and for the global minimum value $f^*$ within the input interval vector $[x]$ and tries to prove the local uniqueness of the minimizers within the computed enclosures. The desired accuracy (relative diameter) of the interval enclosures is specified by the input parameter $\varepsilon$. 1 ulp accuracy is chosen if the specified value of $\varepsilon$ is too small (for example 0). The enclosures for the global minimizers of $f$ are returned in the interval matrix $[Opt]$ row by row, the corresponding information on the local uniqueness of the optimizer is returned in the Boolean vector *Info*. The number of enclosures computed is returned in the integer variable $N$.

We use a function called CheckParameters as an abbreviation for the error checks for the parameters of AllGOp which are necessary in an implementation. If no error occurs, AllGOp delivers the $N$ enclosures $[Opt]_i$, $i = 1, 2, \ldots, N$, satisfying

if $Info_i = true$, then $[Opt]_i$ encloses a locally unique minimizer of $f$,
if $Info_i = false$, then $[Opt]_i$ may enclose a local or global minimizer of $f$.

If $N = 0$, then it is guaranteed that there is *no* stationary global minimizer of $f$ in the starting interval $[x]$.

**Algorithm 14.7:** AllGOp $(f, [x], \varepsilon, [Opt], Info, N, [f^*], Err)$

1. $Err$ := CheckParameters;
2. Set $\varepsilon$ to "1 ulp accuracy" if $\varepsilon$ is too small;
3. **if** $Err \neq$ "No Error" **then return** $Err$;
4. GlobalOptimize $(f, [x], \varepsilon, L_{\text{res}}, [f^*])$;
5. $N$ := Length $(L_{\text{res}})$;
6. $([Opt], Info) := L_{\text{res}}$; {Transfer list to result arrays}
7. **for** $i := 1$ **to** $N$ **do**

VerificationStep ($f$, $[Opt]_i$, $Info_i$);

8. **return** $[Opt]$, $Info$, $N$, $[f^*]$, $Err$;

## Applicability of the Algorithm

We have assumed that $f$ is twice continuously differentiable. However, we can apply our algorithm to functions that are only once continuously differentiable or to functions that are not differentiable if we leave out some parts of the algorithm. If we do not use the interval Newton step (Step 3(c)vii in Algorithm 14.4), and replace it by the sequence

$$[V]_1 := [u]_i; \quad p := 1;$$

then our method works for functions that are only once differentiable. Algorithm 14.6 cannot be applied in this case.

If we replace Step 3(c) of Algorithm 14.4 by the sequence

**for** $i := 1$ **to** $2$ **do**
  $[f_u] := f([u]_i)$;
  **if** $\widetilde{f} \geq \underline{f_u}$ **then** $L := L + ([u]_i, \underline{f_u})$;

then we can apply our method, consisting of subdividing and midpoint test, to functions that are *not* differentiable.

The closer the upper bound $\widetilde{f}$ is to the global minimum value $f^*$, the more intervals we can delete in the midpoint test (Step 3(e)ii of Algorithm 14.4). Thus, the method can be improved by incorporating an approximate local search procedure, to try to decrease the value $\widetilde{f}$. See [31] or [78] for the description of such local search procedures.

For a multiple zero $x^*$ of $\nabla f$, the algorithm cannot verify the existence and the local uniqueness of $x^*$ in the enclosing result box. Nevertheless, the zero of $\nabla f$, which is possibly a global minimizer, is bounded to the desired accuracy specified by $\varepsilon$. In this case, the corresponding component of the *Info*-vector has the value *false*.

As a consequence of the bisecting of the boxes, it may happen that a minimizer lying exactly at the splitting boundary is enclosed in several intervals. Furthermore, because of the known overestimation effects of interval arithmetic, the algorithm may also find "near-global" minima when rounding prevents determination of the true minimum of several candidates. Sophisticated supplements to our method avoiding these effects can be found in [31] or [79].

# 14.3 Implementation and Examples

## 14.3.1 C++ Program Code

We begin by describing our implementation of the operations needed for handling lists and list elements (pairs). Then, we describe the implementation of Algorithm 14.7 and its subalgorithms.

### 14.3.1.1 Module lst_ari

The header file of module *lst_ari* supplies the definition of the class *Pair* representing a pair of an interval vector and a real value. The type *PairPtr* is provided to manage lists of such pairs. The constant pointer *EmptyList* denotes an empty list of pairs.

```
//----------------------------------------------------------------------------
// File: lst_ari (header)
// Purpose: Definition of a list arithmetic used in connection with an
//    interval bisection method in global optimization for storing pairs of
//    an interval vector and a real value.
// Class Pair:
//    Pair()          : constructors
//    ~Pair()         : destructor
//    operator =      : assignment operator
//    _Pair()         : for explicit generating a pair
//    GetInt()        : get the interval vector component
//    GetFyi()        : get the function value
// Global type:
//    PairPtr         : list of pairs
// Global constant:
//    EmptyList       : the empty list (NULL pointer)
// Global functions and operators:
//    operator +      : adding a new element to a list
//    Next(), Head(): access functions for lists
//    Length()        : access function to length of list
//    FreeAll()       : free complete list
//    MultiDelete() : deletes several elements in a list
//    DelHead()       : deletes the first element of a list
//    ListToMatrix(): transfer a list into a dynamic interval matrix
//----------------------------------------------------------------------------
#ifndef __LST_ARI_HPP
#define __LST_ARI_HPP

#include <intvecto.hpp>      // Integer vector type
#include <imatrix.hpp>       // Interval matrix/vector arithmetic

#define EmptyList NULL       // The empty list
                             //---------------

struct                       // Structure used for a list of pairs
  PairElmt;                  //-----------------------------------

typedef                      // Pointer to a list of pairs
  PairElmt* PairPtr;         //--------------------------

class Pair {                 // Pair of an interval vector 'intv'
  private:                   // and the real value 'inf(f(intv))'
    ivector intv;            //----------------------------------
    real    fyi;
```

```
  public:
    Pair ( ) { };
    Pair ( Pair& );

    Pair& operator= ( Pair& );

    friend Pair    _Pair  ( ivector&, real );
    friend ivector GetInt ( Pair& );
    friend real    GetFyi ( Pair& );
};

extern void    MultiDelete  ( PairPtr&, const real& );
extern void    ListToMatrix ( PairPtr, imatrix&, intvector& );
extern PairPtr operator+    ( PairPtr, Pair& );

extern void    FreeAll      ( PairPtr& );
extern int     Length       ( PairPtr );
extern Pair    Head         ( PairPtr );
extern void    DelHead      ( PairPtr& );
#endif
```

A list of pairs is realized as a linearly linked list. The static variable *FreeList* is used to hold the list of freed elements. The static functions *NewPP()*, *Free()*, and the global function *FreeAll()* work on that list to prevent the creation of garbage in memory. The friend functions *_Pair()*, *GetInt()*, and *GetFyi()* are transfer and access functions for pairs.

The operator + enters a new list element $p$ in the list *List* according to Condition (14.10). The function *MultiDelete()* deletes all elements $p$ in *List* for which $p.fyi > fmax$. This function assumes that the list elements are ordered according to Condition (14.10). Function *Head()* delivers the first pair of *List*, whereas *DelHead()* deletes the first pair of *List*. Function *Length()* delivers the number of elements in *List*. The void function *ListToMatrix()* transfers a list of pairs to the result matrix *Opt* and to the result vector *Info* according to Step 6 of Algorithm 14.7.

```
//----------------------------------------------------------------------------
// File: lst_ari (implementation)
// Purpose: Definition of a list arithmetic used in connection with an
//    interval bisection method in global optimization for storing pairs of
//    an interval vector and a real value.
// Method: Overloading of functions and operators of class type 'Pair'
//    (list element) and data type 'PairPtr' (list).
// Class Pair:
//    Pair()        : constructors
//    ~Pair()       : destructor
//    operator =    : assignment operator
//    _Pair()       : for explicit generating a pair
//    GetInt()      : get the interval vector component
//    GetFyi()      : get the function value
// Global type:
//    PairPtr       : list of pairs
// Global constant:
//    EmptyList     : the empty list (NULL pointer)
// Global functions and operators:
//    operator +    : adding a new element to a list
//    Head()        : access functions for the head of a list
//    Length()      : access function to length of list
//    FreeAll()     : free complete list
//    MultiDelete() : deletes several elements in a list
//    DelHead()     : deletes the first element of a list
```

```
//    ListToMatrix(): transfer a list into a dynamic interval matrix
//----------------------------------------------------------------------------
#include <lst_ari.hpp>

struct PairElmt {          // Type for a list of pairs
  Pair    P;               //-------------------------
  PairPtr N;
};

static PairPtr             // Local variable storing list of free
  FreeList = EmptyList;    // elements (automatic garbage recycling)
                           //--------------------------------------

//----------------------------------------------------------------------------
// Constructor and assignment operator
//----------------------------------------------------------------------------
Pair::Pair ( Pair& p )
  { intv = p.intv; fyi  = p.fyi; }

Pair& Pair::operator= ( Pair& p )
{
  intv = p.intv;
  fyi  = p.fyi;
  return *this;
}

//----------------------------------------------------------------------------
// Transfer and access functions for pairs
//----------------------------------------------------------------------------
Pair _Pair ( ivector& i, real r )    // Generate a pair
{                                    //----------------
  Pair p;
  p.intv = i;
  p.fyi  = r;
  return p;
}

ivector GetInt ( Pair& p )           // Get int-component
  { return p.intv; }                 //------------------

real GetFyi ( Pair& p )              // Get fyi-component
  { return p.fyi; }                  //------------------

//----------------------------------------------------------------------------
// Functions for generating and freeing of list elements (pairs)
//----------------------------------------------------------------------------
static void NewPP ( PairPtr& pp )     // 'NewPP' generates a new list element
{                                     // or gets one from 'FreeList'.
  if (FreeList == EmptyList)          //------------------------------------
    { pp = new PairElmt; pp->N = NULL; }
  else
    { pp = FreeList; FreeList = FreeList->N; pp->N = NULL; }
}

static void Free ( PairPtr& pp )              // 'Free' enters one element of a
{                                             // list in the 'FreeList'.
  if (pp != NULL)                             //------------------------------
    { pp->N = FreeList; FreeList = pp; pp = NULL; }
}

void FreeAll ( PairPtr& List )                // 'FreeAll' enters all elements
{                                             // of 'List' in the 'FreeList'
  PairPtr H;                                  //-----------------------------
```

```
  if (List != EmptyList) {
    H = List;
    while (H->N != NULL) H = H->N;
    H->N = FreeList; FreeList = List; List = NULL;
  }
}

//----------------------------------------------------------------------------
// Operator + enters the pair 'p' on the list 'List' in such a way that after
// entering, one of the four following condition holds:
//   1) o.fyi <= p.fyi < q.fyi,
//   2)          p.fyi < q.fyi  and  'p' is the first element of 'List',
//   3) o.fyi <= p.fyi          and  'p' is the last  element of 'List',
//   4)                              'p' is the only  element of 'List',
// where 'o' is the preceding and 'q' is the succeeding element of 'p' in
// the resulting list.
//----------------------------------------------------------------------------
PairPtr operator+ ( PairPtr List, Pair& p )
{
  PairPtr H, HN;
  int     ready, alreadyIn;

  if (List == EmptyList) {                            // List is empty, so new
    NewPP(H); H->P = p;                               // list contains only 'p'.
    return H;                                         //------------------------
  }
  else if (GetFyi(List->P) > GetFyi(p)) {              // 'p' becomes new first
    NewPP(H); H->P = p; H->N = List;                   // element of the list.
    return H;                                          //----------------------
  }
  else {
    H = List; HN = H->N; ready = 0;
    alreadyIn = ( GetInt(H->P) == GetInt(p) );

    while ( !(ready || alreadyIn) ) {                  // Search for the right
      if (HN == NULL)                                  // position to enter 'p'.
        ready = 1;                                     //-----------------------
      else if (GetFyi(HN->P) > GetFyi(p))
        ready = 1;
      else {
        H = HN; HN = H->N;
        alreadyIn = ( GetInt(H->P) == GetInt(p) );
      }
    } // while

    if (!alreadyIn)                                       // Enter 'p' between
      { NewPP(H->N); H = H->N; H->P = p; H->N = HN; }     // H and HN.
                                                          //------------------
    return List;
  }
} // operator+

//----------------------------------------------------------------------------
// 'MultiDelete' deletes all elements 'p' in 'List' for which the condition
// 'p.fyi > fmax' holds.  This function assumes that the 'fyi' components of
// the list elements are sorted in increasing order (see operator +).
//----------------------------------------------------------------------------
void MultiDelete ( PairPtr& List, const real& fmax )
{
  PairPtr  DelPrev, Del;
  int      ready;

  if (List != EmptyList) {
    if (GetFyi(List->P) > fmax)                     // All list elements fulfill
```

```
      { Del  = List; List = EmptyList; }         // 'p.fyi > fmax'.
    else {                                       //--------------------------
      DelPrev = List; Del = DelPrev->N; ready = (Del == NULL);

      while (!ready) {
        if (Del == NULL)
          ready = 1;
        else if (GetFyi(Del->P) > fmax)
          { ready = 1; DelPrev->N = NULL; }
        else
          { DelPrev = Del; Del = Del->N; }
      }
    }
    FreeAll(Del);
  }
} // Multidelete

Pair Head ( PairPtr List )                       // Delivers first pair of the
  { return List->P; }                            // list, i.e. the pair P with
                                                 // the smallest value P.fyi.
                                                 //--------------------------

void DelHead ( PairPtr& List )                   // Deletes the first pair of
{                                                // the List
  PairPtr Del;                                   //--------------------------

  Del = List; List = List->N; Free(Del);
}


int Length ( PairPtr List )                      // 'Length' delivers the number
{                                                // of elements in list 'List'
  int     i = 0;                                 //----------------------------

  while (List != EmptyList)
    { i++; List = List->N; }
  return i;
}

//----------------------------------------------------------------------------
// Copies the interval vector components of every pair of the 'List' to the
// interval matrix 'Mat' and allocates the intvector 'bv'. For an empty list
// matrix 'Mat' and vector 'bv' are sized to unit length with undefined
// components.
//----------------------------------------------------------------------------
void ListToMatrix (PairPtr List, imatrix& Mat, intvector& bv)
{
  int      i, n, m;
  PairPtr  H;

  m = Length(List);                              // Resize the arrays:
  if (m == 0)                                    //   - to minimum length with
    { Resize(Mat,1,1); Resize(bv,1); }           //     undefined components
  else {                                         //
    n = Ub(GetInt(List->P));                     // or
    Resize(bv,m);                                //   - to correct dimensions
    Resize(Mat,m,n);                             //--------------------------
  }

  H = List;                                      // Copy vector components of
  for (i = 1; i <= m; i++)                       // the list to matrix 'Mat'
    { Mat[i] = GetInt(H->P); H = H->N; }         //--------------------------
}
```

### 14.3.1.2 Module gop

The header file of module *gop* supplies the interface of the function *AllGOp()* for computing all global minimizers and the global minimum value of a twice continuously differentiable multi-dimensional function and the interface of the function *AllGOpErrMsg()* which can be used to get an error message for the error code returned by *AllGOp()*.

```
//----------------------------------------------------------------------------
// File: gop (header)
// Purpose: Computing enclosures for all global minimizers and for the global
//    minimum value of a twice continuously differentiable multi-dimensional,
//    scalar valued function, assuming that the global minimum is a stationa-
//    ry point. If it is a boundary point of the search area with gradient of
//    the function being different from zero, the method fails in its form
//    presented here.
// Global functions:
//    AllGOp()      : computes enclosures for all zeros
//    AllGOpErrMsg(): delivers an error message text
//----------------------------------------------------------------------------
#ifndef __GOP_HPP
#define __GOP_HPP

#include <intvecto.hpp>     // Integer vector type
#include <interval.hpp>     // Interval arithmetic
#include <hess_ari.hpp>     // Hessian differentiation arithmetic

// Additional header files to get access to interval standard functions and
// interval utility functions which often are needed for implementing user
// defined functions of type 'HessType'
//---------------------------------------------------------------------------
#include <i_util.hpp>       // Interval utility functions
#include <imath.hpp>        // Interval mathematical functions

const int MaxCount = 10000; // Maximum count of result components

extern char* AllGOpErrMsg ( int );
extern void  AllGOp ( HTscalar_FctPtr, ivector, real, imatrix&, intvector&,
                      int&, interval&, int&, int = MaxCount );
#endif
```

The function *AllGOp()* is an implementation of Algorithm 14.7. The static functions *MonotonicityTest()*, *ConcavityTest()*, *NewtonStep()*, *GlobalOptimize()*, *PosDef()*, and *VerificationStep()* are implementations of the Algorithms 14.1 to 14.6. All derivative evaluations use the differentiation arithmetic *hess_ari* described in Chapter 12.

The function *AllGOp()* uses the *HessType* function *f()* and the starting interval *Start* as input parameters and stores all computed enclosures in the interval matrix *OptiVector*, a vector of interval vectors. The information about the uniqueness of the optimizers is stored in the integer vector *InfoVector*. When the result list of pairs is transferred to the result arrays by calling function *ListToMatrix()*, the number of rows of the matrix *OptiVector* and the length of the vector *InfoVector* are automatically adapted to the number of optimizers located. The function *AllGOp()* has an optional last parameter *MaxNumberOfOptis* by which you can specify the maximum number of optimizers to be computed. Its default value is given by the local constant *MaxCount* = 10000 which should be big enough to compute *all* optimizers in most cases. *AllGOp()* reports a warning if there are more optimizers than specified by parameter *MaxNumberOfOptis*. The value of the desired accuracy *Epsilon* is checked to be greater than or equal to *MinEpsilon* (the relative machine accuracy). If this is not true, then *Epsilon* is reset to 1 ulp accuracy.

```
//----------------------------------------------------------------------------
// File: gop (implementation)
// Purpose: Computing enclosures for all global minimizers and for the global
//    minimum value of a twice continuously differentiable multi-dimensional,
//    scalar valued function, assuming that the global minimum is a stationa-
//    ry point. If it is a boundary point of the search area with gradient of
//    the function being different from zero, the method fails in its form
//    presented here.
// Method: Bisection method combined with midpoint, monotonicity, concavity
//    test and extended interval Newton step.
// Global functions:
//    AllGOp()      : computes enclosures for all zeros
//    AllGOpErrMsg(): delivers an error message text
//----------------------------------------------------------------------------
#include <string.h>          // String handling
#include <xi_ari.hpp>        // Extended interval arithmetic
#include <lst_ari.hpp>       // List arithmetic
#include <mv_util.hpp>       // Real matrix/vector utility functions
#include <mvi_util.hpp>      // Interval matrix/vector utility functions
#include <matinv.hpp>        // Inversion of real matrices
#include <hess_ari.hpp>      // Hessian differentiation arithmetic
#include <gop.hpp>

static int MaxOptiNo;        // Maximum number of optimizers to be computed

//----------------------------------------------------------------------------
// Error codes used in this module.
//----------------------------------------------------------------------------
const int
  NoError      = 0,  // No error occurred
  IllMaxOptiNo = 1,  // Illegal value for the maximum number of optimizers
  NotAllOptis  = 2,  // Not all optimizers computed
  NoStatOpti   = 3;  // No stationary point is global optimizer

//----------------------------------------------------------------------------
// Error messages depending on the error code.
//----------------------------------------------------------------------------
char* AllGOpErrMsg ( int Err )
{
  static char Msg[160] = "";

  switch (Err) {
    case NoError:
      break;  // Blank message
    case IllMaxOptiNo:
      sprintf(Msg,"Error: Parameter for maximum number of optimizers "
                  "must lie in 1,...,%1d!",MaxCount); break;
    case NotAllOptis:
      sprintf(Msg,"Warning: Not all optimizers found due to the user "
                           "limit of %1d optimizer(s).\n"
                  "         The enclosure of the global minimum value "
                           "could not be optimal!",MaxOptiNo); break;
    case NoStatOpti:
      strcpy(Msg,"Warning: No global optimizer (stationary point!) found!");
      break;
    default:
      strcpy(Msg,"Error: Code not defined!");
  }
```

```
  return(Msg);
}

static int MaxDiamComp ( ivector& iv )        // Determine the component 'mc'
{                                             // with maximum interval diameter
  int mc, k;                                  // of 'iv[mc]'.
  rvector d(Lb(iv), Ub(iv));                  //-------------------------------

  d = diam(iv);  mc = Lb(iv);
  for (k = Lb(iv)+1; k <= Ub(iv); k++)
    if (d[k] > d[mc]) mc = k;
  return mc;
}

static int MonotonicityTest ( ivector& gradY )                // returns true if
{                                                             // 'f' is monotone
  int i, n, Delete;                                           //----------------

  Delete = 0; i = 1;  n = Ub(gradY);
  while (i <= n && !Delete) {
    if (0.0 < Inf(gradY[i]) || 0.0 > Sup(gradY[i]) )   // 'f' is monotone
      Delete = 1;                                      //----------------
    i++;
  }
  return Delete;
}

static int ConcavityTest ( imatrix& HessY )                 // returns true if
{                                                           // 'f' is not convex
  int i, n, Delete;                                         //------------------

  Delete = 0;  i = 1;  n = Ub(HessY,1);
  while (i <= n && !Delete) {
    if (Sup(HessY[i][i]) < 0.0)      // 'f' is not convex
      Delete = 1;                    //------------------
    i++;
  }
  return Delete;
}

//----------------------------------------------------------------------------
// Purpose: Execution of one single interval Newton Gauss-Seidel step for the
//    interval vector 'Y' and the gradient of 'f'.
// Parameters:
//    In : 'f'    : must be declared for the type 'HessType' to enable the
//                  internal use of the differentiation arithmetic 'hess_ari'
//         'Y'    : specifies the starting interval
//         'HessY': Hessian matrix of 'f(Y)', already computed outside of
//                  'NewtonStep'
//    Out: 'V'    : stores the enclosures 'V[i]' for the splittings generated
//                  by the Newton step
//         'p'    : number of non-empty interval vectors 'V[i]'
// Description:
//    'NewtonStep' executes the extended interval Newton Gauss-Seidel step
//    for 'Y' with result interval vector(s) 'V[i]' which can be empty.
//    'p' gives the number of non-empty interval vectors stored as vectors
//    'V[1]',...,'V[p]'.
//----------------------------------------------------------------------------
static void NewtonStep ( HTscalar_FctPtr f,
                         ivector         Y,
                         imatrix&        HessY,
                         imatrix&        V,
                         int&            p )
{
```

```
  const int      n = Ub(Y);

  rvector        c(n);
  ivector        GradC(n), b(n), Y_minus_c(n);
  rmatrix        R(n,n);
  imatrix        A(n,n);
  int            i, i0, InvErr, j;
  interval       h, fC;
  ivector        z(2);
  idotprecision dd;

  c = mid(Y);                                 // Midpoint gradient evaluation
  fgEvalH(f,_ivector(c),fC,GradC);            //-----------------------------

  MatInv(mid(HessY),R,InvErr);                   // Invert the midpoint matrix
  if (InvErr) R = Id(R);                         //---------------------------

  A = R * HessY; b = R * GradC;           // Compute data for Gauss-Seidel step
  Y_minus_c = Y - c;                      //----------------------------------

  p = 0;  i0 = 0;                     // Initializations, A[i0,i0] contains zero
                                      //----------------------------------------
  for (i = 1; i <= n; i++ ) {                  // Interval-Gauss-Seidel-step for
                                               // non-zero A[i,i] elements
    if ( in(0.0,A[i][i]) ) {                   //------------------------------
      i0 = i;        // Largest i with 0 in A[i,i]
      continue;      // Next i
    }

    dd = b[i];
    for (j = 1; j < i; j++) accumulate(dd,A[i][j],Y_minus_c[j]);
    for (j = i+1; j <= n; j++) accumulate(dd,A[i][j],Y_minus_c[j]);
    rnd(dd,h);
    h = c[i] - h / A[i][i];

    if ( Disjoint(Y[i],h) ) { p = 0; return; }          // No solution possible
                                                        //---------------------
    Y[i] = Y[i] & h;  Y_minus_c[i] = Y[i] - c[i];
  } // for (i ...

  for (i = 1; i <= i0; i++ ) {                 // Interval-Gauss-Seidel-step for
                                               // zero A[i,i] elements
                                               //------------------------------
    if ( !in(0.0,A[i][i]) ) continue;                                // Next i
                                                                     //-------
    dd = b[i];
    for (j = 1;j < i; j++) accumulate(dd,A[i][j],Y_minus_c[j]);
    for (j = i+1;j <= n; j++) accumulate(dd,A[i][j],Y_minus_c[j]);
    rnd(dd,h);
    z = Y[i] & (c[i] - h % A[i][i]);              // Extended interval division
                                                  //---------------------------
    if (z[1] == EmptyIntval())                  // z[1] == z[2] == EmptyIntval()
      { p = 0; return; }                        // i.e. no solution possible
                                                //------------------------------
                                                             // Compute new 'Y'
    Y[i] = z[1];  Y_minus_c[i] = Y[i] - c[i];                //----------------
    if (z[2] != EmptyIntval())                     // Store further splittings
      { p++; V[p] = Y; V[p][i] = z[2]; }           //-------------------------
  } // for (i ...

  p++; V[p] = Y;
} // NewtonStep

//----------------------------------------------------------------------------
```

```
// Purpose: Execution of the global optimization method including a bisection
//    method, midpoint test, monotonicity test, concavity test, and extended
//    interval Newton steps.
// Parameters:
//    In : 'f'          : must be declared for type 'HTvector' to enable
//                        the internal use of the differentiation arithmetic
//                        'hess_ari'
//         'Start:      : specifies the starting interval
//         'Epsilon'    : specifies the desired relative accuracy
//    Out: 'ResultList': stores the candidates for enclosure of a global
//                        minimizer
//         'ListLength': length of the result list, that is the number of
//                        candidates in the result list
//         'Minimum'    : stores the enclosure of the global minimum value
// Description:
//    The function manages the list 'L' of pending subintervals that may
//    contain global minimizers. Subintervals are removed from the list and
//    placed in the accepted list 'ResultList' when they satisfy relative
//    error acceptance criteria. Subintervals are also removed from the list
//    by the midpoint, monotonicity, concavity tests, or by the interval
//    Newton steps. Subintervals are added to the pending list when an element
//    from the list is bisected or when the extended interval Newton step
//    yields two candidate intervals. 'ResultList' returns the list of
//    enclosures of the global minimizers. The maximum number of enclosures
//    to be computed is given by the value of the global variable 'MaxOptiNo'.
//    'Minimum' returns the enclosure of the global minimum value.
//----------------------------------------------------------------------------
static void GlobalOptimize ( HTscalar_FctPtr f,
                             ivector&        Start,
                             real            Epsilon,
                             PairPtr&        ResultList,
                             int&            ListLength,
                             interval&       Minimum )
{
  const int  n = Ub(Start);

  Pair       PairY;                     // For the pair (Y, inf(f(Y))
  ivector    Y(n);
  imatrix    U(2,n);                    // Subboxes of Y
  imatrix    V(n+1,n);                  // Subboxes of U
  interval   fY, fU, fV, fC, fCV;       // Function evaluations of f
  ivector    GradU(n), GradV(n);        // Gradient evaluations of f
  imatrix    HessU(n,n);                // Hessian evaluation of f
  real       fmax;                      // Upper bound for minimum
  rvector    c(n), cV(n);               // Midpoints of Y and V
  PairPtr    WorkList;                  // List of pairs
  int        i, j, p, k;                // Control variables
  int        Bisect;                    // Flag for iteration

  c = mid(Start);                            // Compute upper bound for minimum
  fEvalH(f,_ivector(c),fC);                  //--------------------------------
  fmax = Sup(fC);

  ResultList = EmptyList;
  ListLength = 0;

  if ( !UlpAcc(Start,1) ) {                                  // Start iteration
    Y = Start; WorkList = EmptyList;                         //----------------
    do {
      k = MaxDiamComp(Y);                            // d(Y[k] == max_i d(Y[i])
      U[1] = Y;  U[2] = Y;                           // Bisect 'Y' with respect
      SetSup(U[1][k],c[k]); SetInf(U[2][k],c[k]);    // to component 'k'.
      for (i = 1; i <= 2; i++) {                     //------------------------
        fgEvalH(f,U[i],fU,GradU);
```

```
      if (MonotonicityTest(GradU)) continue;    // Next i
                                                // Try centered form to get
      fU = (fC + GradU*(U[i] - c)) & fU;        // better enclosure of 'f(U)'
                                                //---------------------------
      if (fmax < Inf(fU)) continue;             // Next i

      fghEvalH(f,U[i],fU,GradU,HessU);            // Compute interval Hessian
                                                  //-------------------------
      if (ConcavityTest(HessU)) continue;         // Next i

      NewtonStep(f,U[i],HessU,V,p);          // Extended interval Newton step
                                             //------------------------------
      for (j = 1; j <= p; j++) {
        fgEvalH(f,V[j],fV,GradV);                // Compute interval gradient
                                                 //--------------------------
        if (MonotonicityTest(GradV)) continue;          // Next j

        cV = mid(V[j]);                                 // Try centered form
        fEvalH(f,_ivector(cV),fCV);                     // to get better en-
        fV = (fCV + GradV*(V[j] - cV)) & fV;            // closure of 'f(U)'
                                                        //------------------
        if (fmax >= Inf(fV)) {                                   // Store V
          PairY = _Pair(V[j],Inf(fV));                           //--------
          WorkList = WorkList + PairY;
        }
      } // for (j = ...
    } // for (i = ...

    Bisect = 0;                                           // Get next 'Y' of
    while (WorkList != EmptyList && !Bisect ) {           // the work list
      PairY = Head(WorkList);  DelHead(WorkList);         //----------------

      Y = GetInt(PairY);  c = mid(Y);
      fEvalH(f,_ivector(c),fC);                              // Compute f(c)
                                                             //-------------
      if (Sup(fC) < fmax) fmax = Sup(fC);
      MultiDelete(WorkList,fmax);

      Minimum = _interval(GetFyi(PairY),fmax);

      if (RelDiam(Minimum) <= Epsilon || MaxRelDiam(Y) <= Epsilon) {
        ListLength++;                    // Check termination criteria. Stop
                                         // due to user specified maximum
        if (ListLength > MaxOptiNo) {    // for the number of optimizers.
          FreeAll(WorkList);             //---------------------------------
          Minimum = _interval(GetFyi(Head(ResultList)),fmax);
          return;
        }

        ResultList = ResultList + PairY;                    // Store 'PairY'
      }                                                     //--------------
      else
        Bisect = 1;
    } // while

  } while(Bisect);
} // if ( !UlpAcc ...
else {
  fEvalH(f,Start,fY);                               // Store starting interval
  ResultList = EmptyList + _Pair(Start,Inf(fY));    // and interval evaluation
}                                                   //------------------------
                                                // Compute good enclosure of
if (ResultList != EmptyList)                    // of the global minimum value
```

```
    Minimum = _interval(GetFyi(Head(ResultList)),fmax);//--------------------
} // GlobalOptimize

//----------------------------------------------------------------------------
// 'MaxNorm' delivers an upper bound of the maximum norm of a symmetric in-
// terval matrix 'H', i.e. the row sum norm or infinity norm.
//----------------------------------------------------------------------------
static real MaxNorm ( imatrix& H )
{
  real  Nm, MaxNm;
  int   i, j;

  MaxNm = 0.0;
  for (i = Lb(H,1); i <= Ub(H,1); i++) {
    Nm = 0.0;
    for (j = Lb(H,1); j <= Ub(H,1); j++) Nm = addu(Nm,Sup(abs(H[i][j])));
    if (Nm > MaxNm) MaxNm = Nm;
  }
  return MaxNm;
}

//----------------------------------------------------------------------------
// 'PosDef' delivers true if it is guaranteed that all real symmetric matri-
// ces enclosed in 'H' are positive definite.
//----------------------------------------------------------------------------
static int PosDef ( imatrix& H )
{
  const int  kmax = 10;      // Maximum number of iterations

  int       pd;
  real      kappa, eps;
  imatrix   S(Lb(H,1),Ub(H,1),Lb(H,2),Ub(H,2));
  ivector   Z(Lb(H,1),Ub(H,1)),U(Lb(H,1),Ub(H,1));
  int       k;

  kappa = MaxNorm(H);  S = Id(H) - H/kappa;
  for (k = Lb(Z); k <= Ub(Z); k++) Z[k] = _interval(-1.0,1.0);
  k = 0;  eps = 0.25;
  do {
    U = Blow(Z,eps);  Z = S * U;  pd = in(Z,U);
    k++;  eps = 8.0*eps;
  } while( !(pd || k == kmax) );
  return pd;
}

//----------------------------------------------------------------------------
// Purpose: Execution of a verification step including the use of an epsilon
//    inflation.
// Parameters:
//    In    : 'f'      : function of 'HessType'
//    Out   : 'yUnique': returns 'true' if the verification is successful
//    In/Out: 'y'      : interval enclosure to be verified
// Description: This function checks the uniqueness of the local minimizer
//    enclosed in the interval variable 'y' by a verification step including
//    the use of an epsilon inflation of the iterates.
//----------------------------------------------------------------------------
static void VerificationStep ( HTscalar_FctPtr f, ivector& y, int& yUnique )
{
  const int
    kmax = 10,        // Maximum number of iterations
    n = Ub(y);        // Length of vectors in use

  interval  fY;
  ivector   yIn(n), yOld(n), GradY(n);
```

```
  imatrix   HessY(n,n);
  imatrix   yp(n+1,n);
  int       k, p;
  real      eps;

  yUnique = (Inf(y) == Sup(y));                 // y is a point interval vector
  if (yUnique) return;                          //----------------------------

  yIn = y;  yOld = y;  k = 0;  eps = 0.25;                    // Initializations
                                                              //----------------
  while (!yUnique && k < kmax) {          // Do 'kmax' loops to achieve inclusion
                                          //-------------------------------------
    yOld = Blow(y,eps);                              // Epsilon inflation of 'y'
                                                     //-------------------------
    // Perform interval Newton step
    //------------------------------
    k++;
    fghEvalH(f,yOld,fY,GradY,HessY);
    NewtonStep(f,yOld,HessY,yp,p);

    if (p != 1) break;                               // No verification possible
                                                     //-------------------------
    if (yp[1] == yOld)
      eps = eps * 8.0;                                          // Increase 'eps'
    else {                                                      //---------------
      y = yp[1];
      yUnique = in(y,yOld);                 // Inner inclusion ===> uniqueness
    }                                       // of a stationary point
  }  // while                               //--------------------------------

  if (yUnique) yUnique = PosDef(HessY);        // Positive definite ===> local
                                               // minimizer is verified
  if (!yUnique) y = yIn;                       //----------------------------
} // VerificationStep

//----------------------------------------------------------------------------
// Purpose: Computation of enclosures for all global minimizers and for the
//    global minimum value of a twice continuously differentiable multi-di-
//    mensional, scalar-valued function.
// Parameters:
//    In : 'f'                : objective function, must be declared for the
//                              'HessType' to enable the internal use of
//                              the differentiation arithmetic 'hess_ari'
//         'Start',           : specifies the starting interval vector
//         'Epsilon'          : specifies the desired relative accuracy
//                              (interval diameter) of the result boxes
//         'MaxNumberOfOptis': maximum number of optimizers to be computed
//                              (default value: 'MaxCount' = compute all)
//    Out: 'OptiVector'       : stores (row by row) and returns the boxes
//                              (enclosures) for the global optimizers of
//                              'f'. 'OptiVector' is a vector of interval
//                              vectors (ie. an interval matrix).
//         'InfoVector'       : stores the corresponding information on the
//                              uniqueness of the local optimizers in these
//                              enclosures
//         'NumberOfOptis'    : number of enclosures computed
//         'Minimum'     '    : enclosure for the minimum value
//         'Err'              : error code
// Description:
//    The enclosures for the global minimizers of 'f' are computed by calling
//    function 'GlobalOptimize'. Then a verification step is applied.
//    The enclosures (boxes) for the global minimizers of 'f' are stored in
//    the interval matrix 'OptiVector' row by row, the corresponding
//    information on the uniqueness of the local minimizers in these
```

```
//    enclosures is stored in the integer vector 'InfoVector'. The number of
//    enclosures computed is returned in the integer variable 'NumberOfOptis'.
//    The enclosure for the global minimum value is returned in the variable
//    'Minimum'. If an error occurs, the value of 'Err' is different from 0.
//    The result arrays 'OptiVector' and 'InfoVector' are automatically
//    resized to standard index range with lower index bounds 1. The maximum
//    number of optimizers which can be computed is given by
//    'MaxNumberOfOptis'. It is bounded by 'MaxCount'. If your system can
//    handle more than 'MaxCount' optimizers, you can increase its value.
//----------------------------------------------------------------------------
void AllGOp ( HTscalar_FctPtr f,
              ivector         Start,
              real            Epsilon,
              imatrix&        OptiVector,
              intvector&      InfoVector,
              int&            NumberOfOptis,
              interval&       Minimum,
              int&            Err,
              int             MaxNumberOfOptis )
{
  // Check range for the maximum number of optimizers
  //--------------------------------------------------
  if (MaxNumberOfOptis < 1 || MaxCount < MaxNumberOfOptis)
    { Err = IllMaxOptiNo;  NumberOfOptis = 0;  return; }

  int     i;
  PairPtr ResultList;
  real    MinEpsilon;

  SetLb(Start,1);                      // Resize start vector to standard index
                                       // range with lower index bound 1
                                       //--------------------------------------
  Err = NoError;
  MaxOptiNo = MaxNumberOfOptis;          // Initialize global static variable
  MinEpsilon = succ(1.0) - 1.0;          // Relative machine accuracy (1 ulp)
                                         //----------------------------------
  if (Epsilon < MinEpsilon) Epsilon = MinEpsilon;        // Set 'Epsilon' to
                                                         // 1 ulp accuracy
  GlobalOptimize(f,Start,Epsilon,ResultList,NumberOfOptis,Minimum);

  if (NumberOfOptis > MaxNumberOfOptis)
    { Err = NotAllOptis;  NumberOfOptis = MaxNumberOfOptis; }

  if (NumberOfOptis == 0)
    Err = NoStatOpti;
  else                                                   // Store results
    ListToMatrix(ResultList,OptiVector,InfoVector);      // in the matrix
                                                         //--------------
  // Verification step for the enclosure intervals
  //-----------------------------------------------
  for (i = 1; i <= NumberOfOptis; i++)
    VerificationStep(f,OptiVector[i],InfoVector[i]);

  FreeAll(ResultList);
}
```

## 14.3.2 Examples

We illustrate the use of function *AllGOp()* to compute all global minimizers and the global minimum of the functions of Branin [93] and Levy [64] within a given

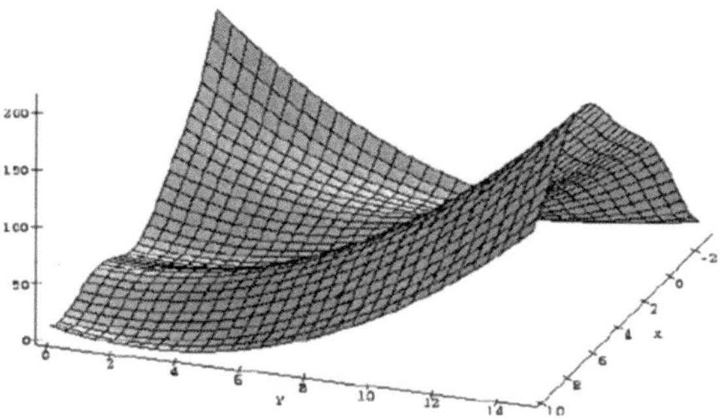

Figure 14.1: Function $f_B(x)$ of Branin

starting box. A plot of Branin's function is shown in Figure 14.1. The function is defined by

$$f_B(x) = \left(\frac{5}{\pi}x_1 - \frac{5.1}{4\pi^2}x_1^2 + x_2 - 6\right)^2 + 10\left(1 - \frac{1}{8\pi}\right)\cos x_1 + 10.$$

It has the global minimizers $(-\pi, \frac{491}{40})^{\mathrm{T}}$, $(\pi, \frac{91}{40})^{\mathrm{T}}$, and $(3\pi, \frac{99}{40})^{\mathrm{T}}$, and the global minimum value $\frac{5}{4\pi}$.

Figure 14.2 shows a plot of Levy's function. The large number of local optimizers, makes it extremely difficult for an approximation method to find the global minimizer $(-1.3068\ldots, -1.424\ldots)^{\mathrm{T}}$. The global minimum value is $-176.13\ldots$ The

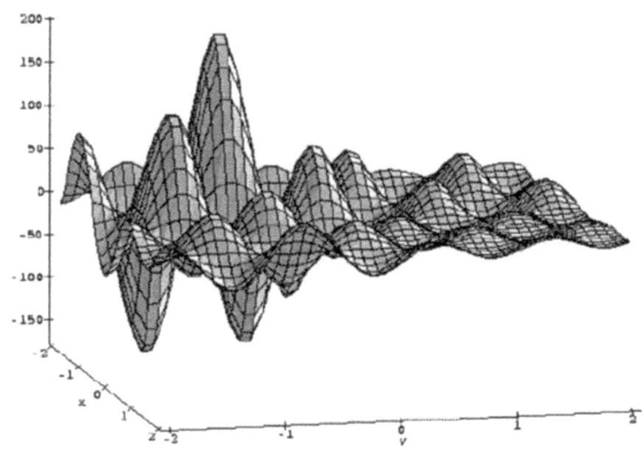

Figure 14.2: Function $f_L(x)$ of Levy with about 700 local minima

function of Levy is defined by

$$f_L(x) = \sum_{i=1}^{5} i\cos((i-1)x_1 + i)\sum_{j=1}^{5} j\cos((j+1)x_2 + j) + (x_1 + 1.42513)^2 + (x_2 + 0.80032)^2.$$

We first define the functions $f_B$ and $f_L$ for the type *HessType*. Then, we use a function *compute()* with a function and a textual description of the function as parameters. This function reads the necessary input data for the call of function *AllGOp()*. If we add additional function definitions in our program, we can apply this function in the same way. For Branin's function we use the starting interval vector $([-5,10],[0,15])^T$. For the function of Levy we take $([-10,10],[-10,10])^T$ as starting box.

```
//----------------------------------------------------------------------------
// Example: Global optimization
// This program uses module 'gop' to compute the global optimizers of the
// function of Branin
//
//    fB(x) = sqr(5/pi*x[1] - 51/(40*sqr(pi))*sqr(x[1]) + x[2] - 6)
//
//             + 10*(1-1/8/pi)*cos(x[1]) + 10
//
// and the function of Levy
//               5                          5
//    fL(x) = sum i cos((i-1)x[1] + i)  sum j cos((j+1)x[2] + j)
//            i=1                         j=1
//
//             + sqr(x[1] + 1.42513) + sqr(x[2] + 0.80032)
//
// A starting interval and a tolerance must be entered.
//----------------------------------------------------------------------------
#include <gop.hpp>         // Global optimization
#include <stacksz.hpp>     // To increase stack size for some
                           // special C++ compilers

HessType fBranin ( HTvector& x )
{
  HessType   hh(2);
  interval   pi = Pi();          // enclosure of pi

  hh = sqr(5.0/pi*x[1] - 51.0/(40.0*sqr(pi))*sqr(x[1]) + x[2] - 6.0)
       + 10.0*(1.0-1.0/8.0/pi)*cos(x[1]) + 10.0;
  return hh;
}

HessType fLevy ( HTvector& x )
{
  HessType   isum(2), jsum(2), hh(2);
  int        i;

  isum = 0.0; jsum = 0.0;
  for (i = 1; i <= 5; i++) {
    isum = isum + double(i)*cos(double(i-1)*x[1] + double(i));
    jsum = jsum + double(i)*cos(double(i+1)*x[2] + double(i));
  }
  hh = isum * jsum +
       sqr(x[1] + _interval(142513.0)/100000.0) +    // Avoid real con-
```

```
      sqr(x[2] + _interval(80032.0)/100000.0);      // version error
  return hh;
}

//----------------------------------------------------------------------------
// Function for printing and reading informations to call the function
// 'AllGOp()'. This function must be called with the function 'f', a string
// 'Name' containing textual description of that function, and an integer
// 'dim' specifying the dimension of the problem.
//----------------------------------------------------------------------------
void compute ( HTscalar_FctPtr f, char* Name, int dim )
{
  ivector   SearchInterval(dim);
  interval  Minimum;
  real      Tolerance;
  imatrix   Opti;
  intvector Unique;
  int       NumberOfOptis, i, Error;

  cout << SetPrecision(23,15) << Scientific;   // Output format

  cout << "Computing all global minimizers of the " << Name << endl;
  cout << "Search interval      : "; cin >> SearchInterval;
  cout << "Tolerance (relative) : "; cin >> Tolerance;
  cout << endl;

  AllGOp(f,SearchInterval,Tolerance,
          Opti,Unique,NumberOfOptis,Minimum,Error);

  for (i = 1; i <= NumberOfOptis; i++) {
    cout << Opti[i];
    if (Unique[i])
      cout << "encloses a locally unique candidate for a global minimizer!";
    else
      cout << "may contain a local or global minimizer!";
    cout << endl << endl;
  }


  if (NumberOfOptis != 0)
    cout << Minimum << endl
         << "encloses the global minimum value!" << endl << endl;
  cout << NumberOfOptis << " interval enclosure(s)" << endl;

  if (Error)
    cout << endl << AllGOpErrMsg(Error) << endl;
  else if (NumberOfOptis == 1 && Unique[1])
    cout << endl << "We have validated that there is "
                    "a unique global optimizer!" << endl;
}

int main ( )
{
  compute(fBranin,"Function of Branin",2);
  cout << endl << endl;
  compute(fLevy,"Function of Levy",2);
  return 0;
}
```

If we execute this program, we get the following runtime output:

```
Computing all global minimizers of the Function of Branin
Search interval      : [-5,10] [0,15]
```

```
Tolerance (relative) : 1e-8

[ 9.424777960768414E+000, 9.424777960770346E+000]
[ 2.474999999999732E+000, 2.475000000000267E+000]
encloses a locally unique candidate for a global minimizer!

[ 3.141592653589792E+000, 3.141592653589795E+000]
[ 2.274999999999997E+000, 2.275000000000004E+000]
encloses a locally unique candidate for a global minimizer!

[-3.141592653589796E+000,-3.141592653589790E+000]
[ 1.227499999999999E+001, 1.227500000000001E+001]
encloses a locally unique candidate for a global minimizer!

[ 3.978873577297381E-001, 3.978873577297418E-001]
encloses the global minimum value!

3 interval enclosure(s)


Computing all global minimizers of the Function of Levy
Search interval      : [-10,10] [-10,10]
Tolerance (relative) : 1e-8

[-1.306853009753609E+000,-1.306853009753535E+000]
[-1.424845041560701E+000,-1.424845041560661E+000]
encloses a locally unique candidate for a global minimizer!

[-1.761375780026126E+002,-1.761375780016284E+002]
encloses the global minimum value!

1 interval enclosure(s)

We have validated that there is a unique global optimizer!
```

Thus, we know that there are three locally unique minimizers of $f_B$ which are good candidates for global minimizers and one global minimizer of $f_L$ within the specified starting boxes.

## 14.3.3 Restrictions and Hints

The objective function $f$ must be expressible in C++ code as a finite sequence of arithmetic operations and elementary functions supported by the differentiation arithmetic module *hess_ari*.

If *AllGOp()* reports a warning that not all optimizers were found due to a user limit for the number of optimizers to be computed, then you may call *AllGOp()* again to find additional optimizers.

The method is not very fast if a very small value of $\varepsilon$ (*Epsilon*) is used, if the interval Newton step does not improve the actual iterates, *and* if the different tests do not discard intervals any more because of rounding and overestimation effects of the machine interval arithmetic. Under these circumstances, the method is equivalent with a bisection method.

In *GlobalOptimize()*, the evaluation of the function with differentiation arithmetic can cause a runtime error if the interval argument of an elementary function does not lie in the domain specified for this interval function or if a division by an interval containing zero occurs. This also may be due to the known overestimation effects of interval arithmetic (see Section 3.1). To get rid of these errors, the user may try to split the starting interval in several parts and call *AllGOp()* for these parts.

The rules for getting true enclosures in connection with conversion errors (see Section 3.7) also apply here. That is why we expressed the real constant 1.42513 as `_interval(142513.0)/100000.0` in the code for the Levy function.

## 14.4 Exercises

**Exercise 14.1** Use our function *AllGOp()* to compute the global minimizers and the global minimum value of the *Six-Hump Camel-Back* function [78]

$$f(x) = 4x_1^2 - 2.1x_1^4 + \frac{1}{3}x_1^6 + x_1x_2 - 4x_2^2 + 4x_2^4$$

shown in Figure 14.3 within the box $([-2.5, 2.5], [-2.5, 2.5])^T$. The global optimizers are $\pm(0.089842\ldots, -0.712656\ldots)^T$ and the global minimum value is $-1.03162\ldots$

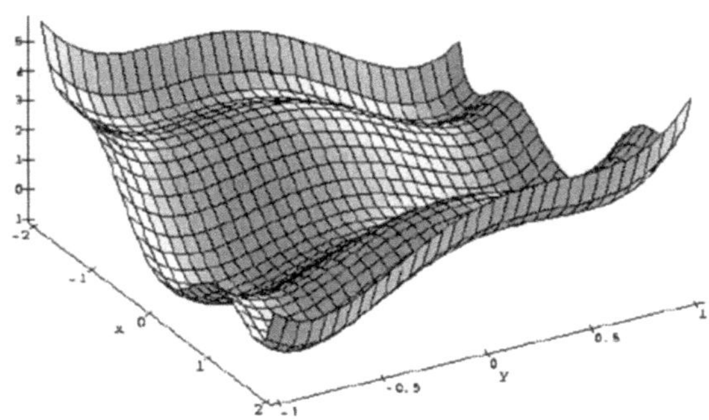

Figure 14.3: Six-Hump Camel-Back function

**Exercise 14.2** Use our function *AllGOp()* to compute the global minimizer and the global minimum value of the function $f : I\!R^n \to I\!R$ with

$$f(x) = \sum_{i=1}^{n} \left( \sum_{j=1}^{i} x_j \right)^2 .$$

Study the behavior of CPU time for our algorithm for increasing values of $n$.

**Exercise 14.3** Use our function *AllGOp()* to compute the global minimizers (a continuum of points) and the global minimum value of the function $f : I\!R^2 \to I\!R$ with

$$f(x) = (x_1 + x_2)^2.$$

**Exercise 14.4** Use our function *AllGOp()* to compute the global minimizer $(1,\ldots,1)^T$ and the global minimum value 0 of $f : \mathbb{R}^n \to \mathbb{R}$ with

$$f(x) = \sum_{i=2}^{n} \left(100(x_i - x_{i-1}^2)^2 + (1 - x_{i-1})^2\right).$$

$f$ is a generalization of Rosenbrock's two-dimensional function [81], which has become accepted as a difficult test for new optimization routines. Study the behavior of CPU time for our algorithm for increasing values of $n = 5, 10, 15, \ldots$.

## 14.5 References and Further Reading

The method we discussed in this chapter is an *a priori* method because the iteration starts with a (possibly large) interval enclosing all the solutions which have to be found. Here, the iterates of the method are subintervals of the previous iterates.

There are also methods for finding (and bounding) one single *local* optimizer called *a posteriori* methods. These methods start with an approximation of a local minimizer of $f$ and apply a test procedure for a neighborhood interval of the approximation to verify that a zero of $\nabla f$ lies within that interval. There are a huge number of approximate optimization methods for local and global optimization available without any verification of the result. For an overview on such approximation methods, see [93] or [96].

The method presented in this chapter can be extended to become more efficient and faster. For more sophisticated extensions, see [13], [29], [31], [40], [78], [79], and [80].

# Appendix A

# Utility Modules

The functions described in the following sections are used by several problem solving routines in this book. They are grouped according to the basic data types of the C–XSC library.

## A.1 Module r_util

The module *r_util* provides the function *Max()* to find the maximum of two real numbers.

```
//----------------------------------------------------------------------------
// File: r_util (header)
// Purpose: Utilities of type 'real'
// Global functions:
//    Max(): delivers the maximum of two real numbers
//----------------------------------------------------------------------------
#ifndef __R_UTIL_HPP
#define __R_UTIL_HPP

#include <real.hpp>     // Real arithmetic

extern real Max ( real, real );
#endif
```

```
//----------------------------------------------------------------------------
// File: r_util (implementation)
// Purpose: Utilities of type 'real'
// Global functions:
//    Max(): delivers the maximum of two real numbers
//----------------------------------------------------------------------------
#include <r_util.hpp>

real Max ( real a, real b )                           // Maximum of two reals
  { return (a > b) ? a : b; }                         //--------------------
```

## A.2 Module i_util

The header file of module *i_util* supplies functions based on the type *interval*.

```
//----------------------------------------------------------------------------
// File: i_util (header)
// Purpose: Utilities of type 'interval'
// Global functions:
//    in()       : contained-in relation for a real
//    in()       : contained-in-the-interior relation for two intervals
//    rnd()      : to round a dotprecision argument to an interval
//    Blow()     : epsilon inflation
//    Disjoint(): test for disjointedness of two intervals
//    AbsMin()   : smallest absolute value of an interval
//    AbsMax()   : greatest absolute value of an interval
//    RelDiam()  : relative diameter of an interval
//    UlpAcc()   : to check whether the width of an interval is less than a
//                 certain number of ulps (ulp = unit in the last place of
//                 the mantissa)
//    Power()    : exponentiation by an integer for intervals
//----------------------------------------------------------------------------
#ifndef __I_UTIL_HPP
#define __I_UTIL_HPP

#include <interval.hpp>       // Interval arithmetic

extern int      in       ( real, interval& );
extern int      in       ( interval&, interval& );
extern void     rnd      ( dotprecision&, interval& );
extern interval Blow     ( interval, real );
extern int      Disjoint ( interval, interval );
extern real     AbsMin   ( interval );
extern real     AbsMax   ( interval );
extern real     RelDiam  ( interval );
extern int      UlpAcc   ( interval, int );
extern interval Power    ( interval, int );
extern interval Pi       ( );
#endif
```

The module provides two implementations of the function *in()* which can be used to check for the contained-in-the-interior relation as defined on page 30. The function *Blow()* performs an epsilon inflation according to Definition (3.24) and the function *Disjoint()* checks two intervals for disjointness. The real-valued functions *AbsMin()*, *AbsMax()*, and *RelDiam()* can be used to compute the smallest absolute value $\langle[x]\rangle$, the greatest absolute value $|[x]|$, and the relative diameter $d_{\text{rel}}([x])$ of a real interval $[x]$, according to the Definitions (3.1) and (3.2). The integer function *UlpAcc()* is used to check for a desired accuracy specified in ulp (see Section 3.6). The interval function *Pi()* delivers an enclosure of $\pi$.

Finally, the module gives an implementation of the interval power function $[x]^n$ with integer exponent $n$ as defined in Example 3.5. The implementation makes use of the static function *Power()* for real argument $x$, positive (!) integer exponent $n$, and a parameter indicating the rounding mode. Depending on the rounding parameter, a lower or an upper bound of $x^n$ is computed using the binary shift method. The binary shift method is used to reduce an exponentiation by an integer to a sequence of multiplications. Therefore in general, it does not yield a result of maximum accuracy. The number of multiplications depends logarithmically on $n$. An algorithmic description of the method is given by Algorithm A.1.

**Algorithm A.1:** Power $(x, n)$

1. {Initialization}
   $p := 1;\quad z := x;$
2. {Binary shift method}
   **while** $(n > 0)$ **do**
     **if** Odd $(n)$ **then** $p := p \cdot z;$
     ShiftRight $(n)$; {Equivalent to $n := n/2$}
     **if** $(n > 0)$ **then** $z := z \cdot z;$
3. {Return function result}
   **return** Power $:= p;$

The method works by binary shifting the exponent $n$. Depending on the digit which was shifted out, the factor $p$ is updated, while $z$ is updated as long as $n$ does not vanish.

**Example A.1** To compute $x^{13}$, the following intermediate results are generated. The binary representation of the exponent is $13_{10} = 1101_2$.

| Digit shifted out | $p$ | $z$ |
|---|---|---|
| 1 | $x$ | $x^2$ |
| 0 | $x$ | $x^4$ |
| 1 | $x^5$ | $x^8$ |
| 1 | $x^{13}$ | – |

To get lower and upper bounds of $x^n$, respectively, Algorithm A.1 is modified by introducing the directed-rounding multiplications *muld()* (downwardly directed) and *mulu()* (upwardly directed). For negative $x$ and odd $n$, we have $\nabla(x^n) = -\Delta(-x^n)$ and $\Delta(x^n) = -\nabla(-x^n)$. The implementation of the module *i_util* is given below.

```
//----------------------------------------------------------------------------
// File: i_util (implementation)
// Purpose: Utilities of type 'interval'
// Global functions:
//    in()       : contained-in relation for a real
//    in()       : contained-in-the-interior relation for two intervals
//    rnd()      : to round a dotprecision argument to an interval
//    Blow()     : epsilon inflation
//    Disjoint() : test for disjointedness of two intervals
//    AbsMin()   : smallest absolute value of an interval
//    AbsMax()   : greatest absolute value of an interval
//    RelDiam()  : relative diameter of an interval
//    UlpAcc()   : to check whether the width of an interval is less than a
//                 certain number of ulps (ulp = unit in the last place of
//                 the mantissa)
//    Power()    : exponentiation by an integer for intervals
//----------------------------------------------------------------------------
#include <imath.hpp>       // Interval mathematical functions
#include <imatrix.hpp>     // Interval matrix/vector arithmetic
#include <i_util.hpp>

int in ( real x, interval& y )                        // Contained-in relation
  { return ( (Inf(y) <= x) && (x <= Sup(y)) ); }      //----------------------
```

```
int in ( interval& x, interval& y )       // Contained-in-the-interior relation
{                                         //-----------------------------------
  return ( (Inf(y) < Inf(x)) && (Sup(x) < Sup(y)) );
}

void rnd ( dotprecision& d, interval& x )    // Round dotprecision to interval
{                                            //--------------------------------
  real Lower, Upper;

  rnd(d,Lower,Upper);          // Rounding to interval bounds
  x = _interval(Lower,Upper);
}

interval Blow ( interval x, real eps )                       // Epsilon inflation
{                                                            //------------------
  x =  (1.0+eps)*x - eps*x;
  return _interval(pred(Inf(x)),succ(Sup(x)));
}

int Disjoint ( interval a, interval b )                // Test for disjointedness
{                                                      //------------------------
  return (Inf(a) > Sup(b)) || (Inf(b) > Sup(a));
}

real AbsMin ( interval x )                                 // Absolute minimum of
{                                                          // an interval
  if ( in(0.0,x) )                                         //--------------------
    return 0.0;
  else if (Inf(x) > 0.0)
    return Inf(x);
  else
    return -Sup(x);
}

real AbsMax ( interval x )                                 // Absolute maximum of
{                                                          // an interval
  real a, b;                                               //--------------------

  a = abs(Inf(x));
  b = abs(Sup(x));

  if (a > b)
    return a;
  else
    return b;
}

real RelDiam ( interval x )                                  // Relative diameter
{                                                            // of an interval
  if ( in(0.0,x) )                                           //------------------
    return diam(x);
  else
    return( divu(diam(x),AbsMin(x)) );
}

//----------------------------------------------------------------------------
// Checks if the diameter of the interval 'x' is less or equal to 'n' ulps.
// Ulp is an abbreviation for units in the last place.
//----------------------------------------------------------------------------
int UlpAcc ( interval x, int n )
{
  int  i;
  real Infimum;
```

```
  Infimum = Inf(x);
  for (i = 1; i <= n; i++) Infimum = succ(Infimum);
  return (Infimum >= Sup(x));
}

//----------------------------------------------------------------------------
// Purpose: The local function 'Power()' is used to compute a lower or an
//    upper bound for the power function with real argument and integer
//    exponent, respectively.
// Parameters:
//    In: 'x'      : real argument
//        'n'      : integer exponent
//        'RndMode': rounding mode,
//                   (-1 = downwardly directed, +1 = upwardly directed)
// Description:
//    This function is used to speed up the interval power function defined
//    below. The exponentiation is reduced to multiplications using the
//    binary shift method. Depending on 'n', this function is up to 40 times
//    as fast as the standard power function for real argument and real
//    exponent. However, its accuracy is less than one ulp (unit in the last
//    place of the mantissa) since about log2(n) multiplications are executed
//    during computation. Since directed roundings are antisymmetric, one
//    gets
//
//       down(x↑n) = -up((-x)↑n)   and   up(x↑n) = -down((-x)↑n)
//
//    for x < 0 and odd n, where 'down' and 'up' denote the downwardly and
//    upwardly directed roundings, respectively.
//----------------------------------------------------------------------------
static real Power ( real x, int n, int RndMode )
{                          // Signals change of the rounding mode
  int  ChangeRndMode;      // for x < 0 and odd n
  real p, z;

  ChangeRndMode = ( (x < 0.0) && (n % 2 == 1) );

  if (ChangeRndMode) {
    z = -x;
    RndMode = -RndMode;
  }
  else
    z = x;

  p = 1.0;
  switch (RndMode) {                                 // Separate while-loops used
    case -1 : while (n > 0) {                        // to gain speed at runtime
                if (n % 2 == 1) p = muld(p,z);       //--------------------------
                n = n / 2;
                if (n > 0) z = muld(z,z);
              }
              break;
    case +1 : while (n > 0) {
                if (n % 2 == 1) p = mulu(p,z);
                n = n / 2;
                if (n > 0) z = mulu(z,z);
              }
              break;
  }

  if (ChangeRndMode)
    return -p;
  else
    return p;
```

```
}

//----------------------------------------------------------------------------
// Purpose: This version of the function 'Power()' is used to compute an
//    enclosure for the power function with interval argument and integer
//    exponent.
// Parameters:
//    In: 'x': interval argument
//        'n': integer exponent
// Description:
//    In general, this implementation does not deliver a result of maximum
//    accuracy, but it is about 30-40 times faster than the standard power
//    function for interval arguments and interval exponents. The resulting
//    interval has a width of approximately 2*log2(n) ulps. Since x↑n is
//    considered as a monomial, we define x↑0 := 1. For negative exponents
//    and 0 in 'x', the division at the end of the function will cause a
//    runtime error (division by zero).
//----------------------------------------------------------------------------
interval Power ( interval x, int n )
{
  int  m;
  real Lower, Upper;

  if (n == 0) return(_interval(1.0,1.0));

  if (n > 0)  m = n;  else  m = -n;

  if ( (0.0 < Inf(x)) || (m % 2 == 1) ) {
    Lower = Power(Inf(x),m,-1);
    Upper = Power(Sup(x),m,+1);
  }
  else if (0.0 > Sup(x)) {
    Lower = Power(Sup(x),m,-1);
    Upper = Power(Inf(x),m,+1);
  }
  else {
    Lower = 0.0;
    Upper = Power(AbsMax(x),m,+1);
  }

  if (n > 0)
    return(_interval(Lower,Upper));
  else                                        // Causes a runtime error
    return(1.0/_interval(Lower,Upper));       // if 0 in 'x'.
}

interval Pi ( )                                       // Enclosure of constant pi
  { return(4.0*atan(_interval(1.0,1.0))); }          //-------------------------
```

## A.3 Module ci_util

The header file of module *ci_util* supplies functions based on the type *cinterval* for complex intervals.

```
//----------------------------------------------------------------------------
// File: ci_util (header)
// Purpose: Utilities of type 'cinterval'
// Global functions:
//    in()  : Contained-in-the-interior relation for two complex intervals
//    Blow(): Epsilon inflation
```

```
//----------------------------------------------------------------------------
#ifndef __CI_UTIL_HPP
#define __CI_UTIL_HPP

#include <cinterva.hpp>     // Complex interval arithmetic

extern int       in   ( cinterval&, cinterval& );
extern cinterval Blow ( cinterval, const real& );
#endif
```

The function *in()* can be used to check if a complex interval is contained in the interior of another complex interval. For complex intervals, the contained-in-the-interior relation is satisfied if it is satisfied for both the real and the imaginary parts (cf. page 30). The function *Blow()* performs an epsilon inflation of a complex interval which is defined componentwise according to Definition (3.24).

```
//----------------------------------------------------------------------------
// File: ci_util (implementation)
// Purpose: Utilities of type 'cinterval'
// Global functions:
//    in()  : Contained-in-the-interior relation for two complex intervals
//    Blow(): Epsilon inflation
//----------------------------------------------------------------------------
#include <i_util.hpp>      // Interval utility functions
#include <ci_util.hpp>

int in ( cinterval& x, cinterval& y )    // Contained-in-the-interior relation
{                                        //----------------------------------
  return ( in(Re(x),Re(y)) && in(Im(x),Im(y)) );
}

cinterval Blow ( cinterval x, const real& eps )          // Epsilon inflation
{                                                        //------------------
  return _cinterval(Blow(Re(x),eps),Blow(Im(x),eps));
}
```

## A.4 Module mv_util

The header file of module *mv_util* supplies utility functions for real and integer matrices and vectors.

```
//----------------------------------------------------------------------------
// File: mv_util (implementation)
// Purpose: Utilities of type 'intvector', 'intmatrix', 'rvector', and
//    'rmatrix'.
// Global functions:
//    VecLen()    : length of an integer or real vector
//    RowLen()    : length of the rows of an integer or real matrix
//    ColLen()    : length of the columns of an integer or real matrix
//    Id()        : identity real matrix
//    transp()    : transposed of a real matrix
//    DoubleSize(): for doubling the size of an integer vector or matrix
//    operator << : output of integer vector
//----------------------------------------------------------------------------
#ifndef __MV_UTIL_HPP
#define __MV_UTIL_HPP

#include <intmatri.hpp>     // Interval matrix/vector type
```

```
#include <rmatrix.hpp>      // Real matrix/vector arithmetic

extern int      VecLen     ( intvector& );
extern int      RowLen     ( intmatrix& );
extern int      ColLen     ( intmatrix& );
extern int      VecLen     ( rvector& );
extern int      RowLen     ( rmatrix& );
extern int      ColLen     ( rmatrix& );
extern rmatrix  Id         ( rmatrix& );
extern rmatrix  transp     ( rmatrix& );
extern void     DoubleSize ( intvector& );
extern void     DoubleSize ( intmatrix& );
extern ostream& operator<< ( ostream&, intvector& );
#endif
```

The functions *RowLen()* and *ColLen()* can be used to get the length of a row and of a column of a real or integer matrix, respectively. *VecLen()* returns the length of a real or integer vector. The function *Id()* returns a real identity matrix of the same shape as its argument and *transp()* returns the transposed of a real matrix. The void functions *DoubleSize()* can be used to dynamically double the size of memory occupied by their arguments. Finally, the module contains the operator << for the output of an integer vector.

```
//----------------------------------------------------------------------------
// File: mv_util (implementation)
// Purpose: Utilities of type 'intvector', 'intmatrix', 'rvector', and
//    'rmatrix'.
// Global functions:
//    VecLen()    : length of an integer or real vector
//    RowLen()    : length of the rows of an integer or real matrix
//    ColLen()    : length of the columns of an integer or real matrix
//    Id()        : identity real matrix
//    transp()    : transposed of a real matrix
//    DoubleSize(): for doubling the size of an integer vector or matrix
//    operator << : output of integer vector
//----------------------------------------------------------------------------
#include <mv_util.hpp>

int VecLen ( intvector& v )                        // Length of an integer vector
  { return Ub(v)-Lb(v)+1; }                        //----------------------------

int RowLen ( intmatrix& A )            // Length of the rows of a integer matrix
  { return Ub(A,2)-Lb(A,2)+1; }        //----------------------------------------

int ColLen ( intmatrix& A )        // Length of the columns of an integer matrix
  { return Ub(A,1)-Lb(A,1)+1; }    //--------------------------------------------

int VecLen ( rvector& v )                               // Length of a real vector
  { return Ub(v)-Lb(v)+1; }                             //------------------------

int RowLen ( rmatrix& A )                 // Length of the rows of a real matrix
  { return Ub(A,2)-Lb(A,2)+1; }           //-------------------------------------

int ColLen ( rmatrix& A )              // Length of the columns of a real matrix
  { return Ub(A,1)-Lb(A,1)+1; }        //----------------------------------------

rmatrix Id ( rmatrix& A )                                  // Real identity matrix
{                                                          //---------------------
  int i,j;
  int lbi = Lb(A,1), ubi = Ub(A,1);
  int lbj = Lb(A,2), ubj = Ub(A,2);
```

```
  rmatrix B(lbi,ubi,lbj,ubj);

  for (i = lbi; i <= ubi; i++)
    for (j = lbj; j <= ubj; j++)
      B[i][j] = (i==j) ? 1.0 : 0.0;
  return B;
}

rmatrix transp ( rmatrix& A )                          // Transposed matrix
{                                                      //------------------
  int      n;
  rmatrix  res(Lb(A,2),Ub(A,2),Lb(A,1),Ub(A,1));

  for (n = Lb(A,1); n <= Ub(A,1); n++) Col(res,n) = Row(A,n);
  return res;
}

// The 'DoubleSize' functions double the number of rows of a matrix
// or double the length of a vector preserving existing components.
//--------------------------------------------------------------------
void DoubleSize ( intvector& x )
{
  int n = Lb(x);
  Resize(x,n,2*Ub(x)-n+1);
}

void DoubleSize ( intmatrix& A )
{
  int n = Lb(A,1);
  Resize(A,n,2*Ub(A,1)-n+1,Lb(A,2),Ub(A,2));
}

ostream& operator<< ( ostream& os, intvector& v )      // Output of integer
{                                                      // vectors
  int i, newline = (Ub(v)-Lb(v) > 15);                 //------------------

  for (i = Lb(v); i <= Ub(v); i++) {
    os << v[i] << ' ';
    if (newline) os << endl;
  }
  if (!newline) os << endl;
  return os;
}
```

## A.5 Module mvi_util

The header file of module *mvi_util* supplies utility functions for interval matrices and vectors.

```
//----------------------------------------------------------------------------
// File: mvi_util (header)
// Purpose: Utilities of type 'ivector' and 'imatrix'.
// Global functions:
//    in()          : contained-in-the-interior relation for two interval
//                    vectors and for an integer in an interval vector
//    Blow()        : epsilon inflation for an interval vector
//    Disjoint()    : test for disjointedness of two interval vectors
//    Id()          : identity interval matrix
//    Zero()        : check for a zero interval matrix
//    mid()         : midpoint vector and matrix
```

```
//    MaxRelDiam(): to get the maximum of the relative diameters of the
//                  components of an interval vector.
//    UlpAcc()    : to check whether all components of an interval vector
//                  have width less than a certain number of ulps
//                  (ulp = unit in the last place of the mantissa).
//    VecLen()    : length of an interval vector
//    RowLen()    : length of the rows of an interval matrix
//    ColLen()    : length of the columns of an interval matrix
//    DoubleSize(): for doubling the size of an interval vector or matrix
//----------------------------------------------------------------------------
#ifndef __MVI_UTIL_HPP
#define __MVI_UTIL_HPP

#include <i_util.hpp>       // Interval utility functions
#include <rmatrix.hpp>      // Real matrix/vector arithmetic
#include <imatrix.hpp>      // Interval matrix/vector arithmetic
#include <mv_util.hpp>      // Utility functions for real matrices, real
                            // vectors, and integer vectors

extern int      in         ( ivector&, ivector& );
extern int      in         ( int,      ivector& );
extern ivector  Blow       ( ivector&, real );
extern int      Disjoint   ( ivector&, ivector& );
extern imatrix  Id         ( imatrix& );
extern int      Zero       ( ivector& );
extern rmatrix  mid        ( imatrix& );
extern rvector  mid        ( ivector& );
extern real     MaxRelDiam ( ivector& );
extern int      UlpAcc     ( ivector&, int );
extern int      VecLen     ( ivector& );
extern int      RowLen     ( imatrix& );
extern int      ColLen     ( imatrix& );
extern void     DoubleSize ( ivector& );
extern void     DoubleSize ( imatrix& );
#endif
```

The module provides two implementations of the function *in()* which can be used to check if a real vector or an interval vector is contained in the interior of an interval vector. For vectors, the contained-in-the-interior relation is defined componentwise (see page 41). The function *Disjoint()* checks two interval vectors for disjointness which is also defined componentwise. The function *Id()* returns an interval identity matrix of the same shape as its argument, *Zero()* can be used to check if an interval vector has zero entries in all its components, and *mid()* delivers the midpoint of an interval matrix or vector. The function *MaxRelDiam()* can be used to get the maximum relative diameter of an interval vector as defined in Section 3.4. The integer function *UlpAcc()* is used to check any component of an interval vector for a desired accuracy specified in ulp. Finally, there are two void functions *DoubleSize()* which can be used to dynamically double the size of memory occupied by their arguments.

```
//----------------------------------------------------------------------------
// File: mvi_util (implementation)
// Purpose: Utilities of type 'ivector' and 'imatrix'.
// Global functions:
//    in()        : contained-in-the-interior relation for two interval
//                  vectors and for an integer in an interval vector
//    Blow()      : epsilon inflation for an interval vector
//    Disjoint()  : test for disjointedness of two interval vectors
```

```
//    Id()         : identity interval matrix
//    Zero()       : check for a zero interval matrix
//    mid()        : midpoint vector and matrix
//    MaxRelDiam(): to get the maximum of the relative diameters of the
//                  components of an interval vector.
//    UlpAcc()     : to check whether all components of an interval vector
//                  have width less than a certain number of ulps
//                  (ulp = unit in the last place of the mantissa).
//    VecLen()     : length of an interval vector
//    RowLen()     : length of the rows of an interval matrix
//    ColLen()     : length of the columns of an interval matrix
//    DoubleSize(): for doubling the size of an interval vector or matrix
//----------------------------------------------------------------------------
#include <mvi_util.hpp>

int in ( ivector& x, ivector& y )          // Inner inclusion for two ivectors
{                                          //---------------------------------
  int i = Lb(x), n = Ub(x), ok = 1;

  while (ok && i <= n) { ok = in(x[i],y[i]); i++; }
  return ok;
}

int in ( int x, ivector& y )     // Inner inclusion of an integer in a ivector
{                                //-------------------------------------------
  int    i = Lb(y), n = Ub(y), ok = 1;
  double d = x;

  while (ok && i <= n) { ok = in(d,y[i]); i++; }
  return ok;
}

ivector Blow ( ivector& x, real eps )                    // Epsilon deflation
{                                                        //------------------
  int     i;
  ivector h(Lb(x),Ub(x));

  for (i = Lb(x); i <= Ub(x); i++) h[i] = Blow(x[i],eps);
  return h;
}

int Disjoint ( ivector& a, ivector& b )            // Test for disjointedness
{                                                  //------------------------
  int al = Lb(a), au = Ub(a), bl = Lb(b);
  int disjointed, i, d;

  d = bl - al;

  i = al;
  disjointed = 0;
  do {
    if (Disjoint(a[i],b[i+d]))
      disjointed = 1;
    else
      i++;
  } while ( !(disjointed || i > au) );
  return disjointed;
}

imatrix Id ( imatrix& A )                         // Interval identity matrix
{                                                 //-------------------------
  int i,j;
  int lbi = Lb(A,1), ubi = Ub(A,1);
  int lbj = Lb(A,2), ubj = Ub(A,2);
```

```
  imatrix B(lbi,ubi,lbj,ubj);

  for (i = lbi; i <= ubi; i++)
    for (j = lbj; j <= ubj; j++)
      B[i][j] = _interval( (i==j) ? 1.0 : 0.0 );
  return B;
}

int Zero ( ivector& x )                                // Check for zero vector
{                                                      //----------------------
  int i, ok;
  for (i = Lb(x), ok = 1; i <= Ub(x) && ok; i++) ok = (x[i] == 0.0);
  return ok;
}

rvector mid ( ivector& v )                              // Vector of midpoints
{                                                       //--------------------
  int i;
  int l = Lb(v), u = Ub(v);
  rvector w(l,u);

  for (i = l; i <= u; i++) w[i] = mid(v[i]);
  return w;
}

rmatrix mid ( imatrix& A )                              // Matrix of midpoints
{                                                       //--------------------
  int i,j;
  int lbi = Lb(A,1), ubi = Ub(A,1);
  int lbj = Lb(A,2), ubj = Ub(A,2);
  rmatrix B(lbi,ubi,lbj,ubj);

  for (i = lbi; i <= ubi; i++)
    for (j = lbj; j <= ubj; j++)
      B[i][j] = mid(A[i][j]);
  return B;
}

real MaxRelDiam ( ivector& v )                    // Maximum relative diameter
{                                                 //--------------------------
  real r;
  int  i, l=Lb(v), u=Ub(v);

  r = 0.0;
  for (i = l; i <= u; i++)
    if (RelDiam(v[i]) > r) r = RelDiam(v[i]);
  return r;
}

//----------------------------------------------------------------------------
// Checks if the diameter of the interval vector 'v' is less or equal to 'n'
// ulps. Ulp is an abbreviation for: units in the last place of the mantissa.
//----------------------------------------------------------------------------
int UlpAcc ( ivector& v, int n )
{
  int k, upper;

  for (k = Lb(v), upper = Ub(v); (k < upper) && UlpAcc(v[k],n); k++);
  return UlpAcc(v[k],n);
}

int VecLen ( ivector& v )                        // Length of a interval vector
  { return Ub(v)-Lb(v)+1; }                      //----------------------------
```

```
int RowLen ( imatrix& A )           // Length of the rows of a interval matrix
  { return Ub(A,2)-Lb(A,2)+1; }     //---------------------------------------

int ColLen ( imatrix& A )        // Length of the columns of a interval matrix
  { return Ub(A,1)-Lb(A,1)+1; }  //------------------------------------------

// The 'DoubleSize' functions double the number of rows of a matrix
// or double the length of a vector preserving existing components.
//------------------------------------------------------------------
void DoubleSize ( ivector& x )
{
  int n = Lb(x);
  Resize(x,n,2*Ub(x)-n+1);
}

void DoubleSize ( imatrix& A )
{
  int n = Lb(A,1);
  Resize(A,n,2*Ub(A,1)-n+1,Lb(A,2),Ub(A,2));
}
```

# Appendix B
# Alphabetical List of Modules

# Appendix C

# List of Special Symbols

| | | |
|---|---|---|
| $I\mathbb{R}$ | Set of real intervals | 30 |
| $I\mathbb{R}^*$ | Set of extended intervals | 40 |
| $J_f$ | Jacobian matrix of $f : \mathbb{R}^n \to \mathbb{R}^n$ | 247 |
| $\mathcal{L}$ | Set of neighboring basic index sets | 213 |
| $L$ | Pending list of pairs $([y], \underline{f_y})$ in global optimization | 113 |
| $m([x])$ | Midpoint of interval data | 31 |
| $M^n$ | Set of $n$-dimensional vectors over $M$ | 30 |
| $M^{n\times m}$ | Set of $n \times m$ matrices over $M$ | 30 |
| $\mathcal{N}$ | Set of nonbasic indices | 211 |
| $N([x])$ | Interval Newton operator for $f : \mathbb{R} \to \mathbb{R}$ | 93 |
| $N_{\text{GS}}([x])$ | Interval Newton Gauss-Seidel operator for $f : \mathbb{R}^n \to \mathbb{R}^n$ | 295 |
| $N'([x])$ | Interval Newton operator for the derivative of $f : \mathbb{R} \to \mathbb{R}$ | 116 |
| $N'_{\text{GS}}([x])$ | Interval Newton Gauss-Seidel operator for the gradient of $f : \mathbb{R}^n \to \mathbb{R}$ | 315 |
| $q^*$ | Coefficients of the deflated polynomial or eigenvector of the companion matrix | 164 |
| $q([x],[y])$ | Distance between two intervals | 32 |
| $R$ | Set of floating-point numbers, floating-point system | 42 |
| $\mathbb{R}$ | Set of real numbers | 30 |
| $\rho(B)$ | Spectral radius of the matrix $B$ | 315 |
| $\mathcal{S}$ | Set of index sets | 212 |
| $\mathcal{V}_{\text{opt}}$ | Set of optimal basic index sets | 212 |
| $X$ | Set of feasible solutions in linear optimization | 210 |
| $x^*$ | Zero or minimizer of a function | 93 |
| $\widetilde{x}$ | Approximation of an exact value $x$ | 58 |
| $\underline{x}$ | Lower bound of an interval | 30 |
| $\overline{x}$ | Upper bound of an interval | 30 |
| $x^{(k)}$ | $k$-th iterate | 30 |
| $[x]$ | Interval | 30 |
| $\langle [x] \rangle$ | Smallest absolute value of an interval | 31 |
| $\|[x]\|$ | Greatest absolute value of an interval | 31 |
| $\|\|[x]\|\|_\infty$ | Maximum norm for interval vectors or matrices | 41 |
| $([x]_i)_{i=1,\ldots,n}$ | Interval vector of length $n$ | 41 |
| $([x]_{ij})_{\substack{i=1,\ldots,n\\ j=1,\ldots,m}}$ | $n \times m$ interval matrix | 41 |

| | | |
|---|---|---|
| $[x] \bowtie \varepsilon$ | Epsilon inflation for an interval | 46 |
| $[x] \mathring{\subset} [y]$ | Contained-in-the-interior relation | 30 |
| $[x] \underline{\cup} [y]$ | Hull of intervals | 30 |
| $[x] \cap [y]$ | Intersection of intervals | 30 |
| $[x_{\mathrm{re}}] + i[x_{\mathrm{im}}]$ | Complex interval | 37 |
| $z^*$ | Zero of a polynomial or eigenvalue of the companion matrix | 164 |
| $z_{\mathrm{opt}}$ | Optimal value | 210 |
| $\emptyset$ | Empty set | 40 |
| $\nabla f$ | Gradient of $f : I\!R^n \to I\!R$ | 244 |
| $\nabla^2 f$ | Hessian matrix of $f : I\!R^n \to I\!R$ | 245 |
| $\bigcirc$ | Rounding $\bigcirc : I\!R \to R$ | 42 |
| $\Box$ | Rounding to the nearest | 42 |
| $\Box(\ldots)$ | Evaluation with maximum accuracy | 44 |
| $\triangledown$ | Downwardly directed rounding | 42 |
| $\triangle$ | Upwardly directed rounding | 42 |
| $\Diamond$ | Interval rounding | 45 |
| $\Diamond(\ldots)$ | Interval evaluation with maximum accuracy | 45 |

# Bibliography


[1] Adams, E., Kulisch, U. (Eds.): *Scientific Computing with Automatic Result Verification*. Academic Press, New York, 1993.

[2] Alefeld, G., Herzberger, J.: *Einführung in die Intervallrechnung*. Bibliographisches Institut, Mannheim, 1974.

[3] Alefeld, G., Herzberger, J.: *Introduction to Interval Computations*. Academic Press, New York, 1983.

[4] American National Standards Institute / Institute of Electrical and Electronics Engineers: *A Standard for Binary Floating-Point Arithmetic*. ANSI/IEEE Std. 754-1985, New York, 1985.

[5] American National Standards Institute / Institute of Electrical and Electronics Engineers: *A Standard for Radix-Independent Floating-Point Arithmetic*. ANSI/IEEE Std. 854-1987, New York, 1987.

[6] Atanassova, L., Herzberger, J. (Eds.): *Computer Arithmetic and Enclosure Methods*. North-Holland, Elsevier, Amsterdam, 1992.

[7] Auzinger, W., Stetter, H. J.: *Accurate Arithmetic Results for Decimal Data on Non-Decimal Computers*. Computing **35**, pp 141–151, 1985.

[8] Bauch, H., Jahn, K.-U., Oelschlägel, D., Süsse, H., Wiebigke, V.: *Intervallmathematik*. Teubner, Leipzig, 1987.

[9] Berz, M.: *Forward Algorithms for High Orders and Many Variables with Application to Beam Physics*. In [23], pp 147–156, 1991.

[10] Böhm, H.: *Berechnung von Polynomnullstellen und Auswertung arithmetischer Ausdrücke mit garantierter maximaler Genauigkeit*. Dissertation, Universität Karlsruhe, 1983.

[11] Buchberger, B., Kutzler, B., Feilmeier, M., Kratz, M., Kulisch, U., Rump, S. M.: *Rechnerorientierte Verfahren*. Teubner, Stuttgart, 1986.

[12] Corliss, G. F.: *Automatic Differentiation Bibliography*. In [23], pp 331–353, 1991.

[13] Csendes, T., Pintér, J.: *The Impact of Accelerating Tools on the Interval Subdivision Algorithm for Global Optimization*. European Journal of Operational Research, **65**, pp 314–320, 1993.

[14] Dantzig, G. B.: *Lineare Programmierung und Erweiterung*. Springer-Verlag, Heidelberg, 1966.

[15] Dennis, J. E., Schnabel, R. B.: *Numerical Methods for Unconstrained Optimization and Nonlinear Equations*. Prentice-Hall, Englewood Cliffs, New Jersey, 1983.

[16] Ellis, M., Stroustrup, B.: *The Annotated C++ Reference Manual.* Addison-Wesley, Reading, Massachusetts, 1991.

[17] Falcó Korn, C.: *Die Erweiterung von Software-Bibliotheken zur effizienten Verifikation der Approximationslösung linearer Gleichungssysteme*. Dissertation, Universität Basel, 1993.

[18] Fischer, H.-C.: *Schnelle automatische Differentiation, Einschließungsmethoden und Anwendungen*. Dissertation, Universität Karlsruhe, 1990.

[19] Fischer, H.-C., Haggenmüller, R., Schumacher, G.: *Auswertung arithmetischer Ausdrücke mit garantierter hoher Genauigkeit*. Siemens Forschungs- und Entwicklungsberichte **16**, Nr. 5, pp 171–177, 1987.

[20] Fischer, H.-C., Schumacher, G., Haggenmüller, R.: *Evaluation of Arithmetic Expressions with Guaranteed High Accuracy*. In [62], pp 149–157, 1988.

[21] Geörg, S.: *Two Methods for the Verified Inclusion of Zeros of Complex Polynomials*. In [95], pp 229–244, 1990.

[22] Griewank, A.: *On Automatic Differentiation*. In: Iri, M., Tanabe, K. (Ed.): *Mathematical Programming: Recent Developments and Applications*, pp 83–108, Kluwer Academic Publishers, 1989.

[23] Griewank, A., Corliss, G. (Eds.): *Automatic Differentiation of Algorithms: Theory, Implementation, and Applications*. SIAM, Philadelphia, Pennsylvania, 1991.

[24] Grüner, K.: *Solving Complex Problems for Polynomials and Linear Systems with Verified High Accuracy*. In [42], pp 199–220, 1987.

[25] Hammer, R.: *Maximal genaue Berechnung von Skalarproduktausdrücken und hochgenaue Auswertung von Programmteilen*. Dissertation, Universität Karlsruhe, 1992.

[26] Hammer, R., Neaga, M., Ratz, D.: *PASCAL-XSC – New Concepts for Scientific Computation and Numerical Data Processing*. In [1], pp 15–44, 1993.

[27] Hammer, R., Hocks, M., Kulisch, U., Ratz, D.: *Numerical Toolbox for Verified Computing I – Basic Numerical Problems*, Theory, Algorithms, and PASCAL–XSC Programs. Springer-Verlag, Heidelberg, 1993.

[28] Hansen, E.: *Global Optimization Using Interval Analysis – The One-Dimensional Case*. Journal of Optimization Theory and Applications **29**, pp 331–344, 1979.

[29] Hansen, E.: *Global Optimization Using Interval Analysis – The Multi-Dimensional Case*. Numerische Mathematik **34**, pp 247–270, 1980.

[30] Hansen, E.: *An Overview of Global Optimization Using Interval Analysis*. In [69], pp 289–307, 1988.

[31] Hansen, E.: *Global Optimization Using Interval Analysis.* Marcel Dekker, New York, 1992.

[32] Hansen, E., Sengupta, S.: *Bounding Solutions of Systems of Equations Using Interval Analysis.* BIT **21**, pp 203–211, 1981.

[33] Heuser, H.: *Funktionalanalysis.* Teubner, Stuttgart, 1986.

[34] Hocks, M.: *Innere-Punkt-Methoden und automatische Ergebnisverifikation in der Linearen Optimierung.* Dissertation, Universität Karlsruhe, to appear in 1995.

[35] IBM *High-Accuracy Arithmetic Subroutine Library* (ACRITH). General Information Manual, GC 33-6163-02, 3rd Edition, 1986.

[36] IBM *High-Accuracy Arithmetic Subroutine Library* (ACRITH). Program Description and User's Guide, SC 33-6164-02, 3rd Edition, 1986.

[37] IBM *High Accuracy Arithmetic – Extended Scientific Computation* (ACRITH–XSC), Reference. SC 33-6462-00, IBM Corp., 1990.

[38] Jansson, C.: *Zur Linearen Optimierung mit unscharfen Daten.* Dissertation, Universität Kaiserslautern, 1985.

[39] Jansson, C.: *A Global Optimization Method Using Interval Arithmetic.* In: [6], pp 259–267, 1992.

[40] Jansson, C., Knüppel, O.: *A Global Minimization Method: The Multi-Dimensional Case.* Bericht **92.1**, TU Hamburg-Harburg, Technische Informatik III, 1992.

[41] Karmarkar, N.: *A New Polynomial-Time Algorithm for Linear Programming.* Combinatorica **4**, pp 373–395, 1985.

[42] Kaucher, E., Kulisch, U., Ullrich, Ch. (Eds.): *Computer Arithmetic – Scientific Computation and Programming Languages.* Teubner, Stuttgart, 1987.

[43] Kaucher, E., Rump, S. M.: *E-Methods for Fixed Point Equations* $f(x) = x$. Computing **28**, pp 31–42, 1982.

[44] Kaucher, E., Markov, S. M., Mayer, G. (Eds.): *Computer Arithmetic, Scientific Computation and Mathematical Modelling.* IMACS Annals on Computing and Applied Mathematics **12**, J. C. Baltzer AG, Scientific Publishing Co., Basel, 1991.

[45] Kearfott, R. B.: *Preconditioners for the Interval Gauss-Seidel Method.* SIAM Journal of Numerical Analysis **27**, 1990.

[46] Kearfott, R. B., Hu, C., Novoa, M.: *A Review of Preconditioners for the Interval Gauss-Seidel Method.* Interval Computations **1**, pp 59–85, Institute for New Technologies, St. Petersburg, 1991.

[47] Kernighan, B., Ritchie, D.: *The C Programming Language*, 2nd Edition. Prentice-Hall, Englewood Cliffs, New Jersey, 1988.

[48] Khachiyan, L. G.: *A Polynomial Algorithm in Linear Programming.* Doklady Akademiia Nauk SSSR **244**, pp 1093–1096 (in Russian), English translation in Sov. Math. Dokl. **20**, pp 191–194, 1979.

[49] Klatte, R., Kulisch, U., Lawo, C., Rauch, M., Wiethoff, A.: *C-XSC - A C++ Class Library for Extended Scientific Computing*, Reference. Springer-Verlag, Heidelberg, 1993.

[50] Klatte, R., Kulisch, U., Neaga, M., Ratz, D., Ullrich, Ch.: *PASCAL-XSC - Sprachbeschreibung mit Beispielen.* Springer-Verlag, Heidelberg, 1991.

[51] Klatte, R., Kulisch, U., Neaga, M., Ratz, D., Ullrich, Ch.: *PASCAL-XSC - Language Reference with Examples.* Springer-Verlag, New York, 1992.

[52] König, S.: *On the Inflation Parameter Used in Self-validating Methods.* In [95], pp 127–132, 1990.

[53] Krämer, W.: *Inverse Standardfunktionen für reelle und komplexe Intervallargumente mit a priori Fehlerabschätzungen für beliebige Datenformate.* Dissertation, Universität Karlsruhe, 1987.

[54] Krämer, W.: *Einschluß eines Paares konjugiert komplexer Nullstellen eines reellen Polynoms.* ZAMM **71**, pp T820–T824, 1991.

[55] Krämer, W.: *Evaluation of Polynomials in Several Variables with High Accuracy.* In [44], pp 239–249, 1991.

[56] Krawczyk, R.: *Newton-Algorithmen zur Bestimmung von Nullstellen mit Fehlerschranken.* Computing **4**, pp 187–220, 1969.

[57] Krawczyk, R.: *Fehlerabschätzungen bei linearer Optimierung.* In: Nickel, K. (Ed.): *Interval Mathematics.* Lecture Notes in Computer Science, No. 29, pp 215–227, Springer-Verlag, Vienna, 1975.

[58] Kulisch, U.: *Grundlagen des Numerischen Rechnens - Mathematische Begründung der Rechnerarithmetik.* Reihe Informatik, No. 19, Bibliographisches Institut, Mannheim, 1976.

[59] Kulisch, U. (Ed.): *Wissenschaftliches Rechnen mit Ergebnisverifikation — Eine Einführung.* Ausgearbeitet von S. Geörg, R. Hammer und D. Ratz. Akademie Verlag, Berlin, and Vieweg, Wiesbaden, 1989.

[60] Kulisch, U., Miranker, W. L.: *Computer Arithmetic in Theory and Practice.* Academic Press, New York, 1981.

[61] Kulisch, U., Miranker, W. L. (Eds.): *A New Approach to Scientific Computation.* Academic Press, New York, 1983.

[62] Kulisch, U., Stetter, H. J. (Eds.): *Scientific Computation with Automatic Result Verification.* Computing Suppl. **6**, Springer-Verlag, Vienna, 1988.

[63] Kulisch, U., Ullrich, Ch. (Eds.): *Wissenschaftliches Rechnen und Programmiersprachen.* Berichte des German Chapter of the ACM, No. 10, Teubner, Stuttgart, 1982.

[64] Levy, A. V., Montalvo, A., Gomez, S., Claderon, A.: *Topics in Global Optimization*. Lecture Notes in Mathematics, No. 909, Springer-Verlag, New York, 1981.

[65] Mayer, G.: *Grundbegriffe der Intervallrechnung*. In [59], pp 101–118, 1989.

[66] Miranker, W. L., Toupin, R. A.: *Accurate Scientific Computations*. Lecture Notes in Computer Science, No. 235, Springer-Verlag, Berlin, 1986.

[67] Moore, R. E.: *Interval Analysis*. Prentice-Hall, Englewood Cliffs, New Jersey, 1966.

[68] Moore, R. E.: *Methods and Applications of Interval Analysis*. SIAM, Philadelphia, Pennsylvania, 1979.

[69] Moore, R. E. (Ed.): *Reliability in Computing: The Role of Interval Methods in Scientific Computations*. Academic Press, New York, 1988.

[70] Neumaier, A.: *Interval Methods for Systems of Equations*. Cambridge University Press, Cambridge, 1990.

[71] Numerik Software GmbH: *C-XSC: A C++ Class Library for Extended Scientific Computing*, Program Product. Numerik Software GmbH, P.O.Box 2232, D-76492 Baden-Baden, Germany, 1993.

[72] Rall, L. B.: *Automatic Differentiation, Techniques and Applications*. Lecture Notes in Computer Science, No. 120, Springer-Verlag, Berlin, 1981.

[73] Rall, L. B.: *Differentiation in PASCAL—SC: Type GRADIENT*. ACM Trans. Math. Softw. **10**, pp 161–184, 1984.

[74] Rall, L. B.: *Optimal Implementation of Differentiation Arithmetic*. In [42], pp 287–295, 1987.

[75] Rall, L. B.: *Differentiation Arithmetics*. In [94], pp 73–90, 1990.

[76] Ratschek, H.: *Die Subdistributivität in der Intervallarithmetik*. ZAMM **51**, pp 189–192, 1971.

[77] Ratschek, H., Rokne, J.: *Computer Methods for the Range of Functions*. Ellis Horwood Limited, Chichester, 1984.

[78] Ratschek, H., Rokne, J.: *New Computer Methods for Global Optimization*. Ellis Horwood Limited, Chichester, 1988.

[79] Ratz, D.: *Automatische Ergebnisverifikation bei globalen Optimierungsproblemen*. Dissertation, Universität Karlsruhe, 1992.

[80] Ratz, D.: *An Inclusion Algorithm for Global Optimization in a Portable PASCAL-XSC Implementation*. In: [6], pp 329–338, 1992.

[81] Rosenbrock, H.: *An Automatic Method for Finding the Greatest and the Least Value of a Function*. Computer Journal **3**, pp 175–184, 1960.

[82] Rump, S. M.: *Kleine Fehlerschranken bei Matrixproblemen*. Dissertation, Universität Karlsruhe, 1980.

[83] Rump, S. M.: *Lösung linearer und nichtlinearer Gleichungssysteme mit maximaler Genauigkeit.* In: [63], pp 147–174, Teubner, Stuttgart, 1982.

[84] Rump, S. M.: *Solving Algebraic Problems with High Accuracy.* In: [61], pp 51–120, 1983.

[85] Rump, S. M.: *New Results on Verified Inclusions.* In [66], pp 31–69, 1986.

[86] SIEMENS: *ARITHMOS (BS 2000) Unterprogrammbibliothek für Hochpräzisionsarithmetik. Kurzbeschreibung, Tabellenheft, Benutzerhandbuch.* BNr.: U2900-J-Z87-1, 1986.

[87] Spaniol, O.: *Die Distributivität in der Intervallarithmetik.* Computing **5**, pp 6–16, 1970.

[88] Stetter, H. J.: *Sequential Defect Correction for High-Accuracy Floating-Point Algorithms.* In: *Numerical Analysis*, Lecture Notes in Mathematics, Vol. 1066, pp 186–202, 1984.

[89] Stoer, J., Bulirsch, R.: *Introduction to Numerical Analysis.* Springer-Verlag, New York, 1980.

[90] Stroustrup, B.: *The C++ Programming Language*, 2nd Edition. Addison-Wesley, Reading, Massachusetts, 1991.

[91] Stroustrup, B.: *The Design and Evolution of C++.* Addison-Wesley, Reading, Massachusetts, 1994.

[92] Syslo, M., Deo, N., Kowalik, J.: *Discrete Optimization Algorithms with Pascal Programs.* Prentice-Hall, Englewood Cliffs, New Jersey, 1983.

[93] Törn, A., Žilinskas, A.: *Global Optimization.* Lecture Notes in Computer Science, No. 350, Springer-Verlag, Berlin, 1989.

[94] Ullrich, Ch. (Ed.): *Computer Arithmetic and Self-Validating Numerical Methods.* Academic Press, San Diego, 1990.

[95] Ullrich, Ch. (Ed.): *Contributions to Computer Arithmetic and Self-Validating Numerical Methods.* J. C. Baltzer AG, Scientific Publishing Co., IMACS, 1990.

[96] Wolfe, M. A.: *Numerical Methods for Unconstrained Optimization – An Introduction.* Van Nostrand Reinhold, New York, 1978.

[97] Zurmühl, R., Falk, S.: *Matrizen und ihre Anwendungen. Teil 2: Numerische Methoden.* Springer-Verlag, Heidelberg, 1984.

# Index

# Order Form

All computer programs presented in this volume are available on diskettes and via ftp (see page ix). To order the programs on diskettes in IBM-PC compatible format, complete the information on the order form and mail it (or a photocopy) to:

Numerik Software GmbH
P.O.Box 2232
D-76492 Baden-Baden
Germany
Fax: +49 721 694418

All orders must be prepaid. Prices are not guaranteed. Ordinary postage is paid by Numerik Software GmbH.

## ORDER FORM

We would like to order

... Toolbox Program Diskette(s) – Volume I (PASCAL–XSC version)

... Toolbox Program Diskette(s) – Volume I (C++ version)

US$ 19.95 (outside Germany)

DM 19.95 (inside Germany)

### Address

Name ..........................................................................

Street ..........................................................................

ZIP, City ......................................................................

### Mode of Payment

□ Check □ Visa □ Eurocard/Mastercard □ American Express

Card No. ................................... Exp. Date ..................

Date/Signature ..............................................................

R. Klatte, U. Kulisch, A. Wieth, C. Lawo, M. Rauch

# C-XSC

**A C++ Class Library for Extended Scientific Computing**

Translated from the German by **G.F. Corliss**, **C. Lawo**, **R. Klatte**, **A. Wiethoff**, **C. Wolff**

1993. XV, 269 pp. 9 figs. Softcover **DM 74,-**; öS 577,20; sFr 71,50
ISBN 3-540-56328-8

C-XSC is an extension of C for solving numerical problems with automatic verification of results. It is immediately usable by C programmers, easy to learn, user-extendable, and may also be combined with other tools.
The book can be used as a textbook and as a reference manual. It consists of an introduction to advanced computer arithmetic, a chapter describing the programming languages C and C++, the major chapter "C-XSC Reference", sample programs, and indices.

R. Klatte, U. Kulisch, M. Neaga, D. Ratz, C. Ullrich

# Pascal-XSC Language Reference with Examples

**Language Reference with Examples**

Translated from the German by **G.F. Corliss**, **R. Klatte**, **U. Kulisch**, **D. Ratz**, **C. Wolff**

1992. X, 344 pp. Softcover **DM 67,-**; öS 522,60; sFr 64,50 ISBN 3-540-55137-9

PASCAL-XSC significantly simplifies programming in the area of scientific and technical computing.
It is easy to learn and ideal for courses in programming. The book can be used as a textbook for lectures on computer programming. It contains a major chapter with sample programs, exercises, and solutions. A complete set of syntax diagrams, detailed tables, and indices complete the book.

Springer

Preisänderungen vorbehalten

Tm.BA94.02.19

MIX
Papier aus verantwortungsvollen Quellen
Paper from responsible sources
FSC® C105338

If you have any concerns about our products,
you can contact us on
**ProductSafety@springernature.com**

In case Publisher is established outside the EU,
the EU authorized representative is:
**Springer Nature Customer Service Center GmbH**
**Europaplatz 3, 69115 Heidelberg, Germany**

Printed by Libri Plureos GmbH
in Hamburg, Germany